PRACTICAL PROCESS CONTROL

PRACTICAL PROCESS CONTROL

TUNING AND TROUBLESHOOTING

Cecil L. Smith

A JOHN WILEY & SONS, INC., PUBLICATION

Published by John Wiley & Sons, Inc., Hoboken, New Jersey
Published simultaneously in Canada

For general information on our other products and services or for technical support, please contact our Customer Care Department within the United States at (800) 762-2974, outside the United States at (317) 572-3993 or fax (317) 572-4002.

Wiley also publishes its books in a variety of electronic formats. Some content that appears in print may not be available in electronic formats. For more information about Wiley products, visit our web site at www.wiley.com.

Library of Congress Cataloging-in-Publication Data:

Smith, Cecil L.
   Practical process control : tuning and troubleshooting / Cecil L. Smith.
      p.   cm.
   Includes index.
   ISBN 978-0-470-38193-9 (cloth)
1. Process control.   I. Title.
   TS156.8.S6145 2009
   670.42′75—dc22

                                                              2008032185

Printed in the United States of America

10 9 8 7 6 5 4 3 2 1

◼ CONTENTS

After working almost exclusively in process control for over 40 years, I have developed a strong opinion that is reflected throughout this book: The process aspects of process control are the key to success; the systems aspects (the bits and bytes) are largely irrelevant. If you can do an application with one commercial system, then you can do it with any of them.

The practice of process control involves two intertwined endeavors:

1. Developing a process and instrumentation (P&I) diagram for the process.
2. Proportional-integral-derivative (PID) controller tuning and troubleshooting.

Deficiencies in the P&I diagram invariably lead to controllers that cannot be successfully tuned. The root problem is with the P&I diagram; tuning difficulties are merely symptoms. This makes troubleshooting an essential component of controller tuning, and vice versa.

Both endeavors are process (as opposed to systems) issues. The PID control equation is implemented in a variety of commercial control products, including a distributed control system (DCS), programmable logic controller (PLC), single-loop (or multiloop) controller, PC-based control system, and so on. A few implementations have a "wart" or two, but in the end, all are largely equivalent.

Automation of any industrial process should be undertaken with a simple objective—achieve a high degree of automatic control. That means controls that deliver the required performance a very high percentage of the time. This capability is a prerequisite to obtaining the greatest economic returns from a continuous process and the minimum product variability from a batch process. The P&I diagram must be correct, and the controllers must be properly tuned.

What does every plant manager know about the controls? The cost! Control system maintenance and support costs usually appear as line items in the budget. But ask the same plant manager how (or even if) these controls are making money. Most managers do not have a clue. The plant manager understands that controls are necessary, but with no understanding of the benefits, he/she does what all good managers do with regard to costs—manage the costs

tion are discussed, but the detailed presentation of cascade has been reserved for a future book. Override controls and split-range controls are given even less treatment in this book, but they will be covered extensively in the future book. Multivariable processes are covered but only to the extent that loop interaction can be recognized as a source of tuning difficulties. The solution to multivariable control is reserved for the future book.

The first 13 years of my career was as a member of the chemical engineering and computer science faculties of Louisiana State University in Baton Rouge. The gulf between the theory as taught in academia and the practice of process control is astounding. Academia is correct to teach theory and basic principles, but in process control, the theory being taught is largely irrelevant to the practice of process control.

This was a factor in my decision to leave academia in 1979. How can one face students knowing that the material being presented is perhaps a prerequisite for admission to graduate school, but it is virtually useless for those electing to pursue a professional career on graduation? While at the university, I worked with some highly capable and motivated Ph.D. students, all of which have done well. LSU was very good to me, and I will always be grateful.

I received my introduction to industrial applications by playing a minor role as a consultant in one of the early efforts to implement dead time compensation on a paper machine at the Westvaco mill in Charleston, SC. I have since enjoyed a long and productive relationship with the fragrances division of International Flavors and Fragrances. Primarily, because of this relationship, this book is relevant to those working with batch processes. And over the years, I have also had the pleasure of meeting and/or working with numerous people kind enough to share some of their insight into the nature of the process control business.

I thoroughly enjoy teaching process control. Starting in 1969, I taught courses on various aspects of process control as part of the continuing education program for the AIChE. In this program I taught over 4,000 attendees, many with some challenging problems back at the ranch. But with rightsizing and outsourcing, the demand for continuing education nosedived. I withdrew from the AIChE program when it was outsourced to the ASME. However, I continue to present courses on process control and related topics for other organizations as well as for in-house courses.

The constant through all of this has been my wife Charlotte. She has endured the late nights, my smart remarks, and the constant traveling. For almost 30 years, I was on the road over 90% of the time. Together we went to some wonderful places (and a few dumps). We still have friends all over the world.

Cecil L. Smith
Homer, AK
July 6, 2008

CHAPTER 1

Introduction

The subject of this book is regulatory control as practiced in the process industries. A good starting point is to explain what constitutes the process industries, and then to explain where regulatory control fits into the larger picture of control functionality.

This introductory chapter will also provide an overview of a control loop and define terms such as "controlled variable" and "manipulated variable." One version of the proportional-integral-derivative (PID) control equation will be presented with a brief explanation of the terms in this equation. All of these topics will be discussed in greater detail in subsequent chapters. The purpose herein is to provide an overview of the subject so that subsequent topics can be understood in the proper context. No doubt there will be unanswered questions at this point, but hopefully these will be answered in the subsequent chapters.

1.1. THE PROCESS INDUSTRIES AND REGULATORY CONTROL

The process industries are generally characterized by the movement of material from one processing unit (or "unit operation") to another. Movement of material (fluids or solids) is via pipes, belt conveyors, pneumatic conveyors, or otherwise. The unit operations are generally one of the following:

Reaction. The entering molecules are combined or broken apart to form new molecules.

Separation. The entering molecules are separated by type into two (or more) product streams.

Segments of the process industries include the following:

- Chemicals and petrochemicals.
  - Commodity.
  - Specialty.
- Food and beverage.
- Gas processing.
- Ore processing (but not metal working).
- Petroleum refining.
- Pharmaceuticals.
- Power generation (but not power distribution).
- Pulp and paper.
- Rubber and plastics.
- Textiles.
- Water and wastewater.

Each segment of the industry has some unique aspect that leads to some differences in the practice and application of the principles of automatic control. But in the end, the similarities outweigh the differences. This observation is supported by the fact that the suppliers of control systems market their products to all segments of the process industries. There have been a few niche suppliers in the past, but most of these have either merged with or been acquired by one of the broad-line control system suppliers.

Continuous Processes. Continuous processes consume raw materials and produce products on a continuous basis. Most commodity products are manufactured in continuous plants. Oil refineries, paper machines, boilers in power plants, and so on, are all continuous processes. They are designed to produce a specific product as efficiently as possible. These businesses are characterized by high production volumes but with small profit margins. Thus, efficiency is paramount.

Ideally, a continuous process would run under the same conditions from one day to the next. In practice, some changes are required, but they are normally small. Occasional adjustments in operating targets are required to compensate for equipment wear, heat transfer surface fouling, decay in catalyst activity, changes in raw material prices, raw material composition, product prices, product demand, and so on. But at least in the short term, the primary objective of the control system is to maintain the same conditions in face of various disturbing influences, whether from raw materials, utilities, ambient conditions, humans, or other. Changes in operating targets are small and relatively infrequent.

Batch Processes. A batch process produces products in a discrete quantity, usually referred to as a batch. Industries such as specialty chemicals and pharmaceuticals are dominated by batch operations. Batch operations are usually appropriate for products that must be produced in relatively small quantities. Multipurpose batch plants consist of very flexible equipment that can be used to produce a variety of products. Most such products have relatively high profit margins, so efficiencies usually take a back seat to other concerns, notably product quality.

The conditions within a batch facility are continually changing. In a sense, a batch facility is in a continual state of startup or shutdown. The most common batch operations are reactors and stills. In each case, the unit is charged with material, processing continues for some time (possibly with the addition of more raw materials and/or removal of some products), and then the contents are discharged. The batch unit must usually be sequenced through an appropriate series of steps, which entail frequent changes in operating targets. Such units are also subject to disturbances, but the first priority of the control system is to translate changes in the targets to changes in process conditions.

Hierarchy of Control Functions. A common way to represent the hierarchy of control functions is via layers in a pyramid. The generic representation in Figure 1.1 contains five layers, including a layer for the process interface. Most companies have developed a hierarchical representation that is specific to their business. These representations often are more detailed, and consequently, they contain more layers.

The pyramid representation is appropriate for the hierarchy of control functions. The functions in a given layer depend on the functions at the next lower layer. The desire is to develop and implement systems for all layers of the hierarchy. But in doing so, the easiest layers to address prove to be at the bottom and the top. The middle layers, specifically production control and supervisory control, are usually the most difficult to define and implement as operational systems. These layers are also the ones that are most likely to change with time.

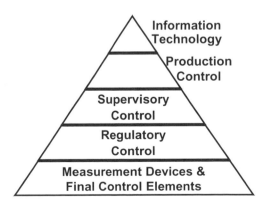

Figure 1.1. Hierarchy of control functions.

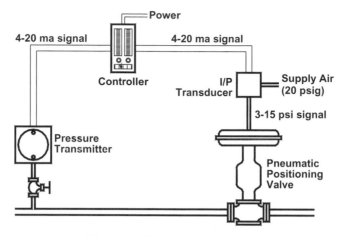

Figure 1.2. Pressure control loop.

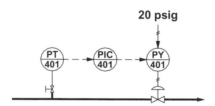

Figure 1.3. Detailed P&I diagram for pressure control loop.

Control configurations are generally represented on drawings referred to as process and instrumentation (P&I) diagrams. As do most process companies, we shall use the Instrumentation Society of America (ISA) symbols (1) for constructing P&I diagrams. However, the fossil fuel electric utility industry generally uses the Scientific Apparatus Makers Association (SAMA) symbols (2).

Detailed P&I Diagram. The P&I diagram in Figure 1.3 includes every component of the pressure control loop:

- The pressure transmitter is designated as PT-401.
- The output of the pressure transmitter is an electronic signal (dashed line).
- The controller is designated PIC-401 (PIC means "pressure indicating controller").
- The output of the controller is an electronic signal (dashed line).
- The electronic signal is converted to pneumatic by the I/P converter PY-401.

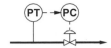

Figure 1.4. General P&I diagram for pressure control loop.

- The control valve is a globe valve equipped with a diaphragm-actuator without a valve positioner.

General P&I Diagram. When the focus is on process issues, most implementation details can be omitted from the P&I diagram, such as the fact that the pressure controller is an indicating controller. We shall generally draw P&I diagrams only to the detail illustrated in Figure 1.4. Only the following details are shown for the pressure loop:

- The pressure transmitter, designated only as PT.
- The controller, designated only as PC.
- The control valve, as a diaphragm-actuated valve.

Do not read implementation details into general P&I diagrams; they are often drawn before many implementation details have been decided.

1.3. REGULATORY CONTROL EXAMPLE

The steam-heated exchanger illustrated in Figure 1.5 heats a liquid stream to a specified temperature. The control configuration in the illustration consists of three components, as follows:

- Temperature transmitter to measure the liquid outlet temperature.
- Control valve on the steam supply to the exchanger.
- Temperature controller to position the control valve so as to attain the desired value for the liquid outlet temperature.

Being the simplest possible control configuration, such configurations are generally referred to as simple feedback control.

Process Variable. Any variable that describes a condition within the process is a process variable. Process variables can be either of the following:

Measured variables. These process variables are the outputs of process measurement devices. For the exchanger, the liquid outlet temperature

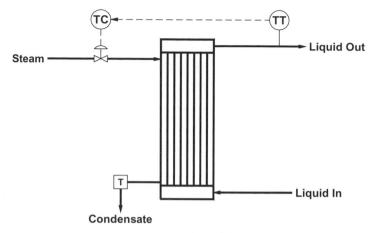

Figure 1.5. Regulatory control example.

is a measured variable. Other possible measured variables include liquid inlet temperature, liquid flow, steam flow, and so on.

Derived variables. Also called computed variables, these process variables are calculated from other process variables (measured variables and possibly other derived variables). For the exchanger, a possible derived variable is the heat transfer rate, which could be computed from the liquid flow, the liquid inlet temperature, the liquid outlet temperature, and the specific heat of the liquid.

Controlled Variable. The controlled variable is the process variable whose value is to be maintained at or near a specified target. For the exchanger, the controlled variable is the liquid outlet temperature.

The term "controlled variable" is routinely used by control engineers and in textbooks on control principles. However, operators and technicians rarely use it, preferring the terms "process variable" or "measured variable" instead.

Manipulated Variable. The manipulated variable is the variable that is adjusted or manipulated by the controller in order to maintain the controlled variable at or near its target. The value of the manipulated variable is entirely at the discretion of the controller.

For the exchanger, the manipulated variable is typically said to be the position of the steam control valve. Technically, the manipulated variable is the signal from the controller to the steam valve. The actual position of the steam valve is unlikely to be exactly equal to the value implied by this signal. However, we tend to ignore this, and refer to the steam valve position as the manipulated variable. Usually the difference between the position indicated by the output signal and the actual valve position is small and has no impact on performance. However, there are exceptions.

Process engineers often prefer to think of the flow through the control valve as the manipulated variable. The steam flow will be one of the variables in the energy balance that describes the exchanger. The steam flow is technically a dependent variable whose value depends on the controller output.

Disturbances (or Disturbance Variables). In process applications, the controlled variable is always influenced by several variables. One of these variables is the manipulated variable. All other variables that influence the controlled variable are considered to be disturbances.

For the exchanger, the liquid outlet temperature depends on the steam valve position (the manipulated variable), the steam supply pressure, the liquid flow, the liquid inlet temperature, the liquid heat capacity, the ambient temperature, and so on. Some of these variables have a greater influence than others. However, all except the steam valve position are considered to be disturbances.

Although many disturbance variables are available, we shall usually represent the disturbances by a single signal.

Load. Load is a term often used interchangeably with disturbances, and we shall largely do so.

For the exchanger, the load is the heat transfer rate. The heat transfer rate can be computed from the liquid flow, liquid inlet temperature, liquid specific heat, and liquid outlet temperature. In most exchangers, changes in the liquid flow have the most impact on the heat transfer rate, so designating the liquid flow rate as the load is common. Any change in throughput usually translates into a major change in the load, and thus, it constitutes a major disturbance to the control system.

1.4. CONTROL LOOP

A block diagram is a pictorial representation of how the various components of a system are interconnected. The block diagram in Figure 1.6 indicates how the process, the measurement device, the controller, and the final control element are interconnected to create a control loop.

In their pure forms, all block diagrams are constructed from only two elements:

Summer/comparator. The output is the sum / difference of the inputs.

Block. The output is related to the input by either an algebraic or a differential equation.

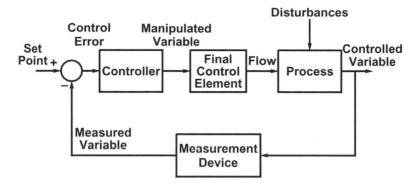

Figure 1.6. Block diagram of a control loop.

But in process control, liberties are commonly taken with the block diagram representations. In Figure 1.6, the block for the process has two inputs and, thus, does not adhere strictly to the conventions for drawing block diagrams.

The Control Loop. The information flow in the block diagram in Figure 1.6 takes the form of a loop:

1. A change in either the flow through the control valve or one of the process disturbances leads to a change in the controlled variable.
2. The change in controlled variable is sensed by the measurement device, which changes the measured variable.
3. The control error is the difference between the set point and the measured variable. Therefore, a change in either the set point or the measured variable leads to a change in the control error.
4. The controller responds to a change in the control error by changing the manipulated variable (or valve position).
5. A change in the valve position leads to a change in the flow through the control valve, which is one of the inputs to the process.

As the measured variable is "fed back" to be compared with the set point, the term "feedback control" is commonly applied to such loops.

Inputs to the Loop. The block diagram in Figure 1.6 indicates that there are two inputs to the loop:

Set point. When a change is made in the set point to the loop, it is expected that the controlled variable will follow the change in an expeditious manner.

Disturbances. A change in a disturbance leads to a change in the controlled variable. The loop is expected to change the manipulated variable in such a manner as to cancel the effect of the disturbance on the controlled variable. In continuous plants, the primary role of most control loops is to respond to changes in disturbances.

One of the factors that distinguishes process control from other control disciplines is that disturbances are present in all loops, and it is essential that the controller effectively respond to changes in the various disturbances as well as to changes in the set point.

Comparator. The comparator determines the difference between the target or set point and the current value of the measured variable. This difference is called the control error.

The term "controller" as used in the block diagram in Figure 1.6 is not consistent with its use in a hardware or software context. As used in Figure 1.6, the "controller" block provides the control equations or control logic. The control equations translate the error signal into the manipulated variable. When used in the context of hardware or software configurations, the "controller" normally includes the comparator.

Control Objective. The objective of the control logic can be stated very simply:

Drive the control error to zero.

A zero control error means the measured variable equals the set point, which is the desired condition.

A major advantage of this approach is that the control system will respond to both changes in the set point and changes in any disturbance. Assume that the controlled variable currently equals the set point. Any change in the set point makes the control error nonzero. The loop responds by changing the manipulated variable until the control error is again zero, which means that the controlled variable has attained the new set point. Any change in a disturbance causes the controlled variable to change, thus making the control error nonzero. The loop responds by changing the manipulated variable until the control error is again zero, which means that the controlled variable has returned to the original set point.

Controller. As represented in the block diagram in Figure 1.6, the function of the controller is to change the manipulated variable in such a manner as to drive the error signal to zero. As applied in the process industries, the control logic in most loops is one of the following:

ON/OFF **control.** The logic is applied in loops where the final control element is a two-state device (motor ON/OFF, valve open/closed, etc.). The hardware to implement this approach is normally the simplest. However, such systems do not line out at the set point, but they cycle about the set point. This behavior is not acceptable for the key controlled variables in a process.

PID control. This logic requires that the final control element be capable of some type of modulation. Valves that can be positioned at any value between fully closed and fully open are the most commonly used final control element. This control logic is capable of determining the valve position that causes the process to line out with the controlled variable equal to the set point.

Controller Function Block. Control engineers often draw loops showing the comparator and the controller as separate entities. They are separate functions, but they are normally implemented in the same hardware or software module.

All single loop controllers, both conventional and microprocessor-based, include the comparator as an integral component of the controller. In multifunction digital control systems, the controller is provided via a software function block that includes both the comparator and the control equations.

While not adhering literally to the conventions for drawing block diagrams, the representation in Figure 1.7 more accurately reflects the hardware and software configurations. The controller is drawn with two inputs, one being the set point (SP) and the other being the process variable (PV).

Percent of Span. When using analog technology for signal transmission, spans are mandatory. Pneumatic transmission uses a signal level of 3 to 15 psi. The current loop electronic transmission uses a signal level of 4 to 20 ma. With digital systems, we tend to use generic spans of 0% to 100%.

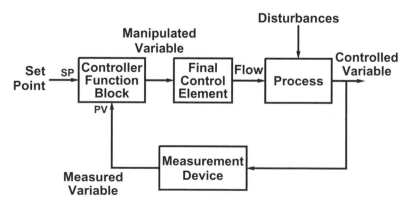

Figure 1.7. Block diagram with controller function block.

Suppose we represent both the measured variable and the manipulated variable as a percent of span. If we also convert the set point to a percent of the span of the measured variable, the error signal is a percent of span. The input (the error signal) to the control equation is in percent of span; the output (the manipulated variable) from the control equation is also a percent of span.

PID Control Equation. The PID control equation can be written in several different ways. In subsequent chapters, we will explain these variations. But until then, we shall write the PID control equation as follows:

$$M = K_C \left[E + \frac{1}{T_I} \int_0^t E\, dt + T_D \frac{dE}{dt} \right] + M_{R,0}$$

where

E = Control error (difference between set point and process variable), %

M = Controller output or manipulated variable, %

$M_{R,0}$ = Initial value of the controller output bias, %

K_C = Controller gain, %/%

T_I = Reset time, min

T_D = Derivative time, min

This equation is the traditional form of the PID control equation, the tuning coefficients being K_C, T_I, and T_D. The coefficient K_C multiplies every term in the equation and is logically referred to as the controller gain.

This control equation is said to have three modes, as follows:

Proportional mode. This term is a coefficient times the current value of the control error. The magnitude of the proportional control action is determined by coefficient K_C, the controller gain. The proportional mode can be used alone, but it is more commonly used in conjunction with the integral mode.

Integral or reset mode. This term is a coefficient times the integral of the control error. The magnitude of the integral control action is determined by coefficient T_I, the reset time. The integral mode can be used alone, but this is rare. The integral is almost always used in conjunction with the proportional mode.

Derivative or rate mode. This term is a coefficient times the rate of change or derivative of the control error (some implementations use the rate of change of the measured variable). The magnitude of the derivative control action is determined by coefficient T_D, the derivative time. The derivative mode is only used in conjunction with the proportional mode.

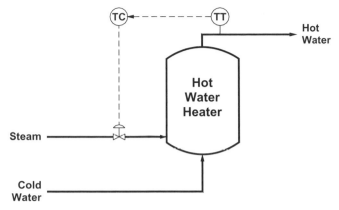

Figure 1.9. Simple feedback control of the hot water process.

The simple feedback control configuration for this process is illustrated in Figure 1.9. A temperature transmitter with a measurement range of 50 °F to 250 °F provides the measured value for the hot water temperature. A control valve on the steam supply is manipulated to obtain the desired hot water temperature.

The following nomenclature will be used:

Quantity	Description	Value and Units	Category
c_P	Water heat capacity	1.0 Btu/lb-°F	Constant
F	Steam flow	32.8 lb/min	Manipulated variable
H_F	Steam enthalpy	1185.3 Btu/lb	Constant
M	Mass of water in tank	2330 lb	Constant
Q	Cold water flow	467.2 lb/min	Dependent variable
T_i	Cold water temperature	75 °F	Disturbance
T_R	Reference temperature	32 °F	Constant
T_w	Hot water temperature	150 °F	Controlled variable
W	Hot water flow	500 lb/min	Disturbance/load

The quantities are divided into four categories, as follows:

Controlled variable. This is the variable that the control system is to maintain at or near a target or set point. For the hot water process, this is the hot water temperature.

Manipulated variable. This is a variable whose value is at the discretion of the control system. Technically the manipulated variable is the steam valve position, but in the model equations, the variable whose value is at the discretion of the controller is the steam flow.

Disturbance/load. The value of a disturbance is determined by external factors. For the control configuration in Figure 1.9, the cold water inlet temperature, the hot water flow, and the steam supply pressure are disturbances.

Dependent variables. A dependent variable is determined by the manipulated variable and/or the disturbances. For the hot water process, the cold water flow is a dependent variable.

Constants. These are either true constants (such as the reference temperature for computing enthalpy) or quantities that vary so little as to have no significant impact on the process controls. For the hot water process, the water heat capacity and the steam enthalpy are treated as constants.

The allocation of certain variables to one of these categories is discretionary. For example, one could classify the cold water flow as the disturbance and the hot water flow as a dependent variable.

The hot water flow can be classified as either a "disturbance" or a "load." When viewed as a load, the term "hot water demand" is often applied. Load is often a throughput. For the hot water process, it is a material flow. But for a heat transfer process, the throughput or load could be the heat transfer rate.

The hot water process is described by two equations:

Material balance: $\quad W = F + Q$

Energy balance: $\quad F H_F + Q c_P (T_i - T_R) - W (T_w - T_R) = M c_P \dfrac{dT_w}{dt}$

The tank is always full and water is incompressible under these conditions; consequently, no dynamics are associated with the material balance. The above expression for the energy balance assumes the contents of the tank are perfectly mixed, which is unlikely. Even for simple processes, analytic expressions have their limits.

Set-Point Changes Versus Disturbances. The hot water temperature controller in Figure 1.9 is typical of many loops in a production facility—its primary purpose is to respond to changes in a load or disturbance. There is a hot water heater in your home–the thermostat is a combined temperature sensor and ON–OFF controller. However, the thermostat has a set point for the hot water temperature. How often is this set point adjusted? Very infrequently, and possibly never changed from the factory setting. The thermostat responds almost exclusively to changes in the hot water demand.

This response is typical of loops in a production facility. The purpose of the loop is to maintain constant operating conditions within the process by responding to the myriad of disturbances. Changes in the set point are infrequent and small. A major advantage of feedback control is that the loop will respond to both disturbances and to set-point changes.

Tuning. For trial-and-error tuning, the performance of the loop is assessed from the response to a change in one of the inputs. For the hot water process, two possibilities exist, as follows:

- Change the set point.
- Change the hot water demand.

The hot water temperature loop is typical of most loops in that the most convenient response to obtain is to a step change in the set point. For trial-and-error tuning, the hot water temperature set point alternates between 150 °F and 160 °F. Starting at 150 °F, change the set point to 160 °F, observe the response, and adjust the tuning parameters. Then change the set point to 150 °F, observe the response, and adjust the tuning parameters. At least conceptually, this can be repeated until the controller is satisfactorily tuned. But in practice, every change in the set point is a disturbance to the process, so one must proceed in an expeditious manner.

Does it really make sense to tune the hot water temperature controller in Figure 1.9 to changes in the set point? The primary purpose of the loop is to respond to changes in hot water demand; set-point changes are small and infrequent. Fortunately, tuning means getting the characteristics of the controller in tune with the characteristics of the process. The input change on which the performance is assessed is of secondary importance.

Suppose the hot water temperature loop is tuned based on alternating the set point between 150 °F and 160 °F, and then the loop is tuned based on alternating the hot water demand between 500 lb/min and 600 lb/min. Would the results be the same? Not exactly. However, the difference for any tuning coefficient rarely exceeds 25%. Just how precisely can a controller be tuned using the trial-and-error approach? In the field, 25% would be doing well. In effect, the difference between tuning to set-point changes and tuning to load changes is about the same as the tolerance to which the loop can be tuned using the trial-and-error approach.

This result is typical of loops in an industrial facility. Consequently, most loops are tuned using the change that is most easily made, which is almost always a change in the set point.

1.6. CASCADE CONTROL

Cascade control is commonly considered for addressing certain deficiencies in simple feedback configurations. For example, the simple feedback configuration in Figure 1.9 for the hot water process has the following deficiencies:

- Changes in the steam supply pressure (a disturbance) have a significant effect on the hot water temperature.

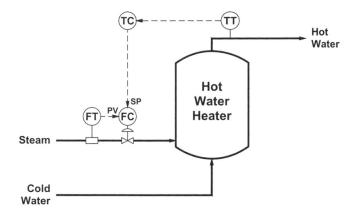

Figure 1.10. Temperature-to-flow cascade control of the hot water process.

- Being mechanical in nature, control valves commonly exhibit behavior that is significantly different from ideal. Such deficiencies in the control valve degrade the performance of the temperature loop.

These issues are addressed by the cascade control configuration in Figure 1.10. Two controllers are available, as follow:

Steam flow controller. This controller provides the position signal to the control valve. The steam flow set point is provided by the hot water temperature controller. The steam flow loop is the inner loop of the cascade.

Hot water temperature controller. The output of this controller is the set point for the steam flow controller. The hot water temperature loop is the outer loop of the cascade.

This configuration addresses both of the above deficiencies, as follows:

- A change in the steam supply pressure affects the steam flow much faster than it affects the hot water temperature. Flow controllers respond very quickly, and they maintain the steam flow very close to its set point. By maintaining a nearly constant steam flow, the flow loop isolates the hot water temperature loop from disturbances in steam supply pressure.
- The flow controller has to contend with any nonidealities exhibited by the control valve. Often a cycle in the steam flow results with a period of a few seconds. The hot water process basically responds to the average of such high-frequency cycles. Thus, the flow loop isolates the temperature loop from all characteristics of the control valve.

Both of these issues depend on the flow loop (the inner loop) being much faster than the temperature loop (the outer loop). For the cascade configuration to function properly, this characteristic is absolutely necessary. The usual desire is for the inner loop of the cascade to be faster than the outer loop by at least a factor of five.

How can the output of one controller be the set point to another? We will start with the following observations:

- The output of the temperature controller is in percent.
- The set point of the flow controller is in flow units, such as lb/min.

In either the flow transmitter or the flow controller (or possibly both), a measurement range is specified for the steam flow. This range permits the flow controller to retrieve the output of the temperature controller and to convert from percent to flow units. In some systems, a range in flow units is specified for the output of the temperature controller, which permits the flow controller to retrieve the temperature controller output in flow units. The mechanics are different, but the net result is the same. In today's digital controls, cascade control is easily implemented.

Enhancing a control configuration from simple feedback to cascade requires an additional measurement, specifically, a measurement for the process variable for the inner loop. For the cascade configuration in Figure 1.10, the steam flow transmitter is required. An additional controller is required for the steam flow controller, but in digital systems, this controller is implemented in software and considered to be "free."

In this book, cascade will be discussed almost exclusively in the context of addressing the limitations of simple feedback configurations. The chapter on P&I diagrams specifically considers two cascade configurations, including the temperature-to-flow cascade in Figure 1.10, to address the limitations of the simple feedback configuration in Figure 1.9. This book will not provide a detailed treatment of cascade. Before implementing cascade, one should understand how to tune cascade configurations, what operational modes are permitted in the various controllers, how to achieve a smooth transition from one mode to another, and so on.

1.7. SUMMARY

As for many subjects, the "chicken or egg" problem with where to start applies to process control. Hopefully this introductory chapter has introduced the terminology and provided an overview of certain concepts, including the PID control equation. With this as the starting point, this book can now proceed to the details on the various topics with process control, or more specifically, regulatory control, as practiced in the process industries.

LITERATURE CITED

1. "Instrument Symbols and Identification", Instrument Society of America, ISA-5.1-1984 (R1992).
2. "Functional Diagramming of Instrument and Control Systems", Scientific Apparatus Makers Association (SAMA), 1981. Declared obsolete; available from the Measurement, Control, and Automation Association (MCAA).

CHAPTER 2

Gain or Sensitivity

You have to understand the process. Those with experience in process control applications love to drop this pearl of wisdom on those just getting started. Unfortunately, most novices are too timid to ask the obvious question: Just how do you do that? Just as well. The usual answer is mumble-jumble anyway.

"Understanding the process" entails the use of parameters to quantify the behavioral characteristics of the process and then understanding the significance of these parameters. If the inputs to a process remain at fixed values, most processes will "line out" at an equilibrium state. One must understand how a change in an input affects the equilibrium state, that is, the steady-state behavior of the process. We will rely on two concepts:

- Process operating line.
- Gain or sensitivity.

Unfortunately, neither concept is completely satisfactory—processes are just too complex.

Most presentations of process control begin with dynamics. This approach reinforces a common misunderstanding that the dynamic characteristics of a process are the primary considerations in designing, implementing, and troubleshooting process control applications. A couple of dynamic characteristics can be very troublesome, but far more problems have their origin in the steady-state characteristics of the process.

2.1. PROCESS DESIGN VERSUS PROCESS CONTROL

Designing a process entails factors such as determining equipment capacities, calculating pipe dimensions, selecting pumps, sizing valves, and so on. All of these factors require that values be known for flow rates, temperatures, pressures, and so on.

For the hot water process described in Section 1.5, design entails determining the flow rate of steam (at a specified pressure) that is required to produce a specified quantity of hot water at a specified temperature, subject to the cold water entering at a specified temperature. This is typical engineering—determining a quantitative solution to a quantitative problem.

Suppose the design objective is to produce 500 lb/min of hot water at a temperature of 150 °F from 75 °F cold water and 90 psig steam (H_F = 1185.3 Btu/lb). The design calculations are as follows:

$$F = \frac{W\, c_P(T_w - T_i)}{H_F - c_P(T_i - T_R)} = \frac{(500)(1.0)(150 - 32)}{1185.3 - (1.0)(75 - 32)} = 32.8\, \text{lb/min}$$

$$Q = W - F = 500 - 32.8 = 467.2\, \text{lb/min}$$

Resolving Design Imperfections. The design calculations suggest that 32.8 lb/min of 90 psig steam produce hot water at 150 °F. But with the process in operation at a hot water demand of 500 lb/min and a cold water inlet temperature of 75 °F, suppose a steam flow of 32.8 lb/min actually gives a hot water temperature of 155 °F. How do we resolve this?

Either the model equations or the values used for the coefficients are slightly in error. The usual design approach to resolve this discrepancy is to calibrate the model to the process and then to repeat the calculations, with the steps being as follows:

1. Adjust one or more parameters in the design model so that the process model calculations agree with the observed operational data for the process. Using a value of 1262.5 Btu/lb for the steam enthalpy H_F in the design calculations would give a hot water temperature of 155 °F.
2. Repeat the design calculations using the new values for the model parameters. Using the revised value for H_F, the steam flow should be 30.8 lb/min.

Process models that are used for online process optimization are sometimes calibrated in this manner. But for regulatory control, a simpler route is normally followed.

Gain or Sensitivity of the Hot Water Process. The gain or sensitivity is always the ratio of a change in the controlled variable or process variable (an "output" from the process) to a change in the manipulated variable (an "input" to the process). For the hot water process, the manipulated variable in the process context is the steam flow F and the output is the hot water temperature T_w. The gain or sensitivity K_E of the hot water process is the amount the hot water temperature would change should the steam flow change by 1.0 lb/min. For example, a process gain or sensitivity of 2.0 °F/(lb/min) means that an increase of 1.0 lb/min in the steam flow will increase the hot water temperature by 2 °F.

Suppose the current value of the hot water temperature is 155 °F, but the desired value is 150 °F. What must we do? Obviously we need to reduce the steam flow. But let us behave like engineers. We have a quantitative problem; we want a quantitative answer. By how much do we need to reduce the steam flow? If the process sensitivity is 2.0 °F/(lb/min), reducing the steam flow by 2.5 lb/min will lower the hot water temperature by 5 °F.

Analytic Expression for the Gain. The process gain is the partial derivative of the controlled variable with respect to the manipulated variable. For the hot water process, this partial derivative is obtained as follows:

1. Rearrange the model equations to obtain an expression for the hot water temperature T_w as a function of the steam flow F:

$$T_w = \frac{H_F - c_P(T_i - T_R)}{W\,c_P} F + T_i$$

2. Take the partial derivative with respect to the steam flow F:

$$K_E = \frac{\partial T_w}{\partial F} = \frac{H_F - c_P(T_i - T_R)}{W\,c_P}$$

3. Substitute values of the variables and coefficients into this equation and compute the value for the gain:

$$K_E = \frac{1185.3 - (1.0)(75 - 32)}{(500)(1.0)} = 2.3°F/(lb/min)$$

In process control, it is not just the value of the gain or sensitivity that is important. Knowing the variables that significantly affect the gain or sensitivity is often more important. From the analytic expression, the following are obvious:

• The hot water flow W significantly affects the gain (K_E is inversely proportional to W).
• The cold water inlet temperature has only a minor effect on the gain (H_F is 1000 or more; $T_i - T_R$ is unlikely to exceed 100 °F).

Figure 2.1 is a graph of the hot water temperature T_w as a function of the steam flow F for three different hot water flows. The lines on such graphs are generally referred to as operating lines. The slope of the lines is clearly a function of the hot water demand W; however, the intercept is always the cold water inlet temperature T_i.

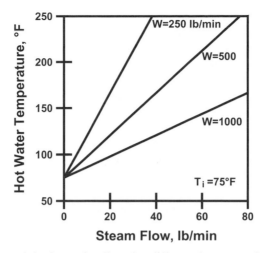

Figure 2.1. Operating lines for different hot water flows.

For most process models, the process gain or sensitivity must be computed numerically. Analytic expressions can be derived only for simple models. Although perhaps an interesting mathematical exercise (at least to some), such approaches have very limited value in practice. But the numerical approaches also have their limitations, one being that from the results of the numerical computations it is difficult to ascertain what variables do and do not significantly affect the value of the gain K_E.

Numerical Value for the Gain by Finite Differences. Regardless of the complexity of the steady-state process model, a numerical value for the gain or sensitivity can be computed using finite differences. To compute the gain for the hot water process, we could use the following solutions:

- Compute the hot water temperature for a steam flow of 30 lb/min, with the result being 143.54 °F. Designate this steam flow as F_1 and the corresponding hot water temperature as $T_{w,1}$.
- Compute the hot water temperature for a steam flow of 35 lb/min, with the result being 154.96 °F. Designate this steam flow as F_2 and the corresponding hot water temperature as $T_{w,2}$.

The finite difference expression for the gain is as follows:

$$K_E \cong \frac{T_{w,2} - T_{w,1}}{F_2 - F_1} = \frac{154.96 - 143.54}{35 - 30} = 2.3°F/(lb/min)$$

When applying finite differences, the value for K_E is actually the average value of the process gain over the interval F_1 to F_2. The smaller the interval, the closer the computed value to the actual value of the gain. But when the model calculations are iterative (as in multicomponent distillation calculations), the results are approximations whose errors are amplified when the difference between F_1 and F_2 is small.

From the analytic expression, it was obvious that the value of K_E is affected in a significant way by the hot water demand W. Such influences are not apparent from the results obtained using finite differences. If you suspect that the hot water flow has a significant influence on the gain, you could verify this by computing the process gain at a high hot water demand and then repeating the calculations for a low hot water demand. Conceptually, you could perform such calculations for every variable, but in practice, you are forced to rely on your understanding of the process to identify the variables that might have a significant influence on the gain, and then use finite differences to confirm your suspicion. Obviously the potential exists to miss an important variable.

2.2. WHAT DO WE MEAN BY "PROCESS GAIN"

The sensitivity K_E computed previously is the sensitivity of the hot water temperature T_w to changes in the steam flow F. For those whose focus is primarily the process, this would logically be the "process gain." The inputs to most process models are the flows through the control valves, which permits values for such sensitivities or gains to be computed from a process model. The term "unit" is often used within a process context (for example, "unit operation"). Hereafter, we will designate the gain or sensitivity K_E as the "process unit gain." To be consistent, the operating lines for the hot water process presented in Figure 2.1 will be referred to as the "process unit operating lines."

Control engineers consider the "process" to be everything between the output of the controller and the input from the measurement device. Therefore, their "process" encompasses three items that are connected as illustrated in the control loop in Figure 2.2:

Component	Input	Output	Gain
Control Valve	Controller output M	Flow through valve F	K_V
Process Unit	Flow through valve F	Controlled variable T_w	K_E
Transmitter	Controlled variable T_w	Process variable PV	K_T

Each component has its own gain. We have already examined K_E, the process unit gain. Shortly we shall examine K_V, the control valve gain, and K_T, the transmitter gain.

To a control engineer, the "process gain" is the sensitivity of the process variable PV to the controller output M. Normally this gain is designated by K:

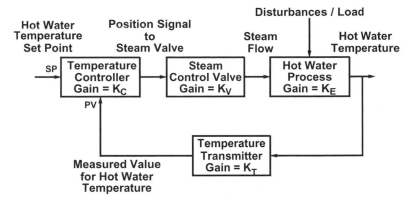

Figure 2.2. Block diagram of simple feedback control configuration.

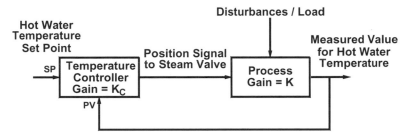

Figure 2.3. Control engineer's view of simple feedback control configuration.

$$K = \frac{\partial PV}{\partial M} = K_V\, K_E\, K_T$$

where

   K = Gain or sensitivity of the process, %/%

   M = Controller output or valve position, %

   PV = Process variable input to controller, %

The process gain K is the product of the gains for each of the three components.

In essence, the control engineer's perspective is reflected in the block diagram in Figure 2.3. The simple feedback loop contains only two components:

Controller. The input is the process variable PV; the output is the control valve position.

Process. The input is the control valve position; the output is the process variable PV, which is the output of the transmitter.

For activities such as controller tuning, this view is logical—the controller has to contend with everything between its output and its PV input.

Operating Line for the Steam Control Valve. The concept of an operating line can be extended to any component of the control loop, specifically, to the control valve, to the measurement device, and even to the PID controller.

The control valve operating line is a plot of the steam flow F as a function of steam control valve position M. This plot depends on the following parameters associated with the control valve and the flow system in which it is installed:

1. Size of the control valve.
2. Inherent characteristics of the control valve.
3. Percent of total flow system pressure drop taken across the control valve.

Figure 2.4 presents the operating line for the steam control valve with the following characteristics:

- Inherent valve characteristics are equal percentage.
- When fully open, the steam flow is 80 lb/min.
- When fully open, 25% of the flow system pressure drop is taken across the valve.

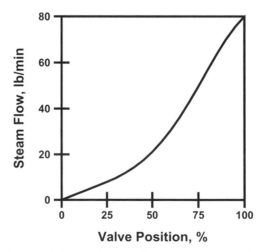

Figure 2.4. Operating line for steam control valve.

These and other issues are addressed in the subsequent chapter devoted to final control elements, so we shall not pursue them at this time.

A gain or sensitivity K_V can be defined by the following partial derivative:

$$K_V = \frac{\partial F}{\partial M}$$

where

K_V = Gain or sensitivity of the control valve, (lb/min)/%
F = Flow through the control valve, lb/min
M = Control valve position, %

The control valve gain is the slope of the operating line in Figure 2.4. The slope of this operating line is not constant, so the gain K_V changes with valve position. However, the region from about 50% open and higher exhibits only a modest departure from linearity, so an average value for K_V over this region could be used and linear behavior could be assumed.

Operating Line for the Hot Water Temperature Transmitter. For digital controls, control engineers typically consider the output of a transmitter to be percent of span, the range for the output signal being 0% to 100%. For the hot water temperature transmitter, the operating line is the graph of the output signal in % as a function of the hot water temperature in °F as illustrated in Figure 2.5. With microprocessors incorporated into the transmitters, all linearizations and compensations are generally performed by the transmitter.

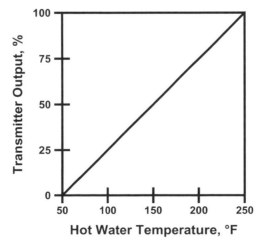

Figure 2.5. Operating line for hot water temperature transmitter.

Controller Operating Line. The equation for the proportional mode of the PID controller is often written as follows:

$$M = K_C(SP - PV) + M_R$$

where

K_C = Controller gain, %/%

M = Controller output (also called the manipulated variable), % of output span

M_R = Controller output bias (output of the integral mode), % of output span

PV = Process variable, % of input span

SP = Set point or target, % of input span

The controller operating line is a plot of the controller output M (the position signal to the control valve) as a function of the process variable PV (the input from the transmitter). Figure 2.7 presents the controller operating line with the PV on the x-axis and M on the y-axis. The following observations apply:

• The point (M = M_R, PV = SP) is on the controller operating line.
• The slope of the operating line is the controller gain K_C.

Actually, the slope of the operating line in Figure 2.7 is the negative of the controller gain K_C. When the control error E is computed as SP – PV, the

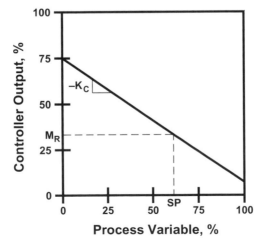

Figure 2.7. Controller operating line.

controller is said to be "reverse acting." Increasing the value of the PV causes the control error E to decrease (either a smaller positive value or a larger negative value). Consequently, the controller output M also decreases.

If the control error E is computed as PV – SP, the controller will be "direct acting." That is, an increase in the PV causes the control error E to increase and, consequently, the controller output M to increase. More on controller action shortly.

Although PV on the x-axis and M on the y-axis are consistent with the customary practice of constructing plots with the input on the x-axis and the output on the y-axis, the controller operating line is sometimes drawn with the axes reversed, that is, PV on the y-axis and M on the x-axis. This permits the process operating line and the controller operating line to be plotted on the same graph.

Loop Gain. In the simplified version of the block diagram in Figure 2.3, there are only two gains or sensitivities:

Controller gain. The customary units for K_C are %/%.
Process gain. If the units for K_C are %/%, the units for K must also be %/%.

In process control, the custom is to express all gains as positive values and use action to specify the directionality of the component. We will examine action shortly; until then, assume that all gains are positive.

The loop gain is the product of the gains of all components of the control loop. For the simplified representation in Figure 2.3, the loop gain is the product of the process gain K and the controller gain K_C:

$$\text{Loop Gain} = K\,K_C$$

The performance of a control loop is influenced in a very significant way by the value of the loop gain $K\,K_C$. Indeed, it is only the product that affects performance; the individual values are immaterial as long as the product remains the same.

Suppose we change the measurement range of the temperature transmitter as follows:

	Range	Span	K_V
Old	50 °F to 250 °F	200 °F	0.5 %/°F
New	100 °F to 200 °F	100 °F	1.0 %/°F

The transmitter gain K_T increases by a factor of 2; the process gain K increases by a factor of 2, and (assuming the controller gain K_C is unchanged) the loop gain $K\,K_C$ increases by a factor of 2.

Doubling the loop gain will have a noticeable effect on the loop performance. But if the controller gain K_C is reduced by a factor of 2, the loop gain returns to its original value and the loop performance remains the same.

When we tune a controller, we establish a value for the controller gain K_C. Basically, this also establishes a value for the loop gain $K K_C$. Unfortunately, we rarely have a numerical value for the loop gain. We know the controller gain K_C, but we do not have a value for the process gain K.

Note that any subsequent change in the loop gain will alter the behavior of the loop. The loop gain will change if either of the following changes:

Controller gain K_C. The value of K_C is at the discretion of the control engineer.

Process gain K. A change in the process gain K will have exactly the same effect as a corresponding change in the controller gain K_C.

2.3. LINEAR VERSUS NONLINEAR PROCESSES

Processes tend to be complex in many respects. Even simple processes such as the hot water process are linear in some respects but nonlinear in other respects. With few exceptions, the control equations, including the PID control equation, used for process control applications are linear. A common situation is the use of a linear control equation for a nonlinear process. Understanding the consequences is very important.

Operating Line for the Hot Water Process. This process unit operating line is a graph of the hot water temperature T_w (the controlled variable or process variable) as a function of steam flow F (the flow through the control valve). The process unit operating lines in Figure 2.1 were constructed for a cold water temperature of 75 °F. Each operating line is linear:

Intercept. The intercept is the cold water inlet temperature T_i.

Slope. The slope is the engineering process gain K_E.

Consequently, the hot water process can be said to be linear with respect to changes in the steam flow.

Figure 2.1 clearly shows that the value of the slope of the process unit operating line (and, consequently, the process unit gain) depends on the hot water demand. This greatly complicates obtaining and using any operating line that includes the unit process. Two questions must be answered.

What variables have a significant influence on the operating line? When analytic expressions are available for the process unit gain K_E, this question is easily answered. But otherwise, one must rely on one's insight into the process to determine which variables influence the operating line.

Do these variables change significantly during process operations?
We must consider both normal process operations and upset conditions.
We want our controls to perform adequately during upsets; controls that
only work when the process is running smoothly are of questionable
value.

Effect of Hot Water Demand on Response to Change In Steam Flow.
Nonlinear effects usually originate within either the control valve or the
process unit, and frequently within both. For modern transmitters, the transmitter is a potential but unlikely source.

To determine whether the nonlinearities originate from within the process
or within the control valve, a measurement of the flow through the control
valve is required. Assume this is available. To determine whether the process
is nonlinear, tests must be conducted under different process operating conditions. Each test normally entails making the same change in an input. For the
hot water process, the same change could be made in either

- The valve position.
- The flow through the control valve.

For the hot water process, let us test by making a change of 1.0 lb/min in the
steam flow. This can be done in two ways:

1. Manually position the steam valve until the steam flow has changed by
   1.0 lb/min from its initial value.
2. Configure a steam flow controller whose measured variable is the steam
   flow and its output positions the steam valve. To obtain the response to
   a 1.0 lb/min, just change the set point of the steam flow controller by
   1.0 lb/min.

We want to examine the process behavior when the hot water demand is low,
specifically, 100 lb/min, and when the hot water demand is high, specifically,
1000 lb/min. For these two hot water demands, the data are obtained as
follows:

1. Determine the valve position and steam flow that gives a hot water
   temperature of 150 °F.
2. Increase the steam flow by 1.0 lb/min and record the hot water temperature when the process lines out.

Figure 2.8 presents the response in hot water temperature, assuming a steam
flow controller is used to implement the change of 1.0 lb/min in the steam flow.
Two cases are presented, one for a low hot water demand and one for a high
hot water demand. From this data, the control valve gain K_V, the process unit

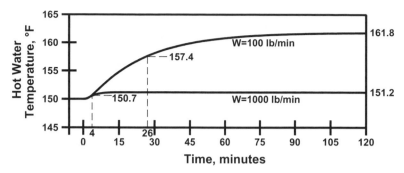

Figure 2.8. Responses at different hot water demands.

gain K_E, and the process gain K can be computed (the transmitter gain K_T is known to be 0.5%/°F). The complete results are as follows:

Hot water demand		100 lb/min	1000 lb/min
Valve position—**initial**		20.7%	86.5%
	final	23.9%	87.3%
	change	3.2%	0.8%
Steam flow—**initial**		6.6 lb/min	65.7 lb/min
	final	7.6 lb/min	66.7 lb/min
	change	1.0 lb/min	1.0 lb/min
Hot water temperature—**initial**		150.0 °F	150.0 °F
	final	161.8 °F	151.2 °F
	change	11.8 °F	1.2 °F
Control valve gain K_V		0.31 (lb/min)/%	1.3 (lb/min)/%
Process unit gain K_E		11.8 °F/(lb/min)	1.2 °F/(lb/min)
Process gain K		3.7 °F/% or 1.85 %/%	1.5 °F/% or 0.75 %/%

The effect of increasing the hot water demand on the gains is as follows:

The process unit gain decreases. The decrease in the above example is by a factor of 10, which is the reciprocal of the increase in hot water demand.

The control valve gain increases. This increase is approximately by a factor of 4.

The process gain decreases. This decrease is approximately by a factor of 2.5, which is the combined effect of the change in the process unit gain and the control valve gain.

In this case, the nonlinearity within the control valve partially compensates for the nonlinearity in the unit process. However, the opposite result can occur.

When we examine process dynamics in the next chapter, we will point out that the time required to attain 63% of the total change is a good measure of the speed of response of the process. As indicated in Figure 2.8, these times are as follows:

Low hot water demand. The total change in hot water temperature is 11.8 °F, 63% of which is 7.4 °F. It takes approximately 26 minutes for the hot water temperature to change from 150.0 °F to 157.4 °F.

High hot water demand. The total change in hot water temperature is 1.2 °F, 63% of which is 0.7 °F. It takes approximately 4 minutes for the hot water temperature to change from 150.0 °F to 150.7 °F.

These results confirm another point made in the next chapter: The speed of response increases as the throughput increases.

2.4. OPERATING LINES AND GAINS FROM PROCESS TESTS

Our capabilities to model our processes continue to improve, and today it can be safely stated that it is technically feasible to develop a model for any process. The obstacle is the economics, especially for the older processes. Once a process has been designed, constructed, and in operation, many incentives to develop a model no longer apply.

An alternative to developing a model is to conduct process tests. Conceptually, process testing is a relatively simple endeavor. However, one should never take any process test lightly. They always prove difficult to execute properly. And they have to be conducted properly; the "garbage in, garbage out" characterization definitely applies.

The Process Test. The procedure consists of a sequence of tests, each of which provides one data point for constructing the operating line. To obtain a point, the procedure for the hot water process is as follows:

1. Set the steam control valve to the desired position.
2. Wait for the process to attain equilibrium conditions (that is, the process lines out).
3. Record the equilibrium values for the hot water temperature, the steam flow (if this measurement is available), and the steam control valve position.

This is repeated until enough points are obtained to construct the operating line.

Figure 2.9 presents an example of a sequence of such tests. The test begins with the steam valve closed (M = 0%) and the hot water temperature equal

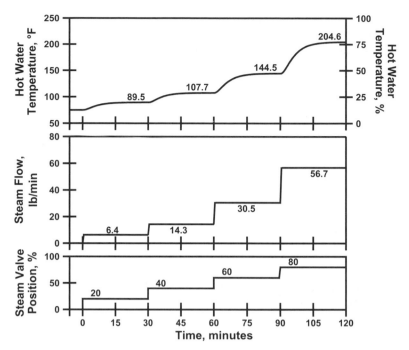

Figure 2.9. Process test to obtain data for the operating lines.

to the cold water temperature ($T_w = 75\,°F$). The steam valve is then set to 20%, then 40%, then 60%, and finally 80%. At each point, the process is allowed to attain equilibrium, giving the following data points:

Valve Position	Steam Flow	Hot Water Temperature
0%	0 lb/min	75.0 °F or 12.5% of span
20%	6.4 lb/min	89.5 °F or 19.8% of span
40%	14.3 lb/min	107.7 °F or 28.9% of span
60%	30.5 lb/min	144.5 °F or 47.3% of span
80%	56.7 lb/min	204.6 °F or 77.3% of span

We could construct several operating lines from this data, including an operating line for the control valve. However, we will only construct the process operating line in Figure 2.10. Since the transmitter is linear, scales are provided in both % of span and °F for the y-axis.

Obstacles to Executing the Test. For even the simple hot water process, three factors complicate conducting a test:

1. The process test is time consuming. The test of the hot water process illustrated in Figure 2.9 required two hours. Although there are a few

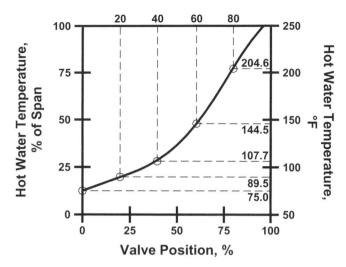

Figure 2.10. Process operating line from test data.

exceptions, industrial processes are notoriously slow, most being even slower than the hot water process.

2. Process variables other than the controlled and manipulated variables must remain constant. For the hot water process, the main concern is the hot water demand. Even for such a simple process, maintaining a constant hot water demand for two hours means telling all users not to change their hot water usage for the next two hours. Good luck on this!

3. Test data can be obtained only over a very restricted range. From a practical standpoint, the process must be operating under safe conditions and must be producing product that meets specifications throughout the duration of the process test (off-spec production is very expensive and will not be tolerated). Suppose we define "hot water" as water between the temperatures of 140°F and 180°F. When the data are restricted to this range, the resulting process operating line is restricted to the very narrow region illustrated in Figure 2.11. This figure also illustrates the fact that even very nonlinear processes will appear linear when the graph is restricted to a narrow range.

Fast processes are easier to test than slow ones. The first two issues are usually manageable for fast processes. As for the third issue, the hot water process permits a much wider range of operating conditions than most.

Why is the behavior of the hot water process outside of 140°F to 180°F of concern? Large swings in hot water demand, equipment malfunctions, and other unusual events will occasionally drive the process outside of the desired operating range. When such events occur, the control system must respond

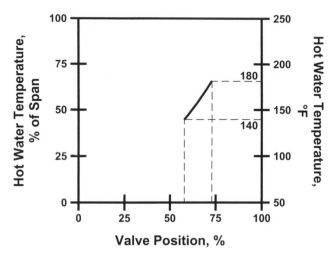

Figure 2.11. Restricted range for test data.

appropriately. A control system that has to be switched to manual on the occurrence of any major upset is of limited value. Instead, control systems are preferably capable of operation in automatic a very high percentage of time and under a wide range of process conditions.

Finite Difference Approximation for Process Gain. For the hot water process, the process gain can be computed from two points using the following finite difference approximation:

$$K \cong \frac{T_{w,2} - T_{w,1}}{M_2 - M_1}$$

where

$T_{w,1}$ = Hot water temperature for a steam valve position of M_1

$T_{w,2}$ = Hot water temperature for a steam valve position of M_2

This finite difference approximation permits the process gain to be calculated from two equilibrium points. Normally one point is obtained with the process operating near the lower end of the acceptable operating range, and the other point is obtained with the process operating near the upper end of the acceptable operating range.

A simple way to obtain two equilibrium points is as follows:

1. Set the steam valve to a position that gives a hot water temperature near the lower end of the acceptable range.
2. Wait until the process lines out and record the valve position, hot water temperature, and steam flow (if available).

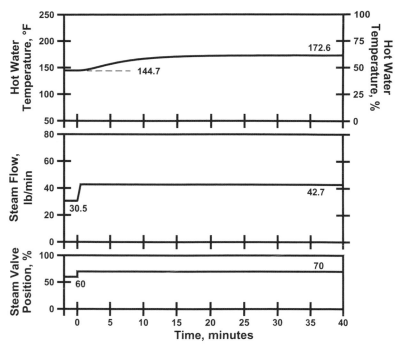

Figure 2.12. Test to obtain two equilibrium points for computing the process gain.

3. Set the steam valve to a position that gives a hot water temperature near the upper end of the acceptable range.
4. Wait until the process lines out and record the valve position, hot water temperature, and steam flow (if available).

The trend in Figure 2.12 illustrates doing this for the hot water process to obtain the following equilibrium points:

Valve Position	Steam Flow	Hot Water Temperature
60%	30.5 lb/min	144.7 °F or 47.4% of span
70%	42.7 lb/min	172.6 °F or 61.3% of span

With the temperatures in percent of span, the process gain is computed from these two points as follows:

$$K \cong \frac{T_{w,2} - T_{w,1}}{M_2 - M_1} = \frac{61.3\% - 47.4\%}{70\% - 60\%} = 1.39\%/\%$$

Alternatively, the finite difference can be computed using the temperature in °F and the result multiplied by the transmitter gain $K_T = 0.5\%/°F$ to obtain the process gain:

$$K \cong \frac{T_{w,2} - T_{w,1}}{M_2 - M_1} K_T = \frac{172.6\,°F - 144.7\,°F}{70\% - 60\%} \times 0.5\%/°F = 1.39\%/\%$$

The control valve gain K_V and the process unit gain K_E can also be computed from the above data, but we will not do so.

Numerical Considerations. In numerical calculations, one is advised to avoid subtracting two large numbers to obtain a small number. A small percentage error in one of the large numbers becomes a larger percentage error in the small number. In the example for the hot water process, 144.7 °F (or 47.4%) was subtracted from 172.6 °F (or 61.3%) to obtain 22.8 °F (or 13.9%). An error of 1% in one of the large numbers translates to an error of more than 5% in the difference. Given the difficulties of conducting process tests, errors in the equilibrium points are unavoidable. These errors will be amplified in the value computed for the process gain.

The greater the separation between the two equilibrium points, the less the errors are amplified. Thus, one equilibrium point should be at one extreme of the acceptable process operating region, and the second equilibrium point should be at the other extreme.

Effect of Disturbance and/or Load Changes. For the hot water process, the hot water demand has a major effect on both the process operating line and the process gain. Any difference in the hot water demand between the two equilibrium points introduces error into the data points and affects the value obtained for the process gain.

For more complex processes, we first have the problem of determining which variables have a major impact on the process gain. Next we face the task of maintaining these variables at the same value for the two equilibrium points. However, this only gives us a value of the process gain under these conditions. We do not have any quantitative measure of the effect of these other variables on the process gain.

At best, the two-point difference approach provides an approximation to the process gain at the conditions under which the two equilibrium points were obtained.

Nonlinear Processes. From only two points, there is no way to determine whether the process is truly linear. The two points used to compute the process gain must lie on the process operating line. The same value for the process gain would be computed for any operating line that passes through these two points. The operating line could be linear, but it could also be highly nonlinear.

At least conceptually, a third equilibrium point could be obtained to determine whether the process exhibits significant nonlinearities. But as noted earlier, process tests are usually conducted within a narrow operating region where the products meet specifications. The two points from which the gain is computed are preferably near the limits of the acceptable region of operation. Consequently, the third point would be approximately in the middle of the allowable operating region. Over any narrow range, the departure from linearity is small. Given that process test data invariably contain errors, detecting a small deviation from linearity using test data is unrealistic.

2.5. ACTION

Action is synonymous with directionality. If the input changes in one direction (increase or decrease), does the output change in the same direction or in the opposite direction? The terminology is as follows:

Direct acting (also called forward acting or increase-increase). An increase in the input leads to an increase in the output; a decrease in the input leads to a decrease in the output.

Reverse acting (also called increase-decrease). An increase in the input leads to a decrease in the output; a decrease in the input leads to an increase in the output.

This behavior can also be represented by the sign of the gain. A positive gain means increase-to-increase; a negative gain means increase-to-decrease. But in process control, the customary approach is to treat all gains as positive and to use action to convey directionality.

Action of the Process. The hot water process is direct acting—an increase in the steam flow results in an increase in the hot water temperature. However, many processes are reverse acting.

In boilers with superheaters, the degree of superheat is often controlled by injecting water in a downstream piece of equipment called an attemporator. In the simplest control configuration, a temperature controller adjusts the amount of water that is injected to give the desired temperature of the superheated steam. Increasing the flow of injected water lowers the temperature of the exit superheated steam. This process is reverse acting—an increase in the water injection rate reduces the steam temperature.

Some vessels are heated by steam, hot oil, and so on; other vessels are cooled by cooling water, chilled water, and so on. A temperature controller manipulates a control valve on the heating or cooling media so as to attain the target for the vessel temperature. A heated vessel is direct acting—increasing the flow of the heating media increases the vessel temperature. A cooled vessel is reverse acting—increasing the cooling media flow decreases the vessel temperature.

The process action is determined by the process design. Control engineers rarely have any ability to influence the action of the process. There is no need for them to do so. The action of the controller is completely at the discretion of the control engineers. In most cases, the only reason a control engineer needs to know the action of the process is to set the action of the controller properly.

Furthermore, control engineers have no reason to prefer a direct-acting process over a reverse-acting process, or vice versa. As long as the action of the controller is properly set, control loop performance is completely independent of the action of the process. That is, a reverse-acting process can be controlled just as effectively as a direct-acting process, and vice versa.

Controller Action. To those just getting into process control, the controller action is defined differently than what seems natural. There are three possible selections for the "input" on which the definition of controller action is based:

- Process variable.
- Set point.
- Control error.

The Instrument Society of America (ISA) publishes a document (1) that defines the key terms used in the industry. This document defines a direct-acting controller as one for which an increase in the measured variable leads to an increase in the controller output. A reverse-acting controller is one for which an increase in the measured variable leads to a decrease in the controller output.

The really key aspect of the definition is that the input for defining controller action is the measured variable or process variable. To control engineers that are not familiar with the process industries, basing action on the control error seems more logical. But there is a problem with this approach: the simple field-mounted regulators for pressure, level, temperature, and so on, do not generate an error signal (as we shall examine in the chapter on the proportional mode). The ISA's definition of action can be applied to these simple regulators as well as the more elegant PID controllers.

Reverse-Acting Controller. When control engineers draw a loop, the comparator to compute the control error is customarily drawn with "+" on the set-point input and "−" on the process variable input. The error signal is

$$E = SP - PV$$

Let us examine the action of this controller:

1. An increase in the process variable PV leads to a decrease in the error signal E (either the error becomes less positive or more negative).

2. As the output is gain K_C times error, a decrease in E leads to a decrease in the controller output.

Consequently, with E = SP – PV, the controller is reverse acting.

Direct-Acting Controller. There are two options for obtaining a direct-acting controller:

1. Compute E = SP – PV, but permit the controller gain to be negative. Most commercial control systems do not accept negative values for the controller gain. But if the controller is configured to be direct acting, the system could internally or "under the hood" treat the controller gain as a negative value.
2. Reverse the inputs to the comparator, as illustrated in Figure 2.13. The control error E is now either SP – PV for reverse acting or PV – SP for direct acting. Many conventional pneumatic and electronic controllers did exactly this.

Either approach can be used. Most manufacturers do not explain exactly how they actually do it, and it is of no concern to the user.

In this book, we shall routinely implement controller action by reversing the signs of the inputs to the comparator, as in Figure 2.13. That is, we shall define the control error E as follows:

E = PV – SP for a direct acting controller.
E = SP – PV for a reverse acting controller.

Setting the Controller Action. There is a simple procedure for setting the action of a controller. The steps are as follows:

1. Suppose the process variable is increasing.
2. Determine whether the appropriate response is an increase or a decrease in the controller output.

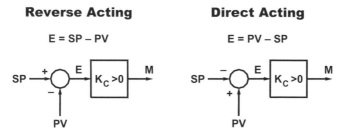

Figure 2.13. Direct- and reverse-acting controllers.

If the conclusion is that the controller output should increase, the controller should be direct acting. If the conclusion is that the controller output should decrease, the controller should be reverse acting.

The procedure for setting the controller action will be illustrated for the hot water process. Assume that the control valve opens on increasing the output signal to the control valve. The logic to determine the controller action is applied as follows:

1. Suppose the hot water temperature is increasing.
2. The steam flow must be reduced.
3. The control valve must be less open.
4. The controller output must decrease.

The conclusion is that an increase in the process variable must lead to a decrease in the controller output. The controller must be reverse acting.

Pneumatic Positioning Valves. Most control valves in industrial plants are equipped with a diaphragm actuator powered by compressed air. Although numerous advantages such as reliability, economy, and so on, are often cited, the major advantage of the diaphragm actuator is its behavior on loss of supply air. Air pressure against the diaphragm generates a force that is opposed by a spring. Two mechanical configurations are available, one that drives the valve fully closed on loss of air and another that drives the valve fully open on loss of air.

Hazards analysis is much simpler when the control valve fails to a known state, with the options being fully closed or fully open. On loss of air (power), this is the case for the diaphragm actuator. On loss of power, an electric actuator, specifically the motor-driven valve, fails in its last position. This is often the major objection to motor-driven valves in process plants.

Three different terms are commonly applied to control valves equipped with a diaphragm actuator:

Fail open/fail closed. This designates the failure state of the valve.

Air-to-open/air-to-close. For an air-to-open valve, an increase in the air signal causes the valve to open. For an air-to-close valve, an increase in the air signal causes the valve to close. An air-to-open valve fails in the closed position; an air-to-close valve fails in the open position.

Normally open/normally closed. In this context, "normal" is used in an equipment context, not in a process context. The normal state is the state of the equipment when all power is removed. Some call this the "shelf" state (a control valve in the warehouse will be in its "normal" state). For the diaphragm actuator, the normal state and the failure state are the same.

The fail open/fail closed choice for a control valve is made by those responsible for process safety, normally as part of the hazards and operability analysis for the facility. The control engineers must know how the valve behaves, but control engineers have no reason to prefer a fail-closed valve over a fail-open valve, or vice versa. As long as the controls are properly configured, control loop performance is completely independent of the failure mode of the control valve. However, these configuration issues require some discussion.

Display Variables for a Control Loop. For each PID controller, the current values of the following variables must be presented to the process operator:

- Process variable PV.
- Set point SP.
- Controller output M.

The initial impression is that these values can be simply displayed. This is always true for the PV and SP, but not necessarily for M. There is one guiding principle in displaying the controller output to the process operator:

Valve positions must be displayed as percent open.

A complication develops when the controller output is the position signal for a fail open/air-to-close/normally open control valve. A position signal of 30% means that such a valve is 30% closed or 70% open.

To repeat, the nature of the position signal for the valve is either percent open or percent closed, depending on the fail-open/fail-closed nature of the control valve:

Percent open. This is the case for a fail-closed/air-to-open/normally closed valve. If the output signal is 60%, the valve is 60% open. This value can be displayed directly.

Percent closed. This is the case for a fail-open/air-to-close/normally open valve. If the output signal is 60%, the valve is 40% open. For this type of valve, the output signal is actually percent closed (the valve is 60% closed or 40% open). This value cannot be displayed directly; instead, the value must be converted to percent open and then displayed.

The next issue is as follows. Assuming the controller action is set properly, the PID controller does not otherwise need to know whether its output is to a fail-open or to a fail-closed valve. However, the nature of the output must be known so that it can always be displayed as percent open.

Output Direction. When the control valve is fail-open/air-to-close/normally open, Figure 2.14 presents two approaches for displaying the value as percent open. In both configurations, the following statements apply:

Output Direction within Display Unit

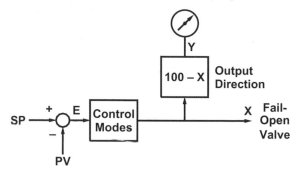

Output Direction within Control Loop

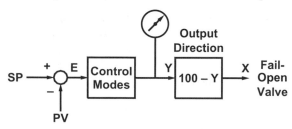

Figure 2.14. Output direction.

- Let X be the valve position as percent closed. The output to the valve must be X.
- Let Y be the valve position as percent open. The value displayed to the process operator must be Y.

These statements are true for both configurations in Figure 2.14. But there are some important differences:

- In the upper configuration in Figure 2.14, the loop is configured so that the controller output is the valve position as percent closed (X), which can be output directly to the control valve. In the display unit, the valve position is converted to percent open (Y) by the simple calculation $Y = 100 - X$.
- In the lower configuration in Figure 2.14, the loop is configured so that the controller output is the valve position as percent open (Y), which can be presented directly to the process operator. To obtain the position signal as percent closed (X), the simple calculation $X = 100 - Y$ must be inserted into the loop.

Although both approaches can be found in commercial control systems, the latter seems to be generally preferred. Those configuring the control functions are told to configure all controllers so that their output is percent open. The mechanics of converting from percent open to percent closed differ from one system to the next, with the possibilities including the following:

- Incorporate the output direction within the PID function block. Output as percent open is available for display; output as percent closed is available to output to the final control element.

- Use a summer function block to configure the calculation $X = 100 - Y$ explicitly within the control logic.

- Incorporate the output direction into the logic that generates the output to the final control element (the "valve block"). The input to the valve block is always percent open; however, the valve block is specifically instructed to output percent closed to the control valve.

- Incorporate the output direction in the "smart valve." With time, this will become the universal approach. However, the use of digital communications is mandatory. When the 4-20-ma current loop is used, the value transmitted via the current loop must be percent closed (a failure in the current loop must be equivalent to 0% closed so that the valve will fail open).

Controller Action and Output Direction. The terms "direct / reverse" and "increase-increase/increase-decrease" are applied to both controller action and output direction. However, do not confuse these two terms. The output direction depends only on the nature of the control valve:

- If the control valve is fail-closed, the output direction will be "direct" or "increase-increase." The "output direction" blocks in Figure 2.14 must be $X = Y$ or $Y = X$ so that they do nothing. Both the output to the control valve and the value to be displayed are percent open.

- If the control valve is fail-open, the output direction will be "reverse" or "increase-decrease." The "output direction" blocks in Figure 2.14 convert between percent closed and percent open.

The controller action depends on two factors:

1. The action of the process. This can be direct or reverse regardless of the behavior of the control valve.
2. The behavior of the control valve and how output direction is implemented.

In setting the controller action, one point must be stressed. The controller action must be based on the controller output, which is not necessarily the same as the output to the control valve. When the output direction is incorporated into the loop itself (as in the lower configuration in Figure 2.14), the controller output is percent open but the output to the control valve is percent closed. The controller action must be based on the controller output as percent open.

For certain loops, direct action is required when implemented in some commercial control systems, but reverse action is required when implemented in others. This only occurs when the control valve is fail-open (output to the valve is percent closed). The reason the controller action is different is because the output direction is implemented differently in the two systems.

Consider the level control application in Figure 2.15 for which the control valve is fail-open (output signal to valve is percent closed). The appropriate controller action depends on which of the approaches in Figure 2.14 is used to implement output direction:

Output Direction within Display Unit (upper configuration in Figure 2.14). The controller output is percent closed. The action is set as follows:

1. Suppose the vessel level is increasing.
2. The discharge flow must increase, so the control valve opening must increase.
3. The controller output (percent closed) must decrease.

A reverse acting controller is required.

Output direction within control loop (lower configuration in Figure 2.14). The controller output is percent open. The action is set as follows:

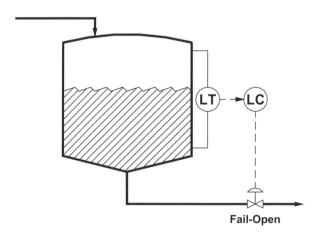

Fail-Open

Figure 2.15. Level control using a fail-open control valve.

1. Suppose the vessel level is increasing.
2. The discharge flow must increase, so the control valve opening must increase.
3. The controller output (percent closed) must decrease.

A direct-acting controller is required. Basically, the combination of the output direction block and the control valve is always equivalent to a fail-closed control valve.

If the control valve is changed from fail-closed to fail-open (or vice versa), the output direction must always be changed. If the output direction is implemented in the display unit (upper configuration in Figure 2.14), the controller action must also be changed.

2.6. IMPACT OF PROCESS NONLINEARITIES ON TUNING

The objective of tuning is to match the characteristics of the controller to the characteristics of the process. If the process is linear, the controller can be tuned for any process operating conditions and the loop will perform equally well under other operating conditions. But when the process is nonlinear, the characteristics of the process change with process operating conditions. When one tunes a controller for a nonlinear process, the resulting controller characteristics match the process characteristics only for the process operating conditions at the time the controller was tuned.

For nonlinear processes, the usual desire is to obtain a set of tuning coefficients that will deliver acceptable performance over the range of operating conditions experienced during production operations. Basically, this is a "one size fits all" approach. The discussion that follows pertains to this objective. But as we shall see, the usual consequence is that the controller performs poorly under the normal process operating conditions because it is tuned to the process behavior exhibited in some fringe region of the operating conditions.

Approaches are available to provide more uniform performance over the range of operating conditions, but applying such techniques requires effort and expense that must be justified by the improved performance. Usually the initial effort is the "one size fits all" approach. Only when this proves inadequate are the alternative approaches considered. One option is scheduled tuning, which will be discussed in the next section.

Hot Water as a Utility Process. Hot water is basically a utility that must change as the process requirements change. Let us start with a simple situation. Our process consists of two lines or trains, say Line A and Line B. Each line requires 500 lb/min of hot water. Normally, both lines are operating, so the requirement for hot water is 1000 lb/min. However, we occasionally have to stop one line, making the hot water requirement 500 lb/min. When we shut down both lines, we also shut down the hot water process.

As utility processes go, this is a relatively simple situation. Nevertheless, we want to show how this complicates the controller tuning effort and leads to a poorly tuned controller during the most frequent mode of process operations.

Only the proportional and integral modes will be used. The loop is to respond to a set-point change with an overshoot of 20% or less and no significant oscillations.

Choice of Process Conditions for Tuning. At what hot water demand should we tune the controller? For our simple example, there are only two choices:

1. Tune when the hot water demand is 1000 lb/min.
2. Tune when the hot water demand is 500 lb/min.

From a process operations perspective, tuning at 1000 lb/min is preferred. The reason is simple: The hot water process usually operates at 1000 lb/min.

Tuning at a Hot Water Demand of 1000 lb/min. Figure 2.16 presents two responses, each to a set-point change from 150 °F to 155 °F and then back to 150 °F:

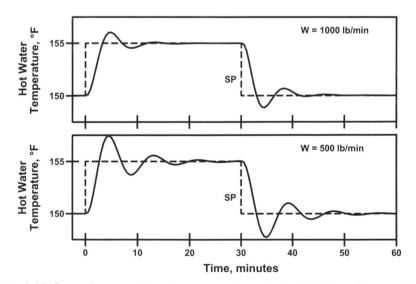

Figure 2.16 Controller tuned for hot water demand of 1000 lb/min ($K_C = 4.0\%/\%$; $T_I = 5.0$ min).

- Hot water demand is 1000 lb/min (upper response). The hot water demand is the same as the demand at the time the controller was tuned. The tuning objectives are met—overshoot of about 20% and negligible cycling.
- Hot water demand is 500 lb/min (lower response). The controller is now operating under process conditions different from those at which it was tuned. The response exhibits an overshoot of almost 50% and a noticeable cycle.

At a hot water demand of 500 lb/min, the process gain K is larger and the response speed is slower. Consequently, the controller gain must be reduced and the reset action made slower (to accommodate the slower process). With the controller tuned at a hot water demand of 1000 lb/min, the tuning coefficients are not acceptable when the hot water demand is 500 lb/min.

Tuning at a Hot Water Demand of 500 lb/min. Figure 2.17 presents two responses, each to a set-point change from 150 °F to 155 °F and then back to 150 °F:

- Hot water demand is 500 lb/min (upper response). The hot water demand is the same as the demand at the time the controller was tuned. The tuning objectives are met—overshoot of about 20% and no cycling.
- Hot water demand is 1000 lb/min (lower response). The controller is now operating under process conditions different from those at which it was tuned. The response approaches the set point much more slowly.

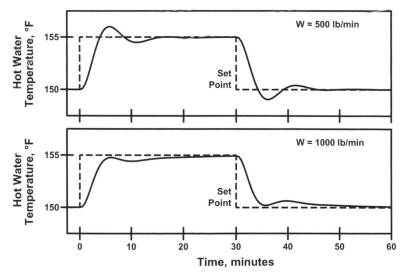

Figure 2.17 Controller tuned for hot water demand of 500 lb/min ($K_C = 2.5\%/\%$; $T_I = 7.5$ min).

For a hot water demand of 1000 lb/min, the process gain K is smaller and the process responds more quickly. The consequence is the slow approach to the set point exhibited by the response in Figure 2.17.

As the process is usually operated at a hot water demand of 1000 lb/min, the following question must be addressed: Is the performance illustrated by the lower response in Figure 2.17 acceptable? From a production perspective, can we make acceptable product with the hot water controller performing in this manner? If the question is yes, then the tuning in Figure 2.17 is deemed to be acceptable. If not, a more sophisticated control configuration must be installed.

Hopefully this example illustrates that a very common consequence of a nonlinear process is a poorly tuned controller for normal process operations. The controller tuning is dictated by process operating conditions that are experienced only occasionally. For this example, the difference in performance is noticeable but not dramatic. In such cases, the plant usually chooses to "live with" the resulting performance. Their other option is to justify the additional expenditure to implement a technique such as scheduled tuning.

The Tuning Experience for a Nonlinear Process. Every process must operate within some envelope of process operating conditions. The normal process operating conditions lie within this envelope. Unless prior experience with the same or similar process is available, the initial tuning is likely to be at the normal process operating conditions.

At some time in the future, the process will be operated at conditions other than the normal operating conditions. If the controller does not perform satisfactorily, a communication (voice mail, e-mail, etc.) is generated usually stating that "it doesn't work." If the process has returned to normal operating conditions, the controller is again working satisfactorily. In such cases the historical data collection features of modern control systems are very useful. From these features, one can ascertain what the process operating conditions were at the time the problem originated.

The key to success is identifying what has changed—the controller works under normal process operating conditions, so the fact that it does not work means that something is different. For the hot water process, the historical data should clearly show that the problem originated when the hot water demand was low. It will be necessary to retune the controller under these conditions, so production must either

- Return the process to these conditions.
- Anticipate when these conditions will occur again so that the appropriate personnel can be present.

Once either of these has been accomplished, the controller is now tuned for the "Low Hot Water Demand" condition. Production is immediately confronted with the issue of whether the controller tuned in this manner functions satisfactorily under normal process operating conditions. Let us suppose the conclusion is "yes."

The potential definitely exists that the controller might not perform satisfactorily under some other condition. Basically, the problem would be approached in the same manner as for the low hot water demand condition. Conceptually, this could be repeated many, many times. But with time, the frequency should decrease. And if a problem occurs in a region where the process is operated very infrequently, one option is to identify these conditions and to stipulate that the controls must be placed on manual when these conditions develop.

2.7. SCHEDULED TUNING

The term "adaptive control" originated in the electrical engineering community, and it was used to describe a control system that somehow measured its own performance and then made adjustments so as to maintain its performance at the desired level. But in the process industries, the term "adaptive control" is frequently applied to any control system that automatically adjusts its parameters with the intent of maintaining its performance at the desired level. This adjustment is often a preprogrammed adjustment based on either other process variables, the controller output, or an artificial variable such as time since the start of a batch. This approach should properly be called "nonlinear control" instead of "adaptive control." The most common implementation of such an approach is scheduled tuning.

Scheduled Tuning. Scheduled tuning involves the following:
1. Divide the possible process operating region into segments, which means that the controller must have some means to determine the segment in which the process is currently operating.
2. Tune the controller in each of these segments. Scheduled tuning can only be successful when the controller can be tuned to deliver acceptable performance in each segment. For loops where the process gain changes with operating conditions, scheduled tuning can be applied subject to one caveat: The process gain must not change to zero or even to a very small value.
3. Create a "tuning table" to provide the values of the tuning coefficients for each segment of the process operating region.
4. In real time, the control system determines the current segment, and then it retrieves the values for the tuning coefficients from the tuning table.

The definition of the segments requires considerable insight into process behavior. The following three examples should illustrate the possibilities:

1. For the hot water process, the hot water demand is the logical basis for defining the segments. The simplest approach would be to define two segments, low flow and high flow. The demarcation between low flow and high flow is a specified value of cold water flow.
2. For many batch processes, time into the batch is the logical basis for defining the segments. Many reactions start with an initiation phase, which might last for 30 minutes. The next four hours might be the main reaction phase. The last two hours might be the "cook-out" phase. The process behavior in these three phases is different, so different tuning is appropriate.
3. Processes such as paper machines produce multiple grades. The process behavior differs from grade to grade, so different tuning is appropriate.

With digital control systems, the number of segments could be very large. However, the practical limit seems to be about three segments.

The simplest configuration for scheduled tuning is where the tuning is changed based on a single variable, say X. Using two cutoff values X_{C1} and X_{C2} defines three segments:

Segment 1: $X \leq X_{C1}$.
Segment 2: $X_{C1} < X \leq X_{C2}$.
Segment 3: $X > X_{C2}$.

The controller must be tuned in each of the three segments, and the values of the tuning coefficients must be stored in a table where they can be retrieved depending on the value of X.

Scheduled Tuning for the Hot Water Process. The P&I diagram in Figure 2.18 illustrates scheduled tuning for the hot water process. The basis for the

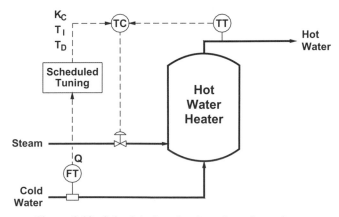

Figure 2.18. Scheduled tuning based on throughput.

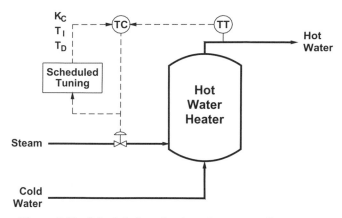

Figure 2.19. Scheduled tuning based on controller output.

scheduled tuning is the measured value of the cold water flow. To implement this configuration, two measurements are required:

1. Hot water temperature is the process variable for the temperature controller.
2. Cold water flow is the basis for the scheduled tuning.

In some situations, the controller output can be used instead of the additional measurement. This is the case for the hot water process. With the hot water temperature controller in automatic, the hot water demand, the steam flow, and the signal to the steam valve are tightly related. To maintain constant temperature, an increase in hot water demand must be quickly followed by an increase in the steam flow, which requires an increase in the signal to the steam valve (the controller output). The existence of such a relationship permits the controller output to be used as the basis for scheduled tuning, as is illustrated in the P&I diagram in Figure 2.19. This eliminates one measurement, permitting scheduled tuning to be implemented with no additional measurements.

2.8. HEAT TRANSFER PROCESSES

One situation that adversely affects the performance of the process controls is a process sensitivity of zero. In practice, the sensitivity usually does not become zero, but it becomes so small that the controller output has little effect on the process variable. Usually this exists only in a certain region of process operations. In other regions, the process controls deliver the required performance.

In heat transfer processes where the heating or cooling media is a utility stream such as hot oil or cooling water, the controls rely on the following relationships:

1. The output of the control system positions a control valve that affects the flow rate of the heating or cooling media.
2. The change in flow rate affects the heat transfer rate.
3. The change in heat transfer rate affects the temperature of interest within the process.

The relationship between the heating or cooling media flow and the heat transfer rate is highly nonlinear. Furthermore, there are situations where changing the flow has only a nominal effect on the heat transfer rate (the heat transfer rate has "maxed out"). This results in a very low gain or sensitivity, which is a problem for the temperature controls.

The process vessel illustrated in Figure 2.20 is equipped with a once-through jacket. The cooling media is cooling water. By "once-through," we mean that the cooling water enters the jacket, makes one pass through it, and then exits to the cooling water return. Although simple in concept, once-through jackets are nonlinear.

Figure 2.20 also presents the customary control configuration for vessel temperature. The vessel temperature is the process variable for a temperature controller whose output is the signal to the cooling water valve. An increase

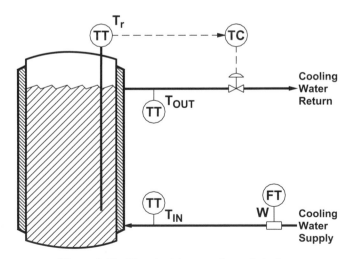

Figure 2.20. Vessel with once-through jacket.

in the valve opening leads to an increase in cooling water flow, which leads to a decrease in vessel temperature.

Heat Transfer Mechanisms. The heat removal is basically a two-step process. The heat that is removed from the contents of the vessel is added to the cooling water that is exiting the jacket. The heat removal proceeds as follows:

1. Heat is first transferred from the vessel contents to the cooling water. This is described by the heat transfer equation:

$$Q = U \, A \, \Delta T_{LM}$$

where

Q = Heat transfer rate, Btu/min

U = Heat transfer coefficient, Btu/hr-ft^2-°F

A = Heat transfer area, ft^2

ΔT_{LM} = Logarithmic mean temperature difference, °F
= $(\Delta T_{IN} - \Delta T_{OUT}) \, / \ln (\Delta T_{IN} \, / \, \Delta T_{OUT})$

ΔT_{IN} = $T_r - T_{IN}$ = Temperature difference at inlet, °F

ΔT_{OUT} = $T_r - T_{OUT}$ = Temperature difference at outlet, °F

T_r = Reactor temperature, °F

T_{IN} = Cooling water supply temperature, °F

T_{OUT} = Cooling water return temperature, °F

2. Heat is removed from the jacket by the cooling water in the form of sensible heat. This is described by the sensible heat equation for the cooling water:

$$Q = W \, c_P (T_{OUT} - T_{IN})$$

where

Q = Rate of heat removal by cooling water, Btu/min

W = Cooling water flow, lb/min

c_P = Cooling water heat capacity, Btu/lb-°F

T_{IN} = Cooling water supply temperature, °F

T_{OUT} = Cooling water return temperature, °F

At steady-state conditions, these two must be equal. That is, the heat transferred from the vessel contents to the jacket must equal the heat being removed from the jacket by the cooling water in the form of sensible heat.

Maximum Heat Transfer Rate. For a once-through jacket, there is a maximum or limiting heat transfer rate. The cooling water exit temperature is a function of cooling water flow. But as the cooling water flow increases, the cooling water exit temperature approaches the cooling water inlet temperature. At very high water flows, the temperature rise from jacket inlet to jacket outlet is very small, and the exit temperature is essentially equal to the inlet temperature. Thus, the driving force for heat transfer is the difference between the vessel temperature and the cooling water inlet temperature. The maximum heat transfer rate Q_{MAX} is

$$Q_{MAX} = U\,A\,(T_r - T_{IN})$$

Heat Transfer as a Function of Cooling Water Flow. Figure 2.21 presents the cooling water return temperature and the heat transfer rate as functions of cooling water flow. As the cooling water flow rate increases from zero, the heat transfer rate initially increases with water flow. But for large cooling water flows, the cooling water return temperature approaches the cooling water supply temperature. The heat transfer rate approaches the maximum possible heat transfer rate, and the sensitivity approaches zero.

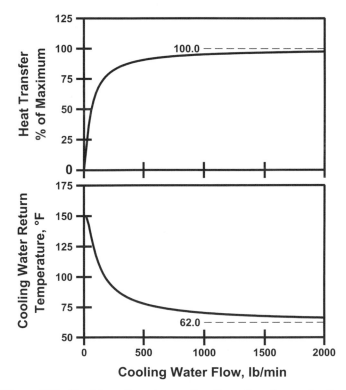

Figure 2.21. Heat transfer rate as a function of cooling water flow.

For very large water flows, the cooling water flow has little effect on the heat transfer rate. Those with little understanding of heat transfer sometimes believe that if the cooling water flow is half the maximum, then we are using only half of the available heat transfer. This is definitely not the case!

Heat Transfer Limited. At high water flows, heat transfer from vessel contents to jacket is the limiting mechanism for heat transfer. The cooling water flow affects the heat transfer by altering the temperature difference for heat transfer. But at high water flows, the temperature rise from jacket inlet to jacket outlet is very small. If the cooling water exit temperature is already close to its inlet temperature, additional increases in the cooling water flow will reduce the temperature rise from jacket inlet to jacket outlet, but they will have little effect on the temperature difference for heat transfer.

Let us illustrate this effect with a numerical example for a heat transfer process with the following characteristics:

A = 100.0 ft².
c_P = 1.0 Btu/lb-°F.
T_{IN} = 62 °F.
T_r = 150 °F.
U = 60.0 Btu/hr-ft²-°F.

For these parameters, the maximum possible heat transfer is

$$Q_{MAX} = U\,A\,(T_r - T_{IN}) = 8800\,\text{Btu/min}$$

The following numerical example computes the heat transfer rates for two different water flow rates:

Cooling water flow	500 lb/min	1000 lb/min
Vessel temperature	150 °F	150 °F
Cooling water supply temperature	62.0 °F	62.0 °F
Cooling water return temperature	78.0 °F	70.4 °F
Temperature rise	16.0 °F	8.4 °F
ΔT_{LM}	79.7 °F	83.7 °F
Heat transfer rate Q	7980 Btu/min	8370 Btu/min
Q / Q_{MAX}	0.907	0.951

Doubling the cooling water flow only increases the heat transfer rate about 5%.

Effect on Process Gain. At high water flows, the heat transfer is the controlling mechanism for removing heat from the vessel. With heat transfer in control, the cooling water flow has little effect on the heat transfer rate. If the cooling water flow has little effect on heat transfer, it also has little effect on vessel temperature. That is, at high cooling water flow rates, the sensitivity of vessel temperature to cooling water flow approaches zero. This spells problems for the temperature controller.

This problem is commonly encountered in practice. Cooling water is an inexpensive commodity. Designers see no need to conserve it. So when it comes to sizing piping, pumps, and other parts of the jacket, oversizing is common. But realize that the oversizing is only in regard to the ability to pump cooling water through the jacket. Increasing the heat transfer area would be productive, but we are not doing that. We are increasing the ability to flow cooling water through the jacket, which has little effect on heat transfer (and vessel temperature) once we enter the heat transfer limited region.

This has undesirable consequences on the vessel temperature control loop. At high cooling water flows, the sensitivity of vessel temperature to cooling water flow approaches zero. This means that the manipulated variable (cooling water valve position) has little effect on the controlled variable (vessel temperature). Controllers cannot function if the process gain is zero. The process gain never goes completely to zero, but it gets so close to zero that the controller cannot accomplish its task.

The consequence on the control loop is a phenomenon called windup. Windup occurs whenever the controller output has little or no effect on the measured variable. If the vessel temperature is above its set point, the controller responds by opening the cooling water valve. The flow increases, but this has little effect on the vessel temperature. So the controller continues to open the valve. This is windup.

The Onset of Heat Transfer Limited.   At the onset of heat transfer limited behavior, there is a significant decrease in the slope of the graph in Figure 2.21 for the heat transfer as a function of cooling water flow. As the slope changes gradually, the "line in the sand" marking the onset of heat transfer limited is somewhat fuzzy. For water flows less than about 250 lb/min, the process is not heat transfer limited. For water flows greater than 500 lb/min, the process is definitely heat transfer limited.

How can one detect the onset of heat transfer limited conditions? Consider using the following two temperature differences:

$T_{OUT} - T_{IN}$. The temperature increase from jacket inlet to jacket outlet.

$T_r - T_{IN}$. The maximum temperature difference for heat transfer, which is the reactor temperature less the cooling water inlet temperature.

The onset of heat transfer limited can be determined by examining the following ratio:

$$\frac{\text{Temperature rise of the cooling water}}{\text{Maximum } \Delta T \text{ for heat transfer}} = \frac{T_{OUT} - T_{IN}}{T_r - T_{IN}}$$

Let us examine this ratio for the heat transfer process for the previous numerical example. The effect of cooling water flow on the outlet temperature, the heat transfer rate, and the ratio is illustrated by the following numerical example:

W lb/min	T_{OUT} °F	$T_{OUT} - T_{IN}$ °F	$T_r - T_{IN}$ °F	Q Btu/min	$\frac{Q}{Q_{MAX}}$	$\frac{T_{OUT} - T_{IN}}{T_r - T_{IN}}$
125	110.5	48.5	88.0	6057	0.688	0.551
250	91.0	29.0	88.0	7253	0.842	0.330
500	78.0	16.0	88.0	7976	0.906	0.181
1000	70.4	8.4	88.0	8374	0.952	0.095
2000	66.3	4.3	88.0	8584	0.975	0.049
∞	62.0	0	88.0	8800	1.000	0.000

This table suggests the following criteria for the onset of heat transfer limited:

$$\frac{T_{OUT} - T_{IN}}{T_r - T_{IN}} < 0.2$$

Again, the demarcation between sensible heat limited and heat transfer limited is somewhat fuzzy, so you will see slightly different values or the onset of heat transfer limited.

2.9. VACUUM PROCESSES

In most installations, the controls position the control valve so as to attain the required flow through the valve. But in reality, the relationships for the control valve involve three variables:

1. Flow through the valve.
2. Pressure drop across the valve.
3. Valve position.

In the short term, changing the valve position always affects the flow through the valve. But in the long term (when the process again attains equilibrium or steady state), we occasionally find that the flow returns essentially to its original value, with the long-term effect being on the pressure drop across the valve.

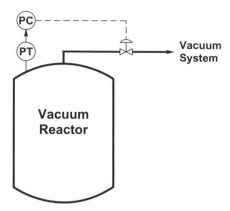

Figure 2.22. Controlling vacuum with control valve in vacuum line.

The vacuum reactor illustrated in Figure 2.22 includes equipment for feeding materials, maintaining the temperature, evacuating the vessel, and so on. We will only be addressing pressure control, so only the equipment pertaining to pressure control is illustrated in Figure 2.22. Also, this discussion is not affected by the type of equipment (vacuum header, dedicated vacuum pump, steam jet ejectors, etc.) used to evacuate the vessel, so this part of the process equipment has been simply designated as the "Vacuum System."

The objective is to maintain the pressure in the reactor vessel at or near the target specified by the reaction chemists. We shall assume that the vacuum system is capable of attaining pressures below the target, so some mechanism to control the vacuum is required. As illustrated in Figure 2.22, one possible approach is to install a control valve between the vessel and the vacuum system. We will show that there are problems with this simple approach.

From a long-term perspective, what is the flow through this valve? There are multiple sources of inert gases within the reactor, including leaks from the outside, inert gases dissolved in the feeds to the reactor, products of some minor side reactions, and so on. If the pressure within the reactor is maintained at a constant value, the flow through the vacuum valve in Figure 2.22 must be the sum of these sources of inert gases. If the control valve is opened slightly, the flow will increase. However, this additional flow comes from the inventory in the vapor space in the reactor, which drops the pressure within the reactor. When equilibrium is again established, the flow is again the sum of the sources of inert gases, but the pressure drop across the valve is lower. The long-term effect of the control valve is on the pressure, not on the flow.

Attaining the Target for Reactor Pressure. Assume for the moment that the pressure control loop illustrated in Figure 2.22 is able to attain the target for the reactor pressure. What statements can we make about the resulting conditions? Assuming that the pressure drop in the vacuum piping can be neglected, the conditions are expressed by the following equation:

$$\Delta P_V = P_R - P_S$$

where

P_R = Reactor pressure
P_S = Pressure at the vacuum system
ΔP_V = Pressure drop across the control valve

The pressure drop across the control valve must be the difference between the reactor pressure and the pressure in the vacuum system. The pressure controller must position the control valve such that this pressure drop is attained across the valve. Is this always possible? One is tempted to respond "Of course." This is one of the lessons to be learned in the process control business. Things that initially seem obvious sometimes turn out not to be true under all regions of process operation.

The pressure drop ΔP_V across a control valve is a function of two variables:

$$\Delta P_V = f(M, F)$$

where

F = Flow through the control valve
M = Valve position

The larger the flow, the more open the valve must be to attain a given pressure drop. But what if the flow through the valve is zero? The pressure drop across the valve is zero. Furthermore, the valve position has no effect on the pressure drop when the flow is zero.

Effect of Off-Gas Flow. The control system illustrated in Figure 2.22 has a deficiency that is serious for many vacuum processes. If the flow through the valve is zero, the controller output (the position signal to the control valve) has no effect on the reactor pressure. That is, the process gain is zero when there is no flow through the control valve.

In practice, the flow through the valve is unlikely to be truly zero. However, for small flows, a very small valve opening is required. The process conditions are such that the control valve must essentially operate "on its seat." Valves tend to exhibit erratic behavior under such conditions. The likely result is cycling in the reactor pressure. As the pressure drops below the target, the controller will fully close the valve. As the reactor pressure rises, the controller increases its output. At some point, the valve comes off its seat and assumes some minimum opening. However, this opening is too large. The pressure drops below the set point, causing the controller to reduce its output and to close the valve.

Except when the flow through the control valve is essentially zero, the pressure control system illustrated in Figure 2.22 will function properly. But should the flow become too small, the loop will cycle. The pressure controller might function quite well most of the time, but not all of the time.

Suppose the vacuum reactor is a batch reactor. During the early portion of the batch, we are feeding materials and the reactions are proceeding. The feeds contain dissolved gases; a few side reactions produce gaseous products. The off-gas flow is sufficient to keep the valve off of its seat. But in the later stages of the batch, we stop the feeds and raise the temperature to drive the reactions to completion. The off-gas flow decreases, and the controller responds by reducing the signal to the control valve. However, as the valve opening becomes too small, the valve behavior becomes erratic, resulting in cycling in the reactor pressure.

Alternatives. The cycling in the reactor pressure is not the fault of the controller. Consequently, it cannot be solved by controller tuning, whether by manual procedures or by automatic tuning systems. We must change the control configuration.

A revised configuration is illustrated in Figure 2.23. The control valve is now on an inert gas stream; the valve between the reactor and the vacuum system is a block valve. The controller now controls the pressure in the reactor by admitting the proper amount of inert gas. Should the process off-gas go to zero, this control system is not affected.

Where dedicated vacuum pumps are used, the inert gas source is often the vacuum pump exhaust, with the result being a recycle around the vacuum

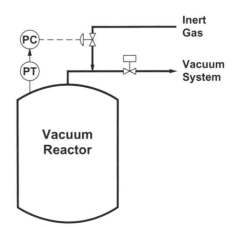

Figure 2.23. Controlling vacuum by bleeding inerts.

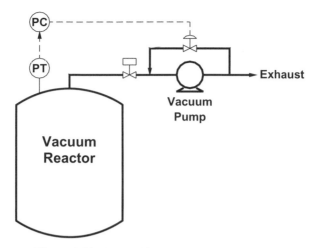

Figure 2.24. Controlling vacuum with a recycle.

pump as illustrated in Figure 2.24. A similar configuration can be used when ejectors provide the vacuum.

2.10. SUMMARY

As noted, the steady-state characteristics of the process have a significant and direct impact on the process controls. Unfortunately, neither the process gain nor the process operating line is completely satisfactory when it comes to characterizing nonlinear processes.

The extent to which nonlinear behavior leads to control problems depends on the plant operating region. Many large commodity processes are designed to operate very efficiently but within a narrow range of operating conditions. For these processes, linear controls generally function quite well. But consider a utility process, such as a steam boiler. Such processes are designed for a maximum production capacity, but they are also designed for sustained operation at a much lower throughput. Over such an operating range, the nonlinear behavior from the process will impact the performance of linear controls. Most batch processes exhibit even stronger nonlinear behavior.

Superficially, scheduled tuning is the easiest way to approach such processes. But in practice, its application proves more difficult. The system issues are indeed easy, but the process issues are not. The alternative is to understand the process better and to develop control configurations that can cope with the nonlinearities. And for those processes where the process gain is essentially zero in certain operating regions, the only approach is to develop a control configuration for which this cannot occur.

LITERATURE CITED

1. "Process Instrumentation Terminology", Instrument Society of America, ISA-51.1-1979 (R1993).

CHAPTER 3

Process Dynamics

The focus of this chapter is process dynamics and related issues, such as the various measures used to characterize the performance of a control loop. But as noted, dynamic issues are not as important in process control as one might initially think. In this respect, two observations are relevant:

1. With few exceptions, the economics of a production process are determined by the long-term averages of key operating variables. The impact of short-term excursions is nil. To affect the economics, the variance must be reduced and the long-term averages must be shifted to more economically favorable values.
2. The use of dynamic simulation in process control is far less than in industries such as aerospace. The first impression is that perhaps the industry is backward. But instead, this is the result of some fundamental differences.

Within the process industries, steady-state simulation is widely practiced. Especially in the commodity production facilities (refineries, paper mills, etc.), efficiencies are crucial, starting with the design of the process. Throughout the life of the process, the focus must be on maintaining operating conditions that result in the most efficient operation of the process. Steady-state models have proven useful in this endeavor.

Dynamic modeling of processes is technically feasible but not widely practiced. Contrast this to the aerospace industry where dynamic simulations are considered essential. This difference is due to one factor. In the aerospace industry, all loops must be in automatic and functioning properly the first time the vehicle leaves the ground. But a process facility can be operated with every loop on manual. For an occasional process, there might be a loop or two that must be in automatic, but for most processes, the answer is truly none.

As a result, it is feasible to start up most process facilities with every loop on manual. Additional personnel are required during the start-up period. And

Practical Process Control: Tuning and Troubleshooting, by Cecil L. Smith
Copyright © 2009 John Wiley & Sons, Inc.

we want to proceed to the automatic mode as quickly as possible, both to reduce the personnel requirements and to achieve the more consistent process operations provided by automatic controls. But look at the consequences on the role of dynamic simulation:

- In the aerospace industry, dynamic simulation is mandatory. Start-up must be on automatic, which can only be achieved through extensive use of dynamic simulation.
- In the process industries, manual start-ups are technically and economically feasible. Dynamic simulation will be pursued only if it can be economically justified. For example, can a dynamic simulation be used to shave enough days off the start-up to pay for the simulation?

To appreciate these differences, envision launching the space shuttle using the approach to plant start-up as practiced within the process industries:

- Move the vehicle out to the launch pad.
- Fuel it up.
- Strike a match.
- As the vehicle ascends, the instrument technicians are hanging on the side frantically tuning loops.

Obviously this will not work. But the counterpart in process plants does work, and most plant start-ups are accomplished in such a manner. Will the process industries ever use dynamic simulation to the extent it is used in the aerospace industry? The day plants must start up with every loop in automatic.

Given such practices within the industry, how much time should be devoted to process dynamics? The focus herein is on the following:

- Understanding the simple dynamic elements (time constants, integrators, dead times) that appear in processes. Also understanding how these elements are combined into simple models that can approximate the behavior of real processes.
- Understanding the key concepts (stability, overdamped/underdamped, performance measures) pertaining to the behavior of control loops.

As process control is practiced today, these are generally sufficient.

3.1. FIRST-ORDER LAG AND TIME CONSTANT

The first-order lag is a first-order differential equation with the following form:

$$\tau \frac{dY}{dt} + Y = K_1 X_1 + K_2 X_2 + \ldots + K_n X_n$$

where

X_i = Input i
K_i = Gain for input X_i
n = Number of inputs
Y = Output
t = Time, min
τ = Time constant, min

In Section 1.5, the energy balance for the hot water process was written as follows:

$$F H_F + Q c_P (T_i - T_R) - W c_P (T_w - T_R) = M c_P \frac{dT_w}{dt}$$

This equation assumes that the contents of the hot water tank are perfectly mixed, which is not very likely. Mixing will be examined later in this chapter. In the meantime, let us proceed with the assumption of perfect mixing.

In the previous chapter, we computed that 32.8 lb/min of 90 psig steam is required to produce 500 lb/min of hot water at 150 °F from cold water entering at 75 °F. The flows are as follows:

W = 500 lb/min
F = 32.8 lb/min
Q = W − F = 467.2 lb/min

As F is relatively small compared with W, the difference between Q and W is small. Now let us make an approximation:

$$Q \cong W$$

With this approximation, the energy balance simplifies to the following:

$$F H_F + W c_P T_i - W c_P T_w = M c_P \frac{dT_w}{dt}$$

Dividing by W c_P and rearranging gives the following equation:

$$\frac{M}{W} \frac{dT_w}{dt} + T_w = \frac{H_F}{W\, c_P} F + T_i$$

This equation is a first-order lag. The output is the hot water temperature T_w. The time constant τ is the ratio M/W. One input is the steam flow F; another input is the cold water temperature T_i.

Holdup and Throughput. The time constant τ is always the ratio of holdup to throughput. For material balances, the holdup is material and the throughput is material flow. For energy balances, the holdup is energy and the throughput is energy flow.

For the above equation, the time constant is M/W. At first, this appears to be the ratio of a material holdup (mass of water in the tank) to a material flow (flow rate of water through the tank). But in rearranging the equation, the heat capacity was canceled from this coefficient. Had this not been done, the time constant would be $(M\, c_P)/(W\, c_P)$. The numerator $M\, c_P$ is the energy holdup for the tank; the denominator $W\, c_P$ is the energy flow through the tank.

The time constant always has the units of time. For example, M is in lbs and W is in lbs/min, so M/W is lbs/(lbs/min) = min. At a hot water demand of 500 lb/min, the time constant of the hot water process is (2330 lb/min)/(500 lb/min) = 4.7 min.

The value of the time constant determines the speed of response of the process. Consider two cases:

> **Small holdup (M); large throughput (W).** Such a process would respond quickly. The value of the time constant τ is small (a "short" time constant).
>
> **Large holdup (M); small throughput (W).** Such a process would respond slowly. The value of the time constant τ is large (a "long" time constant).

Response to a Step Change in the Steam Flow. The graph in Figure 3.1 is the response of the hot water temperature T_w to a step increase of ΔF in the steam flow. Initially, the process is at equilibrium. The parameters are as follows:

	Initial Value	Final Value	Change
Steam flow	F_0	$F_0 + \Delta F$	ΔF
Hot water temperature	$T_{w,0}$	$T_{w,0} + \Delta T_w$	ΔT_w

The output change ΔT_w is related to the input change ΔF by the gain:

$$\Delta T_w = K_F\, \Delta F$$

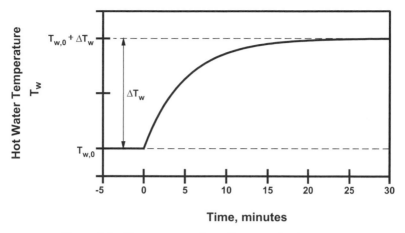

Figure 3.1. Step response for a time constant process.

where $K_F = \dfrac{H_F}{W\,c_P}$ = Sensitivity of hot water temperature to steam flow.

The steady-state sensitivity determines the magnitude of the long-term change in T_w; the time constant τ determines how rapidly the change is implemented. Responses such as in Figure 3.1 are characteristic of a time constant. The most rapid change occurs just after the step increase in F was introduced. The rate of change then decays to zero with time. Such exponentially decaying responses are characteristic of time constant behavior.

Initial Rate of Response. In Figure 3.2, a tangent line has been constructed on the response at time zero. The initial rate of change of the hot water temperature T_w in response to a step change ΔF in steam flow is given by the following expression (obtained by solving the differential equation for the derivative and then evaluating at time zero):

$$\frac{dT_w}{dt} = \frac{(K_F\,F + T_i) - T_w}{\tau} = \frac{K_F\,\Delta F}{\tau} = \frac{\Delta T_w}{\tau} = \frac{\text{Steady-state change in } T_w}{\text{Time constant}}$$

The slope of the tangent line constructed on the step response must be $\Delta T_w/\tau$. Note that the tangent line intersects $T_w = T_{w,0}$ at $t = 0$ and intersects $T_w = T_{w,0} + \Delta T_w$ at $t = \tau$. If the initial rate of change could be sustained, the hot water temperature would attain the final value in one time constant. However, the rate of change decreases as the response approaches the final value.

The tangent line constructed at $t = 10$ in Figure 3.2 shows that this characteristic is true at any point on the response. That is, if the rate of change at time t could be sustained, the final value would be attained at time $t + \tau$.

The 63% Point. As a process described by a time constant approaches the equilibrium state (or final value), the rate of change becomes progressively

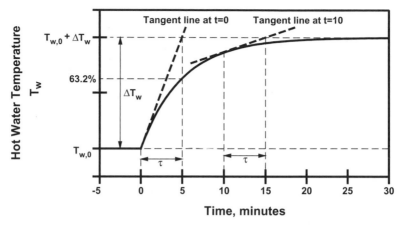

Figure 3.2. Tangent lines and the 63% point.

slower. The rate of change at a point in time is not sustained, but changes in proportion to how far the process is from the equilibrium state.

At any point in time, the response is changing at a rate that would attain the equilibrium value in one time constant. But since the response is becoming progressively slower, the response actually attained in an interval of time equal to one time constant is 63.2% of the response remaining at the beginning of the interval.

The response in Figure 3.2 illustrates this for the time interval starting at $t = 0$. However, this is also true at any point on the response. That is, let ΔX be the response remaining at time t. The response actually attained between time t and time $t + \tau$ is 63.2% of ΔX.

Line-out Time. By line out, we mean that all variables assume constant values. If you want to be theoretically precise, a time constant process never lines out; its rate of change gets progressively slower, but it never reaches zero. But in practice, all measurement devices have limited precision, so the rate of change will eventually become so slow that the output of the measurement device ceases to change. But even before that, we may conclude that the process has essentially attained equilibrium.

The percent of the response attained as a function of time, measured in number of time constants (or essentially in normalized time), is as follows:

Time	Response, %
τ	63.2
2τ	86.5
3τ	95.0
4τ	98.2
5τ	99.3
6τ	99.8

How many time constants are required for the process to line out? Some may consider the process to be lined out when it is within 5% of its equilibrium value. For them, the answer is three time constants. Most engineers would agree that the process is lined out when it is within 1% of its equilibrium state, which requires five time constants. Thus, the lineout time is not a precise number, but it is between three and five time constants.

First-Order Lag. In addition to describing the behavior of some processes, the differential equation for the first-order lag appears frequently in process control:

Measurement device dynamics. A temperature probe is described by a first-order lag. The thermowell provides an additional first-order lag.

Filter. The exponential filter is based on the first-order lag. The time constant of the first-order lag is the filter time constant.

Reset mode implementation. The reset mode of a PI or PID controller can be implemented via a first-order lag, giving the feedback configuration for the PI controller. The time constant of the first-order lag is the reset time.

3.2. INTEGRATING PROCESS

The process in Figure 3.3 is an example of an integrating process. Liquid (reflux) flows from the packing into the tower bottoms. A steam-heated internal coil produces the vapor flow (boilup). As the bottoms flow B is the reflux L less the boilup V, the liquid level H is controlled via the bottoms valve.

A flow controller is provided for the steam. This provides a constant steam flow, which in turn provides essentially a constant boilup.

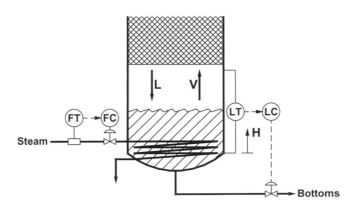

Figure 3.3. Bottoms level.

The major driving force for bottoms fluid flow is the pressure head within the tower. The hydrostatic head has a negligible effect on fluid flow. That is, the change in hydrostatic head caused by a change in bottoms level has no effect on the bottoms flow. If the bottoms valve opening is fixed (level controller on manual), a change in bottoms level does not result in a change in the bottoms flow.

The liquid level must be above the steam coils at all times. Although not illustrated in Figure 3.3, level switches are provided to ensure that this is the case. Above these coils, the cross-sectional area of the tower is A.

Unsteady-State Total Material Balance for Tower Bottoms. The total material balance can be stated as

Flow in – Flow out = Time rate of change of the amount of material in the bottoms of the tower

The terms are as follows:

- The flow in is the reflux flow L.
- The flow out is the boilup V (a constant) plus the bottoms flow B.
- The amount of material in the bottoms of the tower is the density ρ times the volume. The volume of the dished end is V_0; above the dished end, the cross-sectional area A is constant. The level H is measured from the top of the dished end, so the volume of the tower bottoms is $V_0 + A\,H$.

Combining these expressions gives the total material balance for the tower bottoms:

$$L - (V + B) = \frac{d}{dt}\{\rho(V_0 + A\,H)\} = \rho\,A\,\frac{dH}{dt}$$

The expression for the derivative could be simplified because V_0, A, and ρ are constants.

Steady-State Material Balance for the Bottoms Level. At steady state, all variables assume constant values. The derivative dH/dt is zero, so the steady-state material balance is

$$L - [V + B] = 0 \quad \text{or} \quad L = V + B$$

The following points apply to the steady-state material balance:

- The bottoms level H does not appear at all in the steady-state material balance.

- The bottoms level H does not affect any term (L, V, or B) in the steady-state material balance.

The solution to the steady-state material balance does not yield a value for the bottoms level H. Indeed, the process can be at equilibrium for any value of H. That is, the process can be at equilibrium with the level just above the steam coils. For the same flows in and out, the process can be at equilibrium with the level near the upper limit of its acceptable range. Equilibrium means that the flows in and out must sum to zero; however, this can occur at any value of the bottoms level. Such statements apply to all integrating processes.

Integrated Form of the Material Balance. Note the following:

- The reflux flow L is not a function of the level H.
- For this process, the hydrostatic head is not a significant component of the driving force for fluid flow in the bottoms stream. Therefore, the bottoms flow B is not a function of H when the level controller is on manual.

This permits the unsteady-state material balance to be integrated with respect to time, giving the integrated form of the material balance equation:

$$H = \frac{1}{\rho \, A} \int [L - (V + B)] dt$$

The level H is the integral of $L - (V + B)$, which is the imbalance in the steady-state material balance. The process is basically described by an integrator, and hence, the term "integrating process" is most descriptive of such processes. Any imbalance in the total material balance results in a change in bottoms level.

Response to a Step Decrease in Bottoms Flow. Suppose the bottoms level H is initially at an equilibrium state, with the level being H_0. The initial reflux flow is L_0. The initial bottoms flow B_0 must satisfy the steady-state material balance $B_0 = L_0 - V$.

With the bottoms level controller on manual, let us introduce a step decrease of ΔB in the bottoms flow. As the input terms in the material balance now exceed the output terms, the level will increase. As the response in Figure 3.4 indicates, it continues to increase. Neither L nor V nor B is affected by the bottoms level H. The imbalance in the material balance is constant, and the integral of a constant is a ramp.

In practice, the ramp continues until some limiting condition is attained. This makes level switches essential. For the packed tower, the presence of

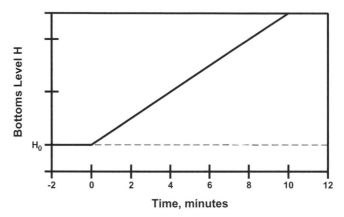

Figure 3.4. Response to a step decrease in B.

liquid in the packed section must be avoided, so a high-level switch is normally provided to shut off the steam on high level. The steam coils must be completely submerged, so a low-level switch is normally provided to shut off the steam on low level.

The response behavior illustrated in Figure 3.4 is the basis for such processes being referred to as ramp processes. This term is used interchangeably with integrating processes.

3.3. SELF-REGULATED VERSUS NON-SELF-REGULATED PROCESSES

We have examined the dynamic behavior of a time constant process and an integrating process. Now let us consider the consequences of the differences in their behavior.

Non-Self-Regulated Process. A non-self-regulated process will not attain the equilibrium state on its own; appropriate control actions are required. The bottoms level process behaves in this manner. The terms "integrating processes," "ramp processes," and "non-self-regulated processes" are equivalent.

For the bottoms level process in Figure 3.3, let us assume that we stay within all design limitations of the process. Let the bottoms flow B and the reflux flow L be set to arbitrary values. The direction of the change in bottoms level depends on the closure of the steady-state material balance. If $L - (V + B)$ is positive, the bottoms level H will increase. If $L - (V + B)$ is negative, the bottoms level will decrease.

Furthermore, note that neither L nor V nor B is affected by the bottoms level. That is, if the level is currently increasing (or decreasing), the level continues to increase (or decrease) until something causes L, V, or B to change.

This "something" is either the process operator taking manual corrective actions or a level controller that takes automatic corrective actions.

Self-Regulated Process. A self-regulated process will attain the equilibrium state on its own and without the need for control actions of any form. The hot water process is a self-regulated process. The hot water flow W, the steam flow F, and the cold water temperature T_i can be set to any values within reason, and the hot water process will seek an equilibrium on its own. By "within reason," we mean that no design limitations are exceeded (such as a hot water temperature in excess of the boiling point of water at the pressure within the tank). All time constant processes are self-regulated processes.

The commonly used PID controller can be applied to both self-regulated and to non-self-regulated processes. However, control engineers seem to pay more attention to self-regulated processes than non-self-regulated processes. Self-regulated processes include the following:

- Flow.
- Temperature (one example is the hot water process).
- Composition.
- Physical properties (density, viscosity, etc.).

In most production facilities, control of these variables is crucial, and thus, they get more attention. Non-self-regulated processes are generally level (except for gravity flow level processes) or occasionally pressure.

Process Operators. For the bottoms level process, the following control issues are of importance to the process operators:

- Control action of some type is required to maintain the bottoms level within its acceptable range. Without an automatic level controller, this task falls on the operators.
- To control manually, the operators must regularly check the bottoms level. Small vessels require frequent checks. Larger vessels require less frequent checks, but the level can never be completely ignored.
- On any lapse of attention, imbalances in the material balance will cause the level to change in some direction. Being an integrating process, the change continues.
- There is no conservative mode of operation in which adverse consequences are very unlikely. High levels lead to adverse consequences, but so do low levels.

Plant Management. For the bottoms level process, the following control issues are of importance to plant management:

- What effect does the bottoms level have on economics? As long as the level is within its acceptable range, the answer is simple: "None."
- There is no economic advantage to operating with a high level as opposed to a low level, or vice versa.
- Economic penalties occur only on the extreme level excursions that precipitate shutdowns.

Assuming shutdowns are not occurring with manual control, there are no economic incentives to automate.

Justification of an Automatic Controller for the Bottoms Level. Let us assume that the process operators are manually controlling the bottoms level. Modest variations in level are occurring, but not a major excursion that would cause a shutdown on high or low level. A proposal to install an automatic level controller is being considered.

The need for automatic control of bottoms level is viewed very differently by two interested parties:

   Process operators. The automatic controller would definitely make their job easier.
   Plant management. There is no economic benefit to be derived from improved bottoms level control.

Loops such as the bottoms level can rarely be economically justified on an individual loop basis. Their justification has to be made on a larger scale. The current number of operations personnel is predicated on a certain degree of automation within the facility. Loops such as bottoms level fall within this expected degree of automation. In most plants, one level controller more or less would not impact staffing requirements, but if you remove all level controllers, additional personnel would be required to operate the facility.

Justification of an Automatic Controller for the Hot Water Temperature. Assume that the plant is currently operating satisfactorily with the steam valve being manually positioned by the process operators. If changes in the hot water demand are small and infrequent, adjustments in the steam valve position are also infrequent, making manual control feasible. Low hot water temperatures must be avoided, but high hot water temperatures are unlikely to have adverse consequences. The operators quickly learn to operate with the hot water temperature on the high side. In this way, upsets are less likely to result in low hot water temperatures. Of course, there is a price for this—a higher consumption of steam than necessary.

A proposal to install an automatic temperature controller is being considered. The need for automatic control of temperature is viewed very differently by two interested parties:

Process operators. The temperature controller does not make their job significantly easier. They just make sure the steam flow is on the high side. However, if changes in hot water demand become larger and/or more frequent, their views might change.

Plant management. The economic benefit is to maintain the hot water temperature close to the design value and not waste steam.

The plant manager wants to control the hot water temperature so that the steam utilization can be minimized. This minimization leads to more efficient plant operations, but it makes the operators' job more difficult. Upsets that previously caused no problems now have the potential to result in low hot water temperatures. The operators must now be alert for such situations.

Process Control and Plant Optimization. The situation described for the hot water process is common. In the manual mode of operation, a chemical, energy, or some other resource is being intentionally consumed in excess. When operated in such a conservative manner, the effect of most upsets on the product is nominal; the plant runs very smoothly. Of course, this comes at a price. Something is being consumed in excess of what is needed. This is reflected on the cost side of the profit-and-loss statement. Optimizing the process entails identifying what is being used in excess and stopping this practice. Backing out the excess has an adverse effect on how smoothly the process operates, thus, the need for automatic control.

3.4. DEAD TIME

The process in Figure 3.5 is an example of a dead-time process. Material is being metered from a feed hopper through a rotary valve onto a belt conveyor. The material is transported by the belt conveyor at velocity V ft/min for a distance L ft.

Material is metered onto the belt conveyor at rate W_1. Material falls off the other end of the conveyor at rate W_2. At equilibrium, W_1 and W_2 must be equal. We want to examine their relationship under transient conditions.

Transportation Time or Dead Time. The time required to transport material from one end of the conveyor to the other is the length L of the conveyor divided by the speed V of the conveyor. Although transportation time is a more descriptive term, control engineers usually call this time the dead time and designate it by θ. When dead time is the result of the transport of material, its value is

$$\theta = \frac{L}{V}$$

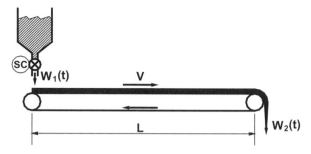

Figure 3.5. Belt conveyor.

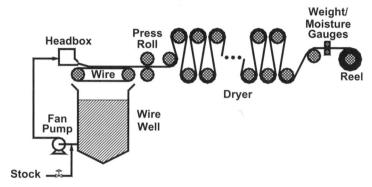

Figure 3.6. Paper machine.

The relationship between W_1 and W_2 is simple. The current value for W_2 is the value of W_1 at one dead time in the past. Mathematically, this is expressed as follows:

$$W_2(t) = W_1(t - \theta)$$

Sheet Processing Industries. From a process perspective, the paper machine illustrated in Figure 3.6 is relatively simple. Fresh stock consisting of about 2% fiber is fed into a large recirculation loop, where it is diluted to about 0.25% fiber before entering the headbox. The headbox uniformly distributes this large flow over the moving wire, which is no simple task (most paper machines are over 20 feet in width).

The dilute slurry jets out of the headbox lip at approximately the same speed as the moving wire. Most water drains through the wire into the wire well, from which it is recirculated. Most fiber is retained to form the sheet. Some water is removed by pressing the wet sheet; more is removed by the steam-heated rotating cylinders that comprise the dryer section of the paper machine.

The dynamics of the paper machine consist largely of dead time. The length of the paper path through the machine is known. The speed of the machine is known. The dead time can be computed as the length divided by the speed.

Paper machines, plastic film lines, carpet lines, and so on, are referred to as the sheet processing industries. In these industries, the dead time dominates. In the paper machine, the dry sheet weight (or fiber per unit area of the sheet) at the press rolls will be the dry sheet weight at the reel one dead time later. The dead time will be a minute or so; the major time constant rarely exceeds 10 seconds.

Dead Time From Fluid Flow in Pipes. Belt conveyors and pipes have one thing in common: Both transport material from point A to point B. Belt conveyors are used for solids; pipes are used for fluids.

Being a transport system, pipes contribute dead time to the process behavior. As compared with a belt conveyor, there is usually a little more mixing within a pipe, but this is small relative to the dead time. However, pipes contribute dead time to some variables but not to all. Specifically, dead time is contributed to temperature and composition, but not to flow. Variables are divided into two classes:

Intensive. An intensive property does not depend on how much material is present. Intensive properties include temperature, composition, and physical properties such as density, viscosity, and so on. Dead time applies to intensive properties.

Extensive. An extensive property depends on how much material is present. Extensive properties include flow. Dead time does not apply to extensive properties.

Step Response of a Dead Time Process. Figure 3.7 presents the response of a dead time process to a step change in its input. There is no change in the output for one dead time (which is 1.0 minute for Figure 3.7). Then the output immediately changes to the final value. Basically, the step change in the input is delayed by one dead time.

The dead time response shares one aspect in common with the step response of a time constant process. The dead time process attains 63% of its final value in time equal to the dead time. Figure 3.7 also presents the response for a time constant process with a time constant of 1.0 min. Both responses attain 63% of their final value in one minute.

But from the perspective of control, a dead time process presents more difficulties. When a change occurs in the input, a time constant process begins to respond immediately, giving the control system some indication that something is occurring. But when a change occurs in the input to a dead time process, nothing occurs in the response until one dead time has elapsed.

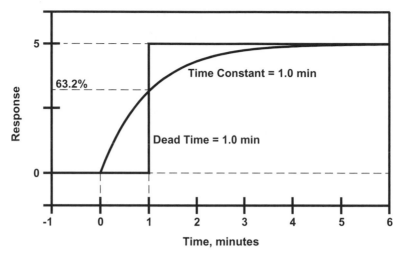

Figure 3.7. Step responses of a time constant process and of a dead time process.

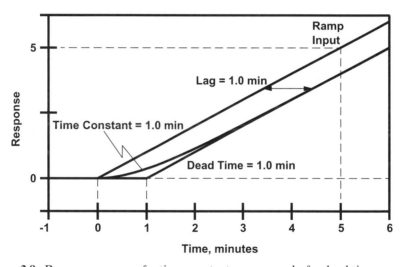

Figure 3.8. Ramp responses of a time constant process and of a dead time process.

First-Order Lag versus Transportation Lag. In process control, the term "lag" is sometimes used with no qualifier. Does this mean "first-order lag," does it mean "transportation lag," or does it matter?

Figure 3.8 presents the ramp response of two processes:

1. A time constant process with a time constant of 1.0 min.
2. A dead time process with a dead time of 1.0 min.

The difference is only in the initial transient response, and it is not very dramatic. Both responses are eventually ramps that are delayed by one minute with respect to the input ramp. When one merely observes lag in a process, it is often not apparent whether the lag is a first-order lag, a transportation lag, or some of each. Consequently, one tends to just call it "lag."

From a control perspective, it does matter. A transportation lag or dead time always degrades the performance of PID controls. PID can cope with time constants and integrators far better than with dead time. From the perspective of PID control, dead time is the "difficult element to control."

3.5. MEASUREMENT ISSUES

In a previous discussion on fluid flow in pipes, it was noted that dead time develops for variables such as temperature and composition (intrinsic properties), but not for flow (an extensive property). We need to examine the implications for the location of measurement devices.

Introduction of Dead time via Location of the Measurement Device. One way to introduce dead time into temperature loops and composition loops is by locating the measurement some distance downstream of the process unit. Figure 3.9 illustrates this for the hot water process. Instead of locating the temperature transmitter at the hot water tank, it has been moved downstream by some distance. The amount of dead time depends on this distance and on the hot water demand:

$$\theta = \frac{D}{V} = \frac{D}{W/(\rho\,A)}$$

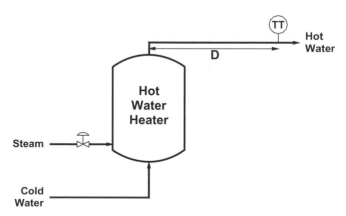

Figure 3.9. Temperature measurement device located downstream.

where

A = Cross-sectional area of the pipe, ft^2
D = Distance downstream, D
V = Fluid velocity, ft/min
W = Hot water demand, lb/min
ρ = Fluid density, lb/ft^3
θ = Dead time, min

At a low hot water demand, the dead time from the sensor location is larger than at a high hot water demand. This will further complicate controller tuning.

As dead time degrades the performance of the control loop, we should avoid this practice. However, several considerations go into sensor location, and sometimes we have no alternative.

What if we install a hot water flow measurement at a distance D from the tank (same location as the temperature transmitter in Figure 3.9)? For incompressible fluids, no dead time is introduced by locating the flow measurement downstream. For a compressible fluid, a small time constant and an even smaller dead time is introduced. These are generally negligible except when the distances are extremely large, such as in pipelines.

Analyzer Sample System. Analyzers such as chromatographs are normally installed at a location some distance from where the sample is withdrawn from the process. To minimize the dead time associated with the sample system, the dual loop arrangement illustrated in Figure 3.10 can be used.

Fast loop. This is a pump-around loop in which a substantial volume of fluid can be recirculated. By keeping the velocity as high as practical in

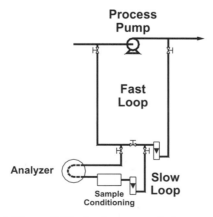

Figure 3.10. Analyzer sample loops.

the fast loop, the time required to transport the material from the sample point to the analyzer location is minimized. However, since a large quantity of material is recirculated, no sample conditioning is provided.

Slow loop. The flow through this loop is small. The sample is conditioned as required and supplied to the analyzer. Since the flow rate is very low, the distances must be kept to a minimum so as to not insert a significant amount of dead time.

There are potentially three sources of dead time in an analyzer system. The total dead time is the sum of the following three components:

Sample transport time θ_T. Provided the sample system is equipped with a flow indicator, the transport time can be computed from the flow rate, the pipe size, and the distance.

Analysis time θ_A. For some analyzers, this is practically zero. For analyzers such as the chromatograph, it could be as short as 15 seconds or as long as 15 minutes.

Effect of sampling θ_E. Analyzers such as chromatographs accept a sample on a time interval called the sampling time. The effect of sampling on control performance is approximately equivalent to adding a dead time equal to half the sampling time to a continuous system.

3.6. EFFECT OF DEAD TIME ON LOOP PERFORMANCE

It has already been observed that dead time has a very detrimental effect on the performance of a PID controller. We need to examine this in more detail.

Analogy to Driving. Let us develop this analogy by locating the hot water temperature measurement device at some distance from the tank, as illustrated in Figure 3.9. As the mixing of steam and cold water to produce hot water occurs in the tank, the temperature of the water in the tank is the true indicator of the state of the process. When the hot water temperature transmitter is located downstream, the temperature controller does not know the current value of the temperature of the water in the tank. Instead, the controller only knows what the temperature was at some time in the past. In essence, the controller is taking actions based on old information, which is never a good idea.

Suppose you try to drive an automobile in this fashion. The equivalent would be to drive down the road looking out the back window. You do not know where you are; you only know where you have been. Your driving performance is seriously impaired. Furthermore, if you insist on driving this way, take some advice: slow down!

Actually, we have to give this same advice to the temperature controller for the hot water process. We do this by reducing the controller gain.

Effect of Dead Time on the Tuning Coefficients. When dead time is present in a loop, the PID controller tends to overcorrect. Suppose locating the sensor downstream results in a dead time of one minute When an error appears in the loop, the controller takes an action. However, the results of this action will not be observed for at least one minute But during this period of time, the PID controller continues to take control actions.

In dead time situations, the only way to prevent the PID controller from overcorrecting is to tune the controller so that it responds more slowly. This is done by reducing the controller gain. Except when the dead time is very small, the controller gain is basically inversely proportional to the dead time. If the dead time doubles, the controller gain must be reduced by approximately a factor of two.

When the dead time is large, the gain must be reduced substantially, resulting in a very slow loop. The usual complaint in large dead time processes is that the controller is so slow that it is useless.

Effect of Dead Time on Overshoot and Oscillations. As the dead time increases, more overshoot is generally observed, and if the controller gain is sufficiently high, oscillations will become pronounced. This is illustrated by the three responses in Figure 3.11. All are for an increase in hot water demand from 500 lb/min to 600 lb/min, and the same tuning coefficients are used in all (the loop is tuned to behave in the desired manner when there is no dead time). The following observations apply to the responses in Figure 3.11:

- Transmitter located at the tank (D = 0). No dead time is inserted because of the transmitter location. The controller is tuned to give a very small

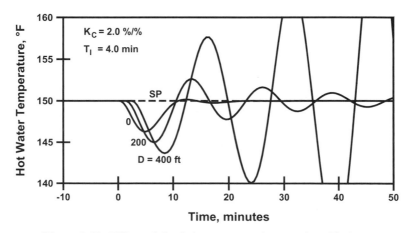

Figure 3.11. Effect of dead time on overshoot and oscillations.

overshoot on its return to the set point. The same tuning coefficients are used for the other two cases.

- Transmitter located 200 ft from the tank (D = 200). The overshoot on return to the set point is significant, and so is the degree of oscillations.
- Transmitter located 400 ft from the tank (D = 400). The amplitude of the oscillations increase with each cycle. Loops that exhibit such behavior are said to be unstable.

Later in this chapter, we introduce the concept of stability margin or degree of stability. Clearly dead time has a negative effect on stability margin.

Effect of Dead Time on the Response Speed. With the tuning parameters held at constant values, increasing the dead time leads to more overshoot and oscillations. To retain the same overshoot (as a percentage) and degree of oscillations, the controller gain must be adjusted.

All responses in Figure 3.12 have approximately the same percentage overshoot. For D = 0, the controller was tuned to give a modest overshoot on returning to the set point. The tuning coefficients for the three responses in Figure 3.12 are as follows:

Distance D	0 ft	200 ft	400 ft
Controller gain K_C	2.0%/%	1.0%/%	0.5%/%
Reset time T_I	4.0 min	4.0 min	4.0 min
Maximum error E_{MAX}	3.7 °F	6.2 °F	8.2 °F
Time at E_{MAX}	5.0 min	7.8 min	10.9 min

Reducing the controller gain retains the same percentage overshoot, but at the expense of a larger maximum departure E_{MAX} from the set point and a

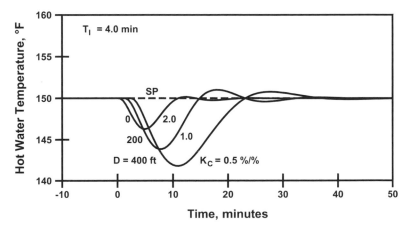

Figure 3.12. Effect of dead time on speed of response.

slower response, as evidenced by the time at which E_{MAX} occurs. This is exactly what happens in practice. The larger the dead time, the lower the controller gain, and the slower the response.

3.7. MIXING

The energy balance for the hot water process was developed under the assumption of the tank contents being perfectly mixed. We also noted that this was unlikely. We could have developed an energy balance under an assumption known as "plug flow," which is the complete absence of mixing. This is also unlikely. The true state is somewhere between these two extremes. Using Figure 3.13 to illustrate the possibilities for mixing, we examine the issues pertaining to imperfect mixing and their consequences on process control.

Residence Time. The residence time for a vessel is the time required to displace the contents of the vessel. Another interpretation is that if the vessel is initially empty, it will completely fill in one residence time.

Let the volume of the vessel be V and the flow through the vessel be F. The residence time is always the volume V divided by the flow F. The residence time does not in any way depend on the nature of the mixing within the vessel.

Perfectly Mixed Vessel. A vessel is perfectly mixed if its contents are uniform in every respect. That is, no composition, temperature, or other variations exist within the vessel. The composition and temperature at one point within the vessel is the same as at all other points within the vessel. The

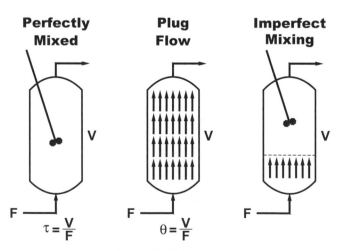

Figure 3.13. Mixing.

composition and temperature of the exit stream are the same as the composition and temperature within the vessel.

The dynamic behavior of a perfectly mixed vessel is described by a time constant whose value is V/F. The time constant is the same as the residence time.

In practice, perfect mixing is never achieved. It is approached in agitated vessels where the energy-to-volume ratio is very high and the fluid in the tank has a low viscosity.

Plug Flow. Plug flow is the complete absence of mixing. The molecules of the fluid leave in exactly the same order that they entered. The term "first-in, first-out" describes such behavior.

The dynamic behavior of plug flow is described by a dead time whose value is V/F. The dead time is the same as the residence time.

The best industrial examples of plug flow are paper machines, plastic film lines, and other sheet-processing operations. For fluid processing, there is always some mixing. However, the flow patterns in tubular reactors approach plug flow, especially if the flow is laminar.

Imperfect Mixing. Industrial processes are neither perfectly mixed nor plug flow; they are somewhere between these two extremes. Describing imperfect mixing is a challenge. To characterize the nature of mixing in a vessel accurately, experimental tests such as tracer studies are required. However, such detailed studies can rarely be justified for control system analysis.

A simpler (but far less accurate) approach is to assume that some fraction of the vessel is plug flow and the remainder is perfectly mixed. The value of the fraction is determined from experimental response data. The temperature or composition of the input stream is varied, and the response of the outlet stream is recorded. The value of the dead time is adjusted until the model behavior matches the experimental response. Another limitation of this approach is that the plug flow fraction is likely to change with the flow F. One would suspect that the plug flow fraction would be smaller at high flows than at low flows.

3.8. PROCESS MODELS

Real processes exhibit complex dynamics that require combinations of the simple elements to approximate their behavior. The term "process model" generally refers to the combination of simple elements used in the approximation. Most process models are constructed as follows:

Time constants. There may be from zero to any number of time constants. Rarely do we identify all of the time constants present in the real process. For an integrating process, we may choose not to include any time constants. For a nonintegrating process, we almost always provide one time constant, but rarely will we go beyond two.

Integrator. Double integrators (and even more) are possible but not in industrial processes. Consequently, the process model is either integrating or it is not.

Dead time. A process may have any number of dead times or transportation lags, but they are additive. That is, a dead time of 1.0 min and a dead time of 0.5 min combine to form a dead time of 1.5 min. Therefore, only one dead time is required. Almost all industrial processes exhibit some dead time. But in addition to the true dead time or transportation lag, we often use the dead time to approximate the contributions of the small time constants within the process.

We now examine some combinations of the simple elements.

Two Time Constants. This model consists of two first-order lags in series, giving a second-order lag. There are three coefficients:

- Process gain K, %/%. The second-order lag is a self-regulated process.
- Major time constant τ_1, min. This is the larger of the two time constants.
- Minor time constant τ_2, min. This is the smaller of the two time constants.

A third-order lag consists of three time constants in series. A fourth-order lag is four time constants in series. Whatever order is desired can be attained by placing the required number of time constants in series. But for process models, rarely are more than two required.

Figure 3.14 presents the response of a second-order lag to a unit step increase in the manipulated variable. Note the following:

1. Since the input is a unit step change, the final value of C is K.
2. The minor time constant τ_2 must be in the range $0 \le \tau_2 \le \tau_1$. We present responses for $\tau_2 = 0$, $\tau_2 = 0.5\tau_1$, and $\tau_2 = \tau_1$.
3. None of the responses in Figure 3.14 exhibit overshoot or oscillations. Mathematically, it is possible for the time constants to be complex conjugates. The resulting response would have a sinusoidal component, and it would exhibit overshoot and oscillations. However, such behavior is unusual in process control applications.
4. For a first-order lag, the 63% point is at time equal to the time constant. For a second-order lag, the 63% point is at time approximately equal to the sum of the time constants. For $\tau_2 = 0.5\tau_1$, the sum of the time constants is $1.5\tau_1$ and the 63% point is at $1.59\tau_1$. For $\tau_2 = \tau_1$, the sum of the time

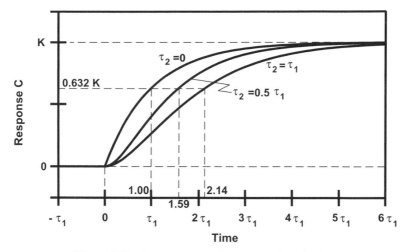

Figure 3.14. Step response for a second-order lag.

constants is $2.0\tau_1$ and the 63% point is at $2.14\tau_1$. Given the realistic precision of process models, these differences are relatively small.

Integrator Plus Dead Time. An integrator-plus-dead-time model has two coefficients:

Process gain K, (%/%)/min or (%/min)/%.
Dead time θ, min.

The main advantage of this model is its simplicity. As will be discussed in the chapter on tuning methods, this model is the basis for both tuning techniques and automatic tuning methods.

Figure 3.15 presents the response of an integrator plus dead time to a unit step increase in the manipulated variable. Note the following:

• There is no change in C until the dead time has elapsed.
• At time equal to the dead time, a ramp immediately appears in C. The rate of change for this ramp is the process gain K.

Clearly this type of model would be required to describe the behavior of an integrating process. One would think that this type of model would not be applicable to a nonintegrating process. But as will be discussed in the chapter on tuning methods, the Ziegler–Nichols tuning equations are based on just such a model, and they have been applied with at least some degree of success to nonintegrating processes.

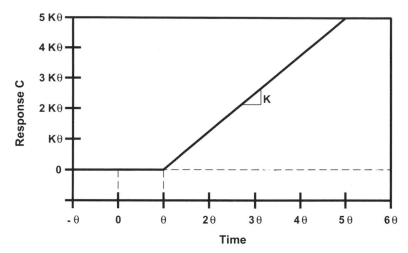

Figure 3.15. Step response for an integrator plus dead time model.

Time-Constant-Plus-Dead Time. This model consists of a first-order lag and a dead time. There are three coefficients:

- Process gain K, %/%. The model is self-regulated.
- Process time constant τ, min.
- Process dead time θ, min.

The time-constant-plus-dead-time model is the simplest practical model that exhibits nonintegrating behavior. Virtually all processes contain some dead time. Furthermore, dead time is usually an acceptable approximation for the small time constants present in most processes. Consequently, dead time is a necessary component of every process model.

Figure 3.16 presents the response C of a time-constant-plus-dead-time model to a unit step increase in the manipulated variable. Note the following:

1. Since the input is a unit step change, the final value of C is K.
2. There is no change in C until the dead time has elapsed. Thereafter, C changes in the characteristic manner of a first-order lag.
3. At t = θ, there is an abrupt change in the rate of change of C. Most processes do not change so abruptly.
4. The 63% point on the response occurs at t = θ + τ.

In the chapter on tuning methods, we will present some simple procedures for obtaining values of the model parameters from process test data. Experience

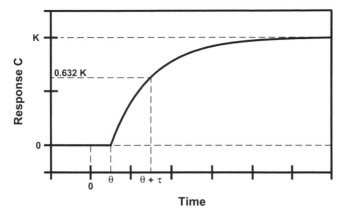

Figure 3.16. Step response for a time-constant-plus-dead-time model.

with this model suggests that it provides an adequate basis for tuning a PI controller. But for PID, questions have been raised about the quality and consistency of the results.

Two Time Constants Plus Dead Time. This model consists of a second-order lag and a dead time. There are four coefficients:

- Process gain K, %/%. The model is self-regulated.
- Major process time constant τ_1, min. This is the larger of the two time constants.
- Minor process time constant τ_2, min. This is the smaller of the two time constants.
- Process dead time θ, min.

The problem with this model is obtaining the model parameters from process test data. Basically, nonlinear regression methods must be applied to the process response data. Packages to do this are available, but the learning curve is steep.

Figure 3.17 presents the response C of a two-time-constants-plus-dead-time model to a unit step increase in the manipulated variable. Note the following:

1. Since the input is a unit step change, the final value of C is K.
2. There is no change in C until the dead time has elapsed. Thereafter, C changes in the characteristic manner of a second-order lag.
3. At $t = \theta$, there is no abrupt change in the rate of change of C. This is more typical of industrial processes.
4. The 63% point on the response occurs at approximately $t = \theta + \tau_1 + \tau_2$.

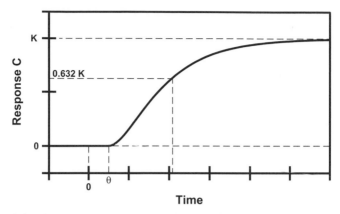

Figure 3.17. Step response for a two-time-constants-plus-dead-time model.

The two-time-constants-plus-dead-time model is capable of adequately representing the dynamic behavior of most industrial processes. Tuning techniques based on a two-time-constants-plus-dead-time model are capable of consistently and effectively tuning a PID controller. However, determining values for the coefficients in this model from process test data is not a trivial problem.

3.9. APPROXIMATING TIME CONSTANTS

Developing a low-order model for a process invariably involves approximations. In some cases, we approximate a dead time by using multiple small time constants. Conversely, we can approximate the contribution of one or more small time constants with a dead time.

Approximating a Dead Time with Multiple Time Constants. The responses in Figure 3.18 illustrate approximating a dead time using multiple time constants. For example, a dead time of 1.0 min could be approximated by 100 time constants of 0.01 min each.

As the number of time constants increases, the approximation becomes better and better. One extreme is using one time constant (n = 1) to approximate the dead time. This is not a very good approximation. Although increasing the value of n improves the approximation, it also increases the complexity of the model that uses the approximation.

We rarely approximate a dead time with a large number of time constants; however, this approximation goes both ways. That is, several small time constants in series can be approximated by a dead time.

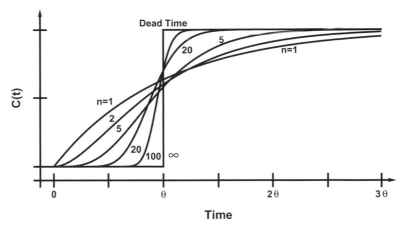

Figure 3.18. Approximating a dead time using time constants.

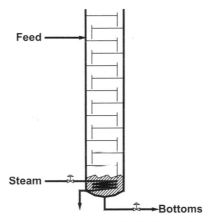

Figure 3.19. Trays in a distillation column.

Apparent Dead Time. True dead time is the result of a transportation delay within the process. But when interpreting process behavior from response data, several time constants in series can give the appearance of dead time.

One such situation involves trays in a distillation column, such as the one in Figure 3.19. The feed is all-liquid, so increasing the feed rate will increase the liquid rate from the bottom tray. But when the feed rate to the column is changed, some time will elapse before the feed rate change is reflected in the bottoms level. When the liquid flow to a tray increases, the amount of liquid retained on the tray (the tray holdup) increases slightly, which results in a lag (called the hydraulic lag) in the liquid flow leaving the tray. The hydraulic time constant for each tray is usually between 5 and 10 seconds. A column with 24 trays below the feed tray will have a total lag between 120 and 240 seconds.

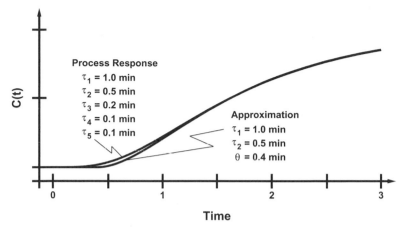

Figure 3.20. Approximating small time constants with a dead time.

The response in the bottoms level will give the appearance of a dead time in the range of 2 to 4 minutes. This dead time is not true transportation lag, but it is the cumulative effect of the small hydraulic time constants associated with the trays.

Approximating High-Order Time Constants with a Dead Time. Although approximating a dead time with multiple time constants is unusual, approximating several small time constants by a dead time is common. The two responses in Figure 3.20 illustrate the use of a dead time to approximate several small time constants.

The original process has five time constants. In the approximation, the three smallest time constants have been replaced with a dead time whose value is the sum of the three smallest time constants. The approximation is very close to the response of the original process.

When developing process models, it is customary to determine the largest (or major) process time constant and possibly the next largest (minor) process time constant. A dead time is then added to represent the remaining time constants plus the true transportation lag, if any, in the process.

3.10. ULTIMATE GAIN AND ULTIMATE PERIOD

In the previous chapter, we introduced the loop gain $K\,K_C$ as the product of the process gain K and the controller gain K_C. The maximum acceptable value of the loop gain depends on the dynamics of the process, especially the value of the dead time. In this section, we want to first define stability and then introduce the ultimate gain.

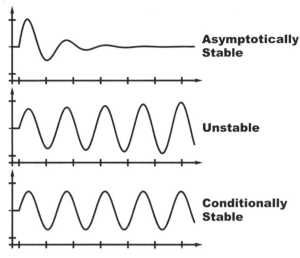

Figure 3.21. Concepts of stability.

Asymptotically Stable. The upper response in Figure 3.21 is from a system that is asymptotically stable. A system is asymptotically stable if its response C eventually lines out at a finite value. This is expressed by the following limit:

$$\lim_{t \to \infty} C = C_\infty, \text{ where } C_\infty \text{ is a finite value}$$

In loops with any version of the PID controller, we require that the loop be asymptotically stable. However, the asymptotically stable definition does not impose any limits on

- The number of cycles in the response before it lines out.
- The time required for the system to line out.

To obtain "reasonable" values for both of these, requirements beyond being merely asymptotically stable must be imposed.

Unstable. The middle response in Figure 3.21 is from a system that is unstable. A system is unstable if its response C is not bounded. In process control, the response of an unstable loop is a cycle whose amplitude increases with each cycle. Clearly unstable systems are not acceptable.

In practice, the cycle does not grow indefinitely. Instead, the cycle grows until the final control element is driven to its limits. The controller then alternates between driving the valve fully closed and driving the valve fully open. If this is allowed to continue, the cycle in C will attain a constant amplitude.

Such cycles are called limit cycles and are caused by a limiting condition being imposed somewhere within the loop.

Conditionally Stable. The lower response in Figure 3.21 is from a system that is conditionally stable. The response C is bounded. This is expressed as follows:

$$\text{As } t \to \infty, C_{min} \le C \le C_{max}, \text{ where } C_{min} \text{ and } C_{max} \text{ are finite values.}$$

In practice, almost all bounded cycles are limit cycles. In noncritical loops, we often use ON–OFF controllers, which take control action by switching their output between two states, say "ON" and "OFF." This leads to a limit cycle in the controlled variable. In noncritical loops, this behavior is acceptable provided the amplitude of the cycle is not excessive.

The thermostat in a residential heating and cooling system is an ON–OFF controller. It switches the air conditioning unit or furnace between ON and OFF, depending on the room temperature. The resulting small limit cycle is usually acceptable.

Relationship of Stability to Controller Gain. Suppose a control loop is tuned so that the loop responds very slowly, making the transition from the original value to the new value with no overshoot or oscillations. What happens if we increase the controller gain?

The responses in Figure 3.22 are for a short-duration pulse or "bump" in the set point to the loop. For starting with a low controller gain and then increasing, the expected results are as follows:

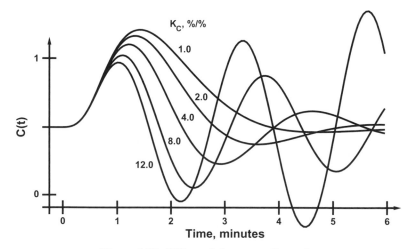

Figure 3.22. Effect of the controller gain.

1. Overshoot and oscillations may appear, but the loop eventually lines out and is asymptotically stable.
2. Continuous cycling is observed in the controlled variable. The loop is conditionally stable.
3. The cycle in the controlled variable grows. The loop is now unstable.

When the observed behavior is other than the above, the loop should be examined very carefully to understand why. Loop interaction is one cause, and if so, this issue has to be addressed.

The Ultimate Gain and Ultimate Period. The ultimate gain K_U is the largest value of the controller gain for which the loop is stable (or not unstable). If the controller gain is set to the ultimate gain, the controlled variable will cycle continuously. The ultimate period P_U is the period of the continuous cycle.

As used in the Ziegler–Nichols tuning technique, the ultimate gain is determined with only the proportional mode in the controller. However, the concept of the ultimate gain applies even if the integral and/or derivative modes are present. For the current integral and derivative settings, the controller gain can be increased until the loop cycles continuously. This is the ultimate gain for this loop with the current integral and derivative settings.

Obtaining a Value for the Ultimate Gain. One approach to obtain numerical values for the ultimate gain K_U and the ultimate period P_U is to conduct a test on the process. The test is conducted with the controller in automatic. At the start of the test, the loop is excited by "bumping" the set point. The controller gain is increased until the loop cycles continuously (but without driving the final control element to its limit). The value of the controller gain that leads to continuous cycling is the ultimate gain K_U. The ultimate period P_U is the period of the continuous cycle. The Ziegler–Nichols tuning technique takes this approach.

Conducting such a test directly on the process usually has serious consequences on process operations, especially if the loop is temperature or composition. Consequently, the direct testing approach is practical only for fast loops.

Some automatic tuners use an alternative test to determine the ultimate gain. The chapter on tuning methods will present such an approach.

Gain Margin. You will sometimes hear questions such as "how stable is this loop?" Technically, stability is a yes/no situation; the loop is either stable or unstable. The question is really inquiring about the proximity to conditions where the loop is unstable. This raises a notion of degree of stability or stability margin. One way to quantify this would be to ask the question, "how much can the controller gain be increased before the loop becomes unstable?"

This is quantified as a parameter called the gain margin, which is the factor by which the controller gain must be multiplied to obtain a conditionally stable loop. This factor is the ultimate gain divided by the current controller gain.

When the gain margin is expressed as a ratio of gains, the following statements apply:

- An asymptotically stable system has a gain margin that is greater than one.
- A conditionally stable system has a gain margin of one.
- An unstable system has a gain margin that is less than one.

As usually tuned, loops have a gain margin of at least 2, and sometimes considerably more.

3.11. DAMPING

The terms "overdamped," "underdamped," and "critically damped" are commonly used to characterize a response.

Overdamped. A system is said to be overdamped when its response approaches the final value without ever crossing to the other side of the final value.

Figure 3.23 illustrates two responses that are overdamped. The upper one is initially below the final value, and it approaches the final value from below without ever going above the final value. The lower one is initially above the final value, and it approaches the final value from above without ever going below the final value.

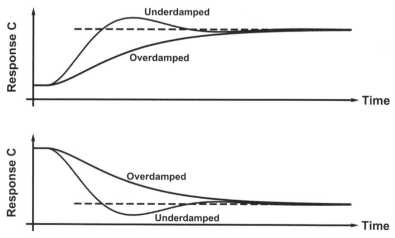

Figure 3.23. Overdamped and underdamped responses.

Underdamped. A system is said to be underdamped when its response crosses the final value one or more times as it approaches the final value. Figure 3.23 also illustrates two responses that are underdamped.

The presence of one or more sinusoidal components in the response leads to underdamped behavior. When the sinusoidal component is not present, the response will be overdamped.

Behavior of Processes. Underdamped behavior is rarely exhibited by the process itself or, more specifically, with the controller on manual. If the pipe is of sufficient length for the inertia of the flowing fluid to be significant, underdamped behavior will be observed in the pressure. How do you close a valve on a pipeline? Very slowly, otherwise the spike in pressure could have adverse consequences. But within process plants, the inertial effects are inconsequential, and most valves can be closed as quickly as possible with no adverse consequences.

Underdamped behavior is so rarely exhibited by a process that the statement is sometimes made that processes are inherently overdamped. If you ever observe underdamped behavior with the controller on manual, first understand the origin of the underdamped behavior.

Behavior of Control Loops. For most control loops, underdamped is the desired behavior. As the process is overdamped, the control loop will exhibit overdamped behavior at low values of the controller gain. However, sufficiently increasing the controller gain will cause most loops to exhibit underdamped behavior.

The responses in Figure 3.24 illustrate the effect of the controller gain on the performance of a typical control loop. The controller contains only the proportional mode. The responses are to a step change in the set point of one

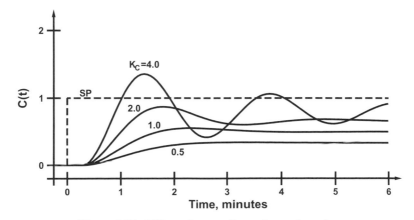

Figure 3.24. Effect of controller gain on damping.

unit. Proportional-only controllers do not line out with the controlled variable equal to the set point (we will leave this explanation for a subsequent chapter). But note that overdamped/underdamped behavior is based on the lineout value, not on the set point.

For a controller gain of 0.5%/%, the behavior is overdamped. Increasing the gain to 1.0%/% gives a small overshoot and underdamped behavior. For controller gains of 2.0%/% and 4.0%/%, the response clearly has a sinusoidal component. For gains less than about 6%/%, the amplitude of the sinusoidal component decreases with each successive cycle (a damped sinusoid). For gains greater than about 6%/%, the amplitude increases with each successive cycle (an undamped sinusoid).

Critically Damped. Critically damped is where the transition from overdamped to underdamped occurs. For the responses in Figure 3.24, critically damped occurs at a controller gain of approximately 1.0%/%. The small overshoot exhibited by this response would have no adverse consequences on the process.

A common definition of "good response" is one that responds as quickly as possible but with "no overshoot." Any overdamped response has no overshoot, but how do you know it is responding as quickly as possible? One approach is to permit the response to exhibit a very small overshoot, perhaps 5% or sometimes as much as 10%. We shall refer to such responses as "minimal overshoot," but they usually pass for "no overshoot" or "critically damped."

3.12. SIMPLE PERFORMANCE MEASURES

This section defines the simple performance measures that are commonly used in process control.

Overshoot. Overshoot is formally defined for the case where the process makes a transition from one operating level to another. Although overshoot can be expressed in engineering units, control engineers most commonly express overshoot in percent.

For the response in Figure 3.25, the controlled variable changes by a units and overshoots its final value by b units. The overshoot in engineering units is b; the overshoot in % is given by the following expression:

$$\text{Overshoot (\%)} = 100 \, \frac{b}{a}$$

In practice, precise calculations of the percent overshoot are rare. Instead, visual estimates usually suffice.

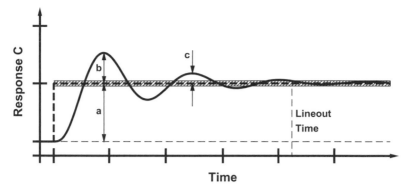

Figure 3.25. Simple performance measures.

Decay Ratio. The decay ratio reflects the rate of decay of the sinusoidal component of the response. The decay ratio is the ratio of the second peak overshoot c to the first peak overshoot b:

$$\text{Decay ratio} = \frac{c}{b}$$

Figure 3.25 defines the overshoots b and c for the response to a step change in the set point. In practice, visual estimates are normally used in lieu of a precise calculation of the decay ratio. The response in Figure 3.25 has a decay ratio of about 1/3.

One advantage of the decay ratio is that it can also be applied to the responses to other changes, including

- Response to a change in a disturbance or load.
- Response to a short-duration pulse input (or "bump") in the set point.

The decay ratios are not identical, but the difference is small.

In process control, the decay ratio is the most commonly applied performance measure. Furthermore, the preferred value is almost always ¼. The term "quarter-wave damping" is sometimes used to refer to responses whose decay ratio is ¼. This response will have a significant first overshoot, followed by a small second overshoot. The third overshoot will be very small, if visible at all.

A response with a quarter decay ratio has a gain margin of approximately 2. That is, the controller gain is about half the ultimate gain.

The quarter decay ratio depends only on the magnitudes of the first two peaks. However, this does not mean that all successive peaks can be ignored. The magnitude of the third peak must be insignificant. Should the third peak become significant before a quarter decay ratio is attained, further increases in the controller gain are unwise.

The performance objective "big hump, little hump" is understood to mean that the response is to exhibit a noticeable first overshoot followed by a small second overshoot. Thereafter, the overshoots, if any, must be insignificant. This is essentially equivalent to a quarter decay ratio.

Line-Out Time or Settling Time. A loop is said to be "lined out" when all variables cease to change. The loop and all of its components are in equilibrium. The lineout time or settling time is the time required for the system to attain equilibrium after a change in one or more of its inputs.

In practice, the lineout time is considered to be the time that the response has attained its final value and no oscillations or other change is visible in the response. To quantify the lineout time mathematically, one has to introduce a tolerance for a variable to be considered as having attained its equilibrium value. When the input is a step change in the set point, the tolerance is typically 2% or 5%.

As indicated in Figure 3.25, the tolerance introduces a band about the final value. The line-out time is the time required for the response to come within that band and to remain within that band thereafter.

Limitations of Simple Performance Measures. Simple performance measures such as decay ratio and percent overshoot are appealing because they are easy to apply. We rarely measure peak heights and compute precise values; we use visual estimates instead.

However, their limitations must be understood. The decay ratio depends on only two points on the response; the percent overshoot depends on only one point. This gives the possibility that two very different responses could have the same value of a simple performance measure.

This has implications for controller tuning. Both responses in Figure 3.26 have a quarter decay ratio, but they are very different. The process is exactly the same; only the controller tuning is different. We would usually prefer the faster of the two responses. The controller tuned with a gain of 2.0%/% results in a response with a quarter decay ratio; however, the controller is capable of delivering better performance.

3.13. THE INTEGRAL CRITERIA

The integral criteria address the deficiencies of the simple performance measures, but at a price. Let us first define the integral criteria and then discuss their advantages and disadvantages.

General Definition. An integral criterion is a performance measure that is based on the integral of some function of the control error and on possibly other variables (such as time). The general expression is

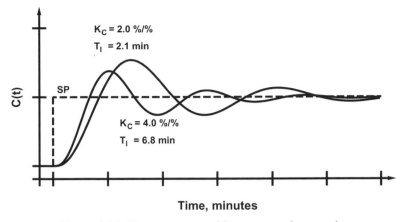

Figure 3.26. Two responses with a quarter decay ratio.

$$\text{Integral criterion} = \int_0^\infty f(E, t)\, dt$$

To be useful as a performance measure, the integrand $f(E,t)$ must be "positive definite"; that is, $f(E,t) \geq 0$ for all values of E.

The three most commonly used integral criteria are as follows:

- Integral of the absolute error (IAE).
- Integral of the square error (ISE).
- Integral of time and absolute error (ITAE).

The smaller the value of the integral criterion, the better the performance of the control loop. Thus, when used as the basis for tuning a PID controller, the objective is to determine the values of the tuning parameters K_C, T_I, and T_D that minimize the selected integral criterion.

Integral of Absolute Error (IAE). This integral is the integral of the control error, with all error treated as positive:

$$\text{IAE} = \int_0^\infty |E|\, dt$$

Integral of Square Error (ISE). The integral of the square error is the counterpart to the sum of squares from linear regression:

$$\text{ISE} = \int_0^\infty E^2\, dt$$

The integral of the square error (ISE) penalizes for large errors more than for small errors. But in attempting to minimize this criterion function, responses with small errors will be accepted provided doing so will reduce the large errors. Often the response has a smaller initial overshoot, but the cycle does not decay rapidly. Basically several small peaks are tolerated to reduce the magnitude of the first peak. This type of behavior is usually not desired in process loops.

Integral of Time and Absolute Error (ITAE). The objective of the integral of time and absolute error (ITAE) is to penalize for even small errors that occur late in time. This is achieved by including time in the integrand:

$$ITAE = \int_0^\infty |E|\, t\, dt$$

Even small errors occurring late in time contribute significantly to the integral. Consequently, minimizing this integral criterion generally leads to responses with a short lineout or settling time. Process engineers generally prefer this type of behavior.

Obtaining a Value for an Integral Criterion. The integral criteria are not "user-unfriendly." Simple performance measures such as decay ratio and percent overshoot are routinely estimated. This is not possible with the integral criteria. The only way to obtain a value is to evaluate the integral numerically, which in practice requires a computer.

We tend to form mental pictures of responses with a certain decay ratio or a certain percent overshoot. However, we cannot do this for the integral criteria. Suppose the response of the loop to a certain change in the set point has an IAE of 27.5. This one number is not especially informative. To some extent, the numbers are only relative. If the tuning parameters are changed such as the IAE for the same set-point change is 24.1, the new tuning coefficients are superior to the old.

Minimizing the Selected Integral Criterion. Determining the values of the tuning coefficients K_C, T_I, and T_D that minimize the selected integral criterion is an iterative endeavor. Given starting values for the tuning coefficients, each iteration consists of the following:

1. Obtain a process response to the input change of choice (such as a step change in set point).
2. Evaluate the integral criterion.
3. Select new values for the tuning coefficients.

This procedure is repeated until the tuning coefficients that minimize the integral criterion of choice are found. Computer routines known as multivariable search techniques are available for efficiently performing such minimizations. But even with a good search technique, 50 to 75 iterations are required to tune a PI controller. For a PID controller, the number is in excess of 200. Such undertakings are only feasible on simulations, never on the real process.

Use of the Integral Criteria. The availability of simulations basically determines the extent to which integral criteria can be used. Process simulations are not routinely done, and even when undertaken, they are often not to the detail required for tasks such as controller tuning.

To date, the use of the integral criteria has been limited to a few tuning methods. For the simple process models used as the basis for the conventional controller tuning techniques, the results of minimizing the integral can be reduced to a set of equations. But for any model, a computer can determine the tuning coefficients that minimize the selected integral criterion.

Advantages of the Integral Criteria. An integral criterion does a far better job of quantifying the nature of the response. Every point on the response contributes to the value of the integral criterion. Simple performance measures such as decay ratio or percent overshoot depend on only one or two points on the response.

Minimizing an integral criterion gives a unique set of values for the tuning coefficients. For PI or PID control of a given process, many sets of tuning coefficients will give a quarter decay ratio. This is never the case for minimizing an integral criterion.

3.14. SUMMARY

Most process control presentations begin with process dynamics, the presumption being that the design and performance of the controls depend primarily, if not exclusively, on dynamic behavior. We wish to challenge this presumption. In practice, two aspects of process dynamics lead to problems:

Dead time. The presence of dead time is one of the distinguishing characteristics of process control. The effectiveness of a PID controller goes down rapidly as the dead time increases. Either the dead time has to be reduced or alternative control approaches have to be pursued.

Integrating processes. The difficulty originates when a short reset time is used in the controller. For integrating processes, the proportional mode has to be the primary mode of control.

The key to a control system that delivers good performance is a P&I diagram that reflects the characteristics of the process. The control structures on the P&I diagram mostly reflect the steady-state characteristics of the process as quantified by the gains or sensitivities. In this regard, problems can surface with regard to process nonlinearities and loop interaction. Control valves as well as the process can contribute to the nonlinearities. In practice, these problems develop more frequently than problems associated with process dynamics.

CHAPTER 4

Controller Modes and Mode Selection

Each mode of control is examined in detail by subsequent chapters. This chapter provides a "warm-up" by addressing the following topics:

- Relationship for each mode. In brief, the contribution of each mode is as follows:

  Proportional. Determines the speed of response of the loop.

  Integral. Forces the controller to line out at its set point.

  Derivative. Enhances stability margin, which permits the gain to be increased.
- Options for the tuning coefficient for each mode.
- Mode combinations that can be considered.
- Special considerations pertaining to flow control and level control.
- Nonlinear control equations.

4.1. MODE CHARACTERISTICS

To begin, let us envision the controller output M to be the sum of the outputs of the individual modes:

$$M = M_P + M_I + M_D$$

where

M_P = Proportional mode output, %

M_I = Integral or reset mode output, %

M_D = Derivative mode output, %

Actual implementations of the control equation do not perform the computations in this manner. As a consequence, the three components cannot be displayed individually in commercial controllers; only the controller output M can be displayed.

Control Error. The traditional PID control equation is based entirely on the control error E, which is the difference between the set point SP and the process variable PV. The sign of the control error depends on the controller action:

$$E = \begin{cases} SP - PV & \text{for a reverse acting controller} \\ PV - SP & \text{for a direct acting controller} \end{cases}$$

Many digital systems permit the proportional mode and/or the derivative mode to be based on the process variable instead of on the control error. However, we will defer this discussion to subsequent chapters.

Proportional Mode. The proportional component M_P of the controller output M is equal to a coefficient times the current value of the control error E plus the controller output bias M_R:

$$M_P = K_C\, E + M_{R,0}$$
$$= \frac{100}{PB}\, E + M_{R,0}$$
$$= K_P\, E + M_{R,0}$$

where

E = Control error, %

K_C = Controller gain, %/%

PB = Proportional band, % $= \dfrac{100\%}{K_C}$

K_P = Proportional gain, %/% $= K_C$

$M_{R,0}$ = Initial value of controller output bias, %
$\quad\quad = M_0 - K_C\, E_0$

E_0 = Initial value of control error, %

M_0 = Initial value of controller output, %

"Proportional plus bias" is a more descriptive term for the proportional mode equation; it is always gain times error plus bias. As usually written, the bias in the proportional mode equation is controller output bias M_R. As the PID equation is normally implemented, the controller output bias M_R is the output of the integral mode, the initial value for the integrator being $M_{R,0}$. We shall examine this formulation shortly, but until then, we must use $M_{R,0}$ instead of M_R in the proportional mode equation.

The value for $M_{R,0}$ is determined when the controller is initialized during the switch from manual to automatic, the objective being to achieve a "bumpless transfer." This will be discussed in detail in the chapter on the proportional mode.

The tuning coefficients K_C and K_P have exactly the same effect on the proportional action. Consequently, there is really no distinction between these two coefficients in the equation for the proportional mode. But as we shall see, their overall effect is very different when other modes are added to the controller. The controller gain K_C multiplies all modes in the controller, whereas the proportional gain K_P is applied to the proportional mode only.

Integral or Reset Mode. The integral component M_I is equal to a coefficient times the integral of the control error:

$$M_I = \frac{K_C}{T_I} \int_0^t E\, dt$$

$$= K_C\, R_I \int_0^t E\, dt$$

$$= K_I \int_0^t E\, dt$$

where

T_I = Integral or reset time, min

R_I = Reset rate, $1/\text{min} = \dfrac{1}{T_I}$

K_I = Integral or reset gain, $(\%/\%)/\text{min} = \dfrac{K_C}{T_I}$

When the controller contains only the proportional mode, the controlled variable is not necessarily equal to the set point when the loop lines out. This behavior, known as droop or offset, will be examined in more detail in the chapter on the proportional mode. When the integral or reset mode is present, the controlled variable must equal the set point when the loop lines out. This type of behavior is desirable in most loops, so the integral or reset mode is usually present.

A commonly held misconception is that the reset mode can be tuned to obtain a fast response. However, the truth is that the speed of response is determined by the proportional mode; the reset mode has little effect on the speed of response.

Derivative Mode. The derivative component M_D of the controller output M is equal to a coefficient times the rate of change of the control error E:

$$M_D = K_C \, T_D \, \frac{dE}{dt}$$

$$= K_D \, \frac{dE}{dt}$$

where

T_D = Derivative time, min

K_D = Derivative gain, (%/%)-min = $K_C \, T_D$

The derivative mode enhances the stability margin for the control loop, which reduces the overshoot and degree of oscillations. When used effectively, the derivative mode permits the controller gain to be increased, thereby providing a faster response to set-point changes and disturbances. But this is achieved only when proper tuning procedures are followed.

4.2. OPTIONS FOR TUNING COEFFICIENTS

The choice of tuning coefficients (for example, controller gain versus proportional band versus proportional gain) differs from one manufacturer to the next, and sometimes from one product line to the next for the same manufacturer. In addition, the choice for the units of time (usually, minutes versus seconds) may also differ.

There is a simple explanation for this: One choice for the tuning coefficients is not clearly superior to another choice. All are numerically related by simple equations.

Many companies have a "preferred vendor" for their control systems. If you are working for such a company, you might think that you only need to know what tuning coefficients that supplier uses. But most companies also purchase packaged units for cooling towers, effluent treatment, compressed air, and so on. The suppliers of such equipment also have their preferred suppliers of control equipment, with the price being a major consideration. They will provide whatever equipment the customer wants but usually at a noticeably higher price. And most purchasers become very flexible when they see the price. So one must be prepared to work with control equipment that uses different choices for the tuning coefficients.

Controller Gain versus Proportional Gain. Although these terms were introduced when the proportional mode equations were presented, their distinction only becomes significant when it is observed that the complete PID control equation can be expressed by either of the following equations:

$$M = K_C \left[E + \frac{1}{T_I} \int_0^t E \, dt + T_D \, \frac{dE}{dt} \right] + M_{R,0}$$

$$M = K_P \, E + K_I \int_0^t E \, dt + K_D \, \frac{dE}{dt} + M_{R,0}$$

These two equations are equivalent provided the values for the tuning coefficients are related as follows:

$$K_P = K_C$$

$$K_I = \frac{K_C}{T_I}$$

$$K_D = K_C \, T_D$$

Although the numerical values for the controller gain K_C and the proportional gain K_P are equal, their overall effect on the control equation is very different. The proportional gain K_P applies to only the proportional term, whereas the controller gain K_C applies to all terms in the control equation.

Until the advent of digital systems, the control equation was always written with the controller gain K_C multiplying all terms in the equation. This form of the PID equation reflected the nature of conventional pneumatic and electronic hardware. The coefficient K_C was the overall sensitivity for the controller.

The flexibility of digital implementations offered the possibility of providing individual gains for each term in the control equation. But when the control system suppliers converted from analog equipment to digital equipment, most (but not all) continued to provide the tuning coefficients as K_C (or PB), T_I (or R_I), and T_D.

There is no inherent advantage of one form over the other. Since the coefficients are algebraically related, the equations are equivalent. Anything that can be accomplished with one form can also be accomplished with the other.

Finally, an observation on terminology. In this book, the term "controller gain" will be applied to K_C (it multiplies all terms in the control equation) and the term "proportional gain" will be applied to K_P (it only appears in the proportional mode). However, this distinction is not universal, and K_C is often referred to as the "proportional gain."

Controller Gain versus Proportional Band. These two coefficients are related very simply. The proportional band PB (in %) is $100/K_C$, or alternatively, the controller gain K_C is $100/$PB. A controller gain of $1\%/\%$ is equivalent to a proportional band of 100%. A controller gain of $2\%/\%$ is a proportional band of 50%.

In the chapter on the proportional mode, the proportional band will be discussed in the context of regulators, which is where the concept originated. For conventional analog controllers, the proportional band was almost universally used. In the United States, the Taylor Instrument Company (now part of ABB) was the major exception. But with digital systems, the use of controller gain is becoming prevalent, with the major exception being Foxboro (now part of Invensys).

In practice, there is no compelling reason to use controller gain over proportional band, or vice versa. The effect on controller design is nil. Consider a conventional controller in which the sensitivity adjustment is implemented via a knob. The adjustment is currently the proportional band in %. What must be done to change the adjustment to the controller gain in %/%? You only need to change the calibration units for the knob. Remove the label with the calibration as proportional band in %, and replace it with a label with the calibration as controller gain in %/%. There is no change to the internal mechanisms of the controller. In digital systems, we simply change the units for the data entry and for displaying the coefficient.

Gains are said to be "high" or "low"; proportional bands are said to be "narrow" or "wide." A controller gain of 5%/% is a high gain or a high sensitivity. The proportional band would be 20%, which is a narrow proportional band. A controller gain of 0.2%/% is a low gain or a low sensitivity. The proportional band would be 500%, which is a wide proportional band.

Reset Time versus Reset Rate. These coefficients are merely the reciprocal of each other. The reset time T_I is in minutes (or occasionally seconds); the reset rate R_I is in reciprocal minutes (or occasionally reciprocal seconds). The term "repeat" is often incorporated into the units for reset time and reset rate, with the units for reset time being "minutes/repeat" and the units for reset rate being "repeats/minute." In the module on the reset mode, we shall explain the origin of "repeats." In the meantime, ignore it.

When the tuning coefficient is the reset rate, the reset action can be completely removed by setting the reset rate to zero. When the tuning coefficient is the reset time, an infinitely large value of the reset time would be required to remove the reset action completely. This necessitates that a "gimmick" be introduced for completely removing the reset action. A reset time of zero makes no sense (it would be an infinitely large reset rate). However, some manufacturers interpret a reset time of zero as meaning no reset action. Another alternative is to interpret reset times greater than some arbitrary value to mean no reset action. For example, a reset time of 99.99 min or greater could be understood to mean no reset action. Unfortunately, there is no universally accepted "gimmick."

The use of zero for the reset time to mean no reset action leads to an anomaly that must be recognized. Most digital systems have a minimum nonzero value for the reset time, such as 0.01 min. If one changes the reset time from 0.0 min to 0.01 min, the reset action changes from none at all to its maximum possible value. To proceed from no reset to the smallest amount of reset action, the reset time must be changed from zero to its maximum possible value (such as 99.99 min).

Minutes versus Seconds. The time units for the integral and derivative modes can be either minutes or seconds. Most systems use minutes, although a few use seconds.

Those with an electronic background usually have some difficulty with the time frame on which processes respond. Their natural inclination is to use seconds. We could certainly measure time constants, dead times, and so on, in seconds. We could certainly plot our variables with the time axis annotated in seconds. With the possible exception of flow loops, minutes are usually far more convenient and appropriate. For some loops, an argument could even be made to measure time in hours.

Choice of Tuning Coefficients. In a sense, the author of a book faces the same dilemma as a control system supplier: what choice to make for the tuning coefficients. The following tuning coefficients will be used from this point on:

- Proportional: controller gain K_C, %/%.
- Integral or reset mode: reset time T_I, min.
- Derivative mode: derivative time T_D, min.

Why choose this combination? Because we had to choose one. There is no compelling reason to choose one combination over another. If such a reason existed, then all of the controls manufacturers would use the same combination.

At least one manufacturer uses this combination, specifically, Honeywell in the TDC product line. But Honeywell's UDC product line of single-loop controllers uses proportional band, reset rate, and derivative time. You might as well become accustomed to different combinations of tuning coefficients.

4.3. COMPUTING THE PID CONTROL EQUATION

Most controllers operate using only the proportional and integral modes (the derivative time is zero). The resulting equation can be expressed using either the definite or the indefinite integral:

$$M = K_C \left[E + \frac{1}{T_I} \int_0^t E \, dt \right] + M_{R,0}$$

$$M = K_C \left[E + \frac{1}{T_I} \int E \, dt \right] \quad \text{Initial condition:} \quad M = M_0$$

An equivalent expression involves two equations, one for the proportional mode and one for the integral or reset mode:

Proportional: $M = K_C \, E + M_R$

Reset: $M_R = \int \dfrac{K_C}{T_I} \, E dt$ Initial condition: $M_R = M_0 - K_C \, E_0 = M_{R,0}$

When expressed in this form, the controller output bias M_R is the output of the reset integrator. The initial value $M_{R,0}$ for the controller output bias is the initial condition for the reset integrator. When the controller is initialized with the set point equal to the process variable, then $E_0 = 0$ and the initial value $M_{R,0}$ of the controller output bias equals the value of the controller output M_0.

Derivative Mode. The effect of derivative is to base the control action on the projected control error $\hat{E}$ instead of on the current control error E. The projected value of the control error is obtained by assuming that the current rate of change of the control error remains constant for one derivative time in the future:

$$\hat{E} = E + T_D \frac{dE}{dt}$$

In effect, a straight line projection is used to obtain the projected control error $\hat{E}$ from the current control error E and its rate of change.

In the proportional mode equation, the projected control error $\hat{E}$ is always used in lieu of the control error E. In the integral mode equation, there are two options:

 • Base the integral mode on the control error E.
 • Base the integral mode on the projected control error $\hat{E}$.

The resulting equations are slightly different and will be examined in detail in the chapter on the derivative mode. For the moment, we will base the integral mode on the control error E.

The resulting PID control equations are as follows:

Derivative: $\hat{E} = E + T_D \dfrac{dE}{dt}$

Proportional: $M = K_C \hat{E} + M_R$

Reset: $M_R = \displaystyle\int \dfrac{K_C}{T_I} E dt$ Initial condition: $M_R = M_0 - K_C E_0 = M_{R,0}$

The computation sequence is as follows:

- Perform the derivative mode calculations to obtain the projected error $\hat{E}$.
- Perform the integral mode calculations to obtain a new value for the controller output bias M_R.
- Perform the proportional mode calculations to obtain the controller output M.

Digital Systems. Virtually all current control systems are digital in nature. For digital systems, the continuous equations must be approximated by discrete equations. Instead of continuous functions such as E, digital systems work on number sequences E_n, where E_n is the value of the control error at sampling instant n. The sampling interval Δt is the time between the values in the number sequences. The derivatives and integrals are replaced by numerical approximations, with the simplest being the following:

$$\frac{dE}{dt} \cong \frac{E_n - E_{n-1}}{\Delta t}$$

$$\int \frac{K_C}{T_I} E \, dt \cong \sum_{i=1}^{n} \frac{K_C}{T_I} E_i \, \Delta t = M_{R,n}$$

For numerical computations, the recursive form of this relationship is usually preferred:

$$M_{R,n} = M_{R,n-1} + \frac{K_C}{T_I} E_n \, \Delta t$$

Using these approximations, the control equations are as follows:

Derivative: $\hat{E}_n = E_n + T_D \dfrac{E_n - E_{n-1}}{\Delta t}$

Proportional: $M_n = K_C \hat{E}_n + M_{R,n}$

Reset: $M_{R,n} = M_{R,n-1} + \dfrac{K_C}{T_I} E_n \, \Delta t$

The simple approximation for the integral is satisfactory for discrete implementations of the PID control equation, especially since the sampling interval is typically 1 second or less. However, the simple finite difference approximation to the derivative is not. We shall explore this in more detail in the chapter on the derivative mode.

With the capabilities of today's digital technology, the sampling intervals for PID control calculations are very fast relative to the response characteristics of most processes. Under these conditions, the behavior of controls that use discrete approximations is indistinguishable from the behavior of controls based on the continuous equations. Consequently, only the continuous equations will generally be presented in this book.

4.4. MODE COMBINATIONS

In the days of conventional controls, the modes to be available were determined when the controller was purchased. If a proportional-integral (PI) was purchased, the controller did not contain the circuits for the derivative mode.

With digital systems, all three modes are provided in every PID controller. But for most loops, a conscious decision is made as to what modes will be used. Although present, the unused modes are essentially removed by setting the respective tuning coefficients to appropriate values. For example, a PI controller is obtained by simply setting the derivative time to zero.

Proportional-Only (P). The control equation is written as follows:

$$M = K_C E + M_R$$

Although commonly called proportional-only, "proportional plus bias" is more descriptive of the control equation: When the control error E is zero, the controller output M is equal to the controller output bias M_R. In a proportional-only controller, the controller output bias is constant; that is, $M_R = M_{R,0} = M_0 - K_C E_0$.

Proportional-only control is infrequently used in loops that reside within the main process controls. However, most field-mounted regulators are proportional-only controllers.

The principal limitation of proportional-only control is that the loop will exhibit droop or offset. When the loop lines out, the process variable is not necessarily equal to the set point, or said another way, the control error is not necessarily zero when the loop lines out. But except for the issue of offset or droop, the performance of proportional-only control would be satisfactory in most applications.

Integral-Only (I). The control equation is written as follows:

$$M = K_I \int_0^t E \, dt + M_0$$

The tuning parameter is the reset gain K_I, with the units being (%/%)/min. In implementing the equation, the initial controller output M_0 becomes the initial condition on the reset integrator (provided the integrand is $K_I E$).

Integral-only control leads to a loop that responds very slowly. Its use is limited to two situations, as follows:

1. Very fast loops (primarily flow loops).
2. Loops with substantial noise on the measured variable.

When substantial noise is present, the use of proportional action causes the controller to translate the noise on the measured variable to the noise on the controller output. Smoothing of the noise is required. However, an integrator is a very effective smoother. Thus, a viable approach is to resort to integral-only control.

Proportional-Integral (PI). The control equation is written as follows:

$$M = K_C \left[E + \frac{1}{T_I} \int_0^t E \, dt \right] + M_{R,0}$$

PI is by far the most commonly used mode combination. When suppliers were selling conventional pneumatic and electronic controllers, sales figures were tabulated on the types of controllers being shipped. The data were approximately as follows:

- PI controllers: 90%.
- PID controllers: 10%.
- All others combined: <1%.

Again, it should be emphasized that these are sales figures. We have no idea how many of the PID controllers spent their lives with the derivative time set to zero. With digital controls, all controllers are supplied with all three modes, so no sales data are available. However, there is little reason to believe that a major shift has occurred.

Proportional-Derivative (PD). The control equation is written as follows:

$$M = K_C \left[E + T_D \frac{dE}{dt} \right] + M_R$$

The following statements for proportional-only also apply to PD:

- The PD controller exhibits offset or droop.
- The controller output bias is constant; that is, $M_R = M_{R,0} = M_0 - K_C E_0$.

With digital systems, the use of PD control is extremely rare. In a few situations where the use of PID control was intended, the limited windup prevention capabilities of conventional controls could not effectively cope with integral mode windup. Removal of the integral mode was the path of least resistance, resulting in PD control. The more extensive windup protection options available with digital controls make addressing the windup issues the preferable option, resulting in PID control.

Proportional-Integral-Derivative (PID). The control equation is written as follows:

$$M = K_C \left[E + \frac{1}{T_I} \int_0^t E \, dt + T_D \frac{dE}{dt} \right] + M_{R,0}$$

Tuning a PID controller is a challenge for both manual and automated tuning. The substantial commitment of time and effort will be made only in the important loops within a process. Many of these loops are temperature loops, most of which respond slowly. A challenging tuning endeavor in a very slow loop should not be taken lightly.

The traditional guideline for mode selection was to use PID controllers in temperature loops, but to use PI everywhere else. But even for temperature loops, the rewards must be noticeable to justify the effort involved in tuning a PID controller.

4.5. FLOW CONTROL

Flow loops are generally the fastest loops within a process. Their dynamics are measured in seconds; the dynamics of most other loops are measured in minutes. When viewed from the time frame of these other loops, flow loops seem to respond instantly.

Most loops are tightly tuned so that they will respond more rapidly to changes to both set-point changes and disturbances. But since flow loops are inherently very fast, rarely is there an incentive to tune a flow loop aggressively. This puts flow loops in a separate category; the considerations pertaining to mode selection and tuning of flow loops are different from that of other loops.

Another issue is that considerable measurement noise sometimes accompanies a flow measurement. If the controller responds rapidly, the measurement noise will be translated to noise on the signal to the control valve. This favors a controller that relies primarily on the integral mode for the control action.

Mode Selection. PI control is used for most flow loops. However, the proportional sensitivity is usually very low, effectively resulting in integral-only control.

With conventional controllers, the tuning procedure was to set the controller gain to its minimum value, and then to tune the reset to attain the desired performance. For conventional controls, the minimum value of the controller gain ranged from about 0.1%/% to 0.5%/%, depending on the manufacturer. With such a low gain, why include the proportional mode at all?

With the older pneumatic and electronic controls, the following considerations led to most flow loops being PI:

1. The cost of a PI controller (a commodity item) was less than an integral-only controller (a specialty item).
2. Since most loops in the plant are PI, spare parts considerations favored PI (most plants had no other requirement for an integral-only controller).

With digital systems, neither of these is an issue. Integral-only control could be used in flow loops, but old habits are hard to break, especially when they get the job done.

Tuning. A flow loop is the one loop for which values can be proposed for the tuning coefficients with reasonable expectation that they will be satisfactory. For loops where the measured flow is the flow through the control valve (such as the hot water flow loop in Figure 4.1), the typical recommendations are as follows:

	Controller Gain K_C %/%	Reset Time T_I sec	Reset Gain K_I (%/%)/sec
Recommendation I	0.1	2	0.05
Recommendation II	0.2	3	0.0667
Recommendation III	0.3	5	0.06

Actually, there is not much difference among the three recommendations. The controller gain is so low that the proportional mode is essentially doing nothing. The reset gain $K_I = K_C/T_I$ for the above three recommendations varies from 0.05 (%/%)/sec to 0.0667 (%/%)/sec. Most flow loops will perform satisfactorily with any of the above recommendations. Recommendation II will be used in the examples in this book, but the results would be the same if either of the other recommendations was used.

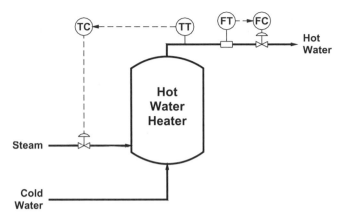

Figure 4.1. Hot water process with hot water flow controller.

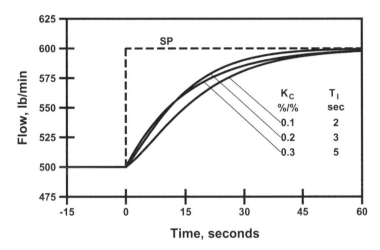

Figure 4.2. Flow loop performance.

Performance. Figure 4.1 illustrates a hot water process with a hot water flow controller. For a change in the flow set point from 500 lb/min to 600 lb/min, Figure 4.2 presents the response for each of the recommended sets of tuning coefficients. The difference in performance is nominal.

The flow controller is certainly capable of a faster response. However, what is the incentive to do so? In most cases, none at all.

The time constant for the temperature responses for the hot water process is about 7.5 minutes, which means that its response time is over 20 minutes. Suppose the responses in Figure 4.2 are plotted using a time axis with a division of 10 minutes instead of 15 seconds for the tic marks. The three responses would be indistinguishable, basically plotting as a step change. Adjusting the

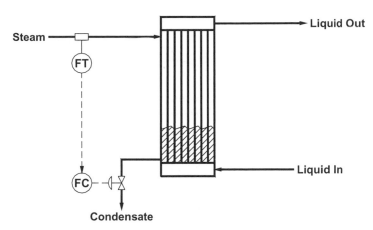

Figure 4.3. Steam-heated exchanger with valve on condensate.

tuning of the flow controller will have no influence on the responses for the hot water temperature.

Exceptions to the Rule. The values suggested for the tuning coefficients apply to flow loops such as the hot water flow loop in Figure 4.1. Specifically, the measured flow is the flow through the control valve. This applies to most, but not all, flow loops.

The example presented in Figure 4.3 does not meet this criterion. The flow of steam to the exchanger is controlled using the control valve on the condensate return. Closing the control valve causes more condensate to back up into the exchanger, reducing the exposed heat transfer surface area. This reduces the heat transfer and, thus, the steam condensation rate and steam flow. The dynamics between a change in the control valve position and a change in steam flow are far slower than in a typical flow loop. Therefore, the usual guidelines for tuning a flow loop do not apply to this case.

Problem Flow Loops. Consider the flow loop for the chlorine vaporizer illustrated in Figure 4.4. On the surface, the flow loop appears to be a typical flow loop, with the measured chlorine flow being the flow through the control valve. However, none of the recommended settings for the flow controller will be satisfactory. The real problem is that the control configuration is inappropriate. The difficulties in controller tuning are merely symptoms of the fact that the P&I diagram does not reflect the characteristics of the process.

With the pressure controller on manual, does the control valve on the chlorine gas have any effect on the chlorine flow? In the short term, the answer is "yes." Opening the valve initially increases the chlorine flow. However, the pressure in the vaporizer also drops, which decreases the chlorine flow. In the long term, the chlorine flow will return to a value only slightly higher than its

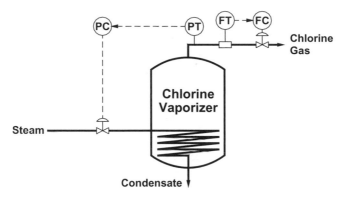

Figure 4.4. Chlorine vaporizer.

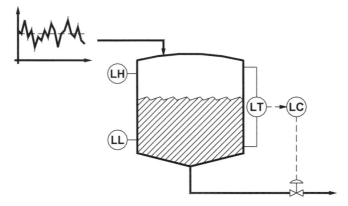

Figure 4.5. Typical level control application with the variance in feed flow.

initial value. The chlorine flow is largely determined by the heat input to the vaporizer. The chlorine gas valve largely affects the pressure; its direct effect on the long-term value of the chlorine flow is minimal.

This example is discussed in more detail in the subsequent chapter on developing P&I diagrams.

4.6. LEVEL CONTROL

The following statements apply to many level loops, including the one illustrated in Figure 4.5:

- There is no need to maintain the vessel level at a fixed set point. The vessel is equipped with level switches that initiate a shutdown on high or low level. As long as the vessel level is between the limits imposed by these switches, the vessel level has no effect on process operations or on plant economics.

- The discharge flow from the vessel is the feed flow to some downstream processing unit. Processing units can usually handle gradual changes in their feed rate. However, a rapid change in feed flow is a major upset to most processing units.

In most level controllers, aggressive tuning is not necessary, and it is often counterproductive. Instead of maintaining a fixed level, the controls should preferably take advantage of the surge capacity within the vessel to smooth any variations in the feed flow(s) into the vessel. It is especially crucial to do so when the downstream unit is significantly affected by rapid changes in its feed rate.

Variance in Feed Flow. For the level process in Figure 4.5, significant variance is present in the feed flow to the vessel. The variance in the feed flow is determined by whatever processing operations are upstream of the vessel. Variance is an issue that can be addressed in two ways:

1. Eliminate (or at least reduce) the variance at its source. This means that the upstream processing units must be analyzed to ascertain the source of the variance, and then to correct whatever is causing the variance in the feed flow to the level process. This approach is generally preferred by those within the statistical quality control community.
2. Implement level control to respond appropriately to the effect of feed flow changes on the vessel level. This approach is generally preferred by those within the process control community.

Eliminating all variance at its source is usually unrealistic. Consequently, level control of some form will be required. But the process design invariably imposes a limit on what the process controls can reasonably accomplish.

No Level Control. Figure 4.6 depicts the situation when the level controller is on manual. The output of the level controller is constant, and we shall assume that this gives a constant discharge flow. The variance in the feed flow propagates to the variance in the vessel level. The variance in the vessel level depends on two factors:

Variance in feed flow. Obviously the greater the variance in feed flow, the greater the variance in liquid level.

Process design. The parameter of importance is the cross-sectional area of the vessel. The larger this area, the smaller the variance in liquid level for a given variance in the feed flow. To reduce variance in level, we prefer a tank with a large cross-sectional area.

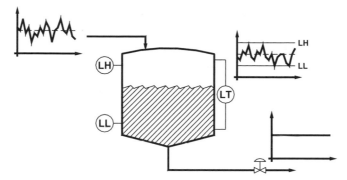

Figure 4.6. Propagation of the variance with level controller on manual.

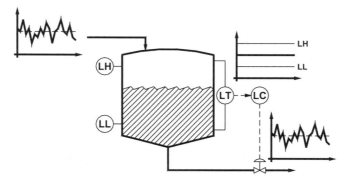

Figure 4.7. Propagation of the variance with level controller on automatic.

With no level control, the feed to the downstream processing unit would be constant. Thus, no feed flow disturbances to that unit would occur. But a major upset in the vessel feed flow is likely to drive the vessel level either above the high level switch or below the low level switch. This activates the shutdown logic, which is usually a major upset to the process.

Perfect Level Control. Figure 4.7 depicts the situation where the level control is perfect. By this, we mean that the level remains at its target despite the variance in the feed flow. When we claim to improve the performance of the level controller, we are claiming that its performance is closer to this ideal or perfect performance. Of course, no feedback controller can accomplish this.

What must the controller do to achieve perfect level control? Any change in the feed flow must be immediately translated to an identical change in the discharge flow. The level controls are propagating the variance in the feed flow to the variance in the discharge flow. Consequently, disturbances to the level process are merely passed through to the downstream process unit.

Most processing operations are adversely affected by disturbances in the feed flow. In most cases, modest variations in vessel level are far preferable to rapid changes in the feed flow to the downstream processing operation. What do we mean by "modest?" Whatever does not cause the level switches to initiate a shutdown.

Regulatory Controls and Variance. In the global sense, regulatory controls do not reduce variance. Regulatory controls merely shift the variance from the controlled variable to the manipulated variable. The final result is bracketed by two extremes:

No control (or controller on manual). The variance in the disturbances leads to the variance in the controlled variable. There are no changes in the manipulated variable, so its variance is zero.

Perfect control. The variance in the disturbances leads to the variance in the manipulated variable. The further consequences of this depend on the nature of the manipulated variable:

Utility variable. Where the manipulated variable is a utility variable such as cooling water flow, steam flow, and so on, the regulatory controls move the variance from the process to the utility system. Most utility systems are designed to cope with such variance.

Process variable. In the level process in Figure 4.5, the controller moves the variance from one process variable (vessel level) to another process variable (discharge flow). The variance remains in the process, so situations exist where this is counterproductive. This issue frequently originates in level control, but it can originate in any regulatory control loop.

Level Controller Behavior. For applications such as in Figure 4.5, conservative tuning of the level controller is desirable under normal process operations. But there is always the possibility of a major process upset that causes a large change in the feed flow to the vessel. The required behavior of the controller differs between these two situations:

Normal operations. The controller should change its output only in a gradual manner that does not upset the downstream processing units.

Major upset. When the level approaches either of the limits imposed by the two level switches, very aggressive action should be taken. In fact, prior to the time the level reaches either limit, the controller output should be at the appropriate extreme, that is, fully open before attaining the high limit and fully closed before attaining the low limit.

The customary implementation of the PID control equation does not provide such a capability. One approach is to tune the controller conservatively and to rely on the process operators to respond to the major upsets. However, some nonlinear control equations can provide such behavior, and one of these should be considered.

4.7. NONLINEAR ALGORITHMS

The customary relationship for the proportional mode is as follows:

$$M - M_R = K_C E$$

The control action is linear with respect to the control error.

Relationships other than linear are sometimes appropriate. The most common requirement is for a low sensitivity when the control error is small but a large sensitivity when the control error is large. In some applications, small deviations from the set point are of no concern. Therefore, there is no need to take control action to correct for such deviations. However, large deviations are very undesirable, so very aggressive control actions should be taken to avoid them. Level control is one application where such behavior is sometimes required.

Error Squared. Although commonly referred to as "error squared," the relationship is actually "signed error squared" in that E^2 is negative if E is negative. The equation is usually written as follows:

$$M - M_R = K_C |E| E$$

Figure 4.8 presents a graph of M as a function of E. The effective sensitivity is $K_C |E|$, or in other words, the sensitivity is proportional to the magnitude of the control error. As the error increases, the sensitivity increases from zero to a very large value.

Error Gap or Error Deadband. As illustrated by the graph of M as a function of E in Figure 4.9, this nonlinear controller applies a small controller gain within the gap and a larger controller gain outside the gap. The equation is as follows:

$$M - M_R = \begin{cases} r\,K_C\,E & \text{if } |E| \leq E_{DB} \\ r\,K_C\,E_{DB} + K_C\,[E - E_{DB}] & \text{if } E > E_{DB} \\ -r\,K_C\,E_{DB} + K_C\,[E + E_{DB}] & \text{if } E < -E_{DB} \end{cases}$$

The adjustable coefficients are the gain K_C, the error gap or deadband E_{DB}, and the ratio r of sensitivity inside the gap to sensitivity outside the gap. A typical value for r is 0.1. However, it is possible for r to be zero, in which case

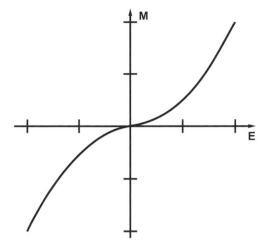

Figure 4.8. Proportional based on error-squared.

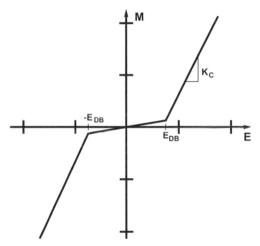

Figure 4.9. Proportional based on error-gap.

no control action is taken if the control error is less than E_{DB}. Such control is often referred to as error deadband.

In Figure 4.9, the gap above the target is the same as the gap below the target. Many implementations permit separate values to be specified for the positive gap and the negative gap.

General Nonlinear Function. Some digital systems provide the possibility of using a user-defined function for the control equation. This can be implemented in two different ways:

$$M - M_R = f[E]$$

$$M - M_R = K_C E \quad \text{where } K_C = g[E]$$

where f and g are functions supplied by the user. They are usually implemented via either function generator blocks or calculation blocks.

Use of the Nonlinear Functions in the Various Modes. The previous discussion presented the nonlinear functions in the context of the proportional mode. Now let us address the other modes.

Reset Mode. Basically there are two options:
1. Use the nonlinear function for the integrand.
2. Use the control error as the integrand.

The pros and cons of each are not entirely clear. Most systems adopt one approach or the other; an option to select the desired approach is not commonly provided.

Derivative Mode. In applications where nonlinear control relationships are applied, the derivative is rarely used. However, if it is used, it should be based on either the control error or the measured variable in the same manner as in the customary PID equations.

Use of Error Gap Algorithm for Level Control. The objective is to achieve the following:

Normal operations. Use a low controller gain so that the controller output changes only in a gradual manner that does not upset the downstream processing units.

Major upset. When the level reaches the limit of the gap, switch to a high controller gain so that the controller will make the rapid changes required to prevent the level from exceeding the limits imposed by the level switches.

The level set point, the error gap, and the controller gain are adjusted so as to achieve the following:

1. Controller output will attain 100% (full discharge flow) before the level attains the location of the high level switch.
2. Controller output will attain 0% (no discharge flow) before the level attains the location of the low level switch.

Tuning Error Gap for Level Control. During normal process operations, the controller output should change as little as possible to minimize the disturbances in the feed flow to the downstream process unit. This suggests a very low value for the sensitivity within the gap, possibly even a sensitivity of zero (giving a true deadband).

However, when the sensitivity within the gap is too low, the level tends to first drift in one direction and then in the other. Suppose the level is slowly increasing. If the sensitivity of the controller is too low to respond effectively, the level continues to increase until the upper extreme of the gap is attained. The controller aggressively reacts, and it usually overcompensates. Normally the result is a slowly decreasing level, with a similar behavior at the lower extreme of the gap.

The controller sensitivity within the gap must be sufficient to respond effectively to the variations in vessel feed rate during normal operations. The level should attain the extremes of the gap rather infrequently, and only when there has been a major process upset.

4.8. LEVEL-TO-FLOW CASCADE

In most control loops, the main objective is to minimize the variance in the controlled variable. Level loops are an exception. A certain degree of control of level is required, but this must be accomplished without significantly increasing the variance in the flow through the control valve. Conservative tuning of the level controller is appropriate. But with a slowly responding level controller, other disturbances can significantly affect the flow through the control valve and increase the variance.

Polymer Reactors. Let us illustrate this using the two polymerization reactors in Figure 4.10 as the example, starting with an explanation of the role of the two reactors:

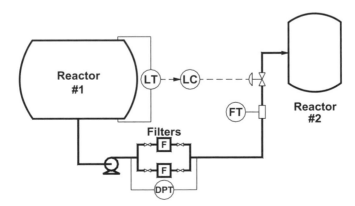

Figure 4.10. Simple level control of polymer reactors.

- Reactor #1. This is the first reactor in a series of reactors for converting monomer feeds (not shown) into a polymeric product. This reactor is very large, having a residence time of approximately eight hours. The large volume assures that the reaction will achieve equilibrium conditions.
- Reactor #2. This reactor operates at a higher temperature than the first reactor, and it is also operated under vacuum. These conditions promote the polymerization reaction. With a residence time of approximately two hours, Reactor #2 is considerably smaller than Reactor #1.

Material is pumped from Reactor #1 to Reactor #2 through a cartridge filter. Two cartridge filters are provided, but only one is in service at a given time. While that filter is in service, the cartridge in the other filter is replaced so that it is available as soon as required.

Switching the filters is initiated based on the pressure drop across the filter that is in service. This occurs on approximately a 12-hour interval. When a new filter is switched into service, the polymer flow increases abruptly because of the lower pressure drop across the filter. The flow then slowly decreases as the pressure drop across the filter increases. The effect on the level in Reactor #1 is minimal; the effect on the polymer flow is the concern.

Simple Feedback Control. Figure 4.10 presents the simple feedback control configuration for Reactor #1. A measurement for the polymer flow is provided but is not used in the control configuration.

For this configuration, switching the filters is a major disturbance to the polymer flow. This change in the flow affects the level in Reactor #1, but it takes some time for the effect to be significant. By then, the change in flow has significantly affected Reactor #2 (the smaller reactor).

Each time the filters are switched, a significant change in the polymer flow occurs. On a switch from a plugged filter to a clean filter, a major pressure drop is removed from the polymer flow system. As the trend in Figure 4.11 clearly shows, this results in a significant change in the polymer flow. The effect on the level in Reactor #1 is nominal, so no response is elicited from the level

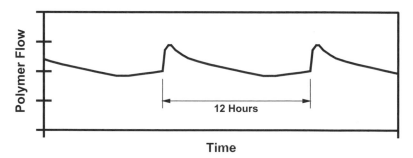

Figure 4.11. Performance of simple level control.

controller. However, the change in polymer flow affects the residence time of Reactor #2, which leads to product variability.

The Reactor #1 level controller is typical of many level controllers. The acceptability of the performance of such control loops cannot be judged solely on its ability to maintain the controlled variable at the target. The effect of the manipulated variable on other process variables must be taken into consideration. When a controller responds slowly, it will not effectively respond to upsets that affect the flow through a control valve. This is a problem when these flow changes are upsets to other processing units. For the polymer reactors, switching the filters leads to a change in polymer flow. The slow responding level controller for Reactor #1 does not effectively react to this, thus permitting the change in polymer flow to upset Reactor #2.

Cascade. The level-to-flow cascade control configuration illustrated in Figure 4.12 consists of two controllers:

> **Polymer flow controller.** This controller is configured as follows:
> - Controlled variable: Polymer flow.
> - Manipulated variable: Polymer control valve position.
> **Reactor #1 level controller.** This controller is configured as follows:
> - Controlled variable: Level in Reactor #1.
> - Manipulated variable: Set point for to the polymer flow controller.

If the level in Reactor #1 is increasing, the level controller for Reactor #1 must increase the set point to the polymer flow controller, which will then open the polymer control valve. This will have very little impact on the performance of the level control in Reactor #1. The major improvement will be to the upsets to polymer flow associated with switching the filters. The flow

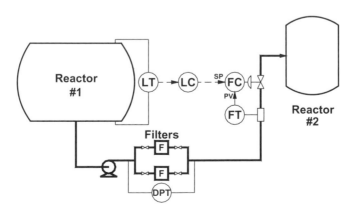

Figure 4.12. Level-to-flow cascade.

CHAPTER 5

Proportional Mode

The discussion of controller modes can only start with the proportional mode. Since we have not discussed integral or derivative, we will be examining proportional-only controllers.

From the perspective of the control loops implemented within the main plant controls, proportional-only control is rare. But additional controls are provided by field-mounted regulators, primarily for pressure control but occasionally for level control. Regulators are attractive for their simplicity, so proportional-only control action is appropriate.

As noted in Chapter 4, the control equation for a proportional-only controller should properly be referred to as "proportional plus bias." The main limitation of a proportional-only controller is that at line out or equilibrium, the controlled variable is not necessarily equal to the set point, a behavior known as either offset or droop. We need to explain why the proportional-only control equation leads to such behavior.

We shall define bumpless transfer and then present the initialization equations required to achieve bumpless transfer. We will also describe the feature generally known as "PV tracking" and explain its impact on operator training.

5.1. CONTROL EQUATION

The proportional-mode equation is as follows:

$$M = K_C E + M_R$$

where

Practical Process Control: Tuning and Troubleshooting, by Cecil L. Smith
Copyright © 2009 John Wiley & Sons, Inc.

E   = Control error, %

 = SP – PV for a reverse acting controller

 = PV – SP for a direct acting controller

K_C = Controller gain, %/%

M   = Controller output or manipulated variable, %

M_R = Controller output bias, %

PV = Process variable or controlled variable, %

SP = Set point, %

To understand the need for the bias M_R in the control equation, consider the hot water process. For a hot water demand of 500 lb/min, we provide the controller with a set point of 150 °F. Let us assume the controller positions the control valve to 62.1%, which gives a hot water temperature of 150 °F. The control error is zero; however, the controller output is 62.1%, not zero. When the control error E is zero, how can the output of the proportional-only controller be 62.1%? Provided the value of the controller output bias M_R is 62.1%.

Equation Relating the Controller Output M to the Process Variable PV.

Although we routinely write the proportional control equation in terms of the control error, the equation can also be expressed in terms of the process variable PV.

Let us start with the proportional-mode control equation and then replace E by SP – PV (the controller for the hot water process must be reverse acting):

$$E = SP - PV \quad \text{(reverse acting controller)}$$
$$M = K_C\,E + M_R$$
$$= K_C(SP - PV) + M_R$$
$$= -K_C\,PV + (K_C\,SP + M_R)$$

Let us put some numbers into this equation. The measurement range of the hot water temperature transmitter is 50 °F to 250 °F. A temperature of 150 °F is 50% of span. The controller gain is 2.0%/%. With these values, the proportional mode equation is

$$M = -(2.0\%/\%)PV + (2\%/\% \times 50\% + 62.1\%)$$
$$= -(2.0\%/\%)PV + 162.1\%$$

If PV = 150 °F = 50%, then M = 62.1%.

Multiple combinations of SP and M_R will give this same equation:

SP	M_R	Equation
150 °F or 50%	62.1%	$M = -(2.0\%/\%)\,PV + (2\%/\% \times 50\% + 62.1\%)$
		$= -(2.0\%/\%)\,PV + 162.1\%$
160 °F or 55%	52.1%	$M = -(2.0\%/\%)\,PV + (2\%/\% \times 55\% + 52.1\%)$
		$= -(2.0\%/\%)\,PV + 162.1\%$
140 °F or 45%	72.1%	$M = -(2.0\%/\%)\,PV + (2\%/\% \times 45\% + 72.1\%)$
		$= -(2.0\%/\%)\,PV + 162.1\%$

The proportional mode equation is exactly the same for all three of the above combinations. For all, the equilibrium is $PV = 150\,°F = 50\%$ and $M = 62.1\%$. However, the controller lines out at the set point for only one combination, specifically, for $SP = 150\,°F$ and $M_R = 62.1\%$. For the second combination ($SP = 160\,°F$ and $M_R = 52.1\%$), the controller lines out with a control error of $10\,°F$ ($SP = 160\,°F$; $PV = 150\,°F$).

Offset is defined as the difference between the set point and the process variable when the controller lines out. For the second combination, the loop lines out at $PV = 150\,°F$ when the set point is $160\,°F$. The controller exhibits an offset of $10\,°F$ or 5% of span. This behavior is a characteristic of any controller that does not contain the integral mode.

That the controller will line out with $E \neq 0$ may initially seem odd. But examine the control equations more carefully:

$$M = K_C\,E + M_R \qquad \text{Proportional mode equation in terms of E.}$$
$$M = -K_C\,PV + (K_C\,SP + M_R) \quad \text{Proportional mode equation in terms of PV.}$$

What in these equations requires that E be zero at steady state? Absolutely nothing. A constant value of E gives a constant value of M. If this value of M gives an $E = SP - PV$ that is this same value of E, then both process and controller are at equilibrium.

Significance of the Controller Output Bias. Perhaps the best way to think of the controller output bias M_R is as the value of the controller output M when the control error E is zero; that is, $PV = SP$.

$$M = K_C\,E + M_R$$
$$= M_R \text{ when } E = 0 \text{ or } PV = SP$$

As we will discuss shortly, the initial value for the controller output bias M_R is determined when the controller is switched from manual to automatic. In a proportional-only controller, the controller output bias remains at its initial value.

In commercial control systems, the term "bias" is used in several different contexts. For example, most implementations of ratio control provide for a bias to the ratio calculation. This has nothing to do with the controller output bias. Furthermore, most commercial controllers will not directly display the value of the controller output bias. In these systems, the only way to determine the current value of the bias is to change the set point to the current value of the process variable. This makes the error zero, so the output of the controller will be the value of the controller output bias.

Offset. Offset is defined as the difference between the set point and the process variable when the controller lines out. For a previous example, the process lined out at PV = 150 °F when the set point was 160 °F. The controller exhibited an offset of 10 °F.

If the controller contains only the proportional mode, the controller output bias M_R is a fixed value, or perhaps said more properly, the normal control calculations will not change the value of the controller output bias M_R. But if the controller contains the integral or reset mode, the value of the controller output bias M_R can be considered to be the output of the integral or reset mode. Consequently, the value of M_R will be changing whenever the control error E is nonzero. Such a controller can only line out with PV = SP or E = 0 (and, consequently, no offset). This will be explained in Chapter 6 on the integral mode.

Alternative Equation. You will frequently see the proportional mode equation written without the controller output bias:

$$m = K_C \, e$$

But note that this equation is expressed using lowercase letters m and e instead of M and E. Indeed, m and e are not the same as M and E:

- M and E are the actual values for the controller output and control error.
- m and e are the changes in the controller output and control error from their equilibrium values (M_R and E_R, respectively).

We have previously noted that when E = 0, then M = M_R. The equilibrium value for the control error is zero; consequently, E_R = 0. The relationships between the lower case and the uppercase variables are as follows:

$$M = m + M_R \quad \text{or} \quad m = M - M_R$$
$$E = e + E_R = e \quad \text{or} \quad e = E - E_R = E$$

Consequently, the following two expressions of the proportional mode equation are equivalent:

$$m = K_C \, e$$
$$M - M_R = K_C \, E$$

We shall only express the control equations in terms of the actual values M and E. However, the proportional mode equation is commonly written as $m = K_C \, e$, with no explanation that m and e are changes from the equilibrium state and not actual values. Nor is it universal or even common practice to use uppercase letters for actual values and lowercase for the changes from the equilibrium state.

Velocity or Incremental Form of the Control Equation. For digital systems, the proportional control equation should be written in a discrete form as follows:

$$M_n = K_C \, E_n + M_R$$

where

E_n = Control error at sampling instant n, %
     = $SP_n - PV_n$ for a reverse acting controller
     = $PV_n - SP_n$ for a direct acting controller
M_n = Controller output at sampling instant n, %
PV_n = Process variable at sampling instant n, %
SP_n = Set point at sampling instant n, %

This equation is referred to as the position form of the control equation. It computes the controller output from the control error, which in turn is computed from the set point and the process variable.

Let us write this equation for sampling instant $n - 1$:

$$M_{n-1} = K_C \, E_{n-1} + M_R$$

Now let us subtract the two equations:

$$\begin{aligned}
\Delta M_n &= M_n - M_{n-1} \\
&= (K_C \, E_n + M_R) - (K_C \, E_{n-1} + M_R) \\
&= K_C (E_n - E_{n-1}) \\
&= K_C \, \Delta E_n
\end{aligned}$$

This equation is referred to as the incremental or velocity form of the control equation. It computes the change in the controller output from a change in the control error.

In digital systems, the control equation can be implemented in either the position form or the velocity form. Provided the controller output is within the limits imposed on the output (normally 0% and 100%), the performance of the two forms of the control equation are exactly the same. The control equation implemented in conventional pneumatic and electronic controls is the position form; the velocity or incremental form of the control equation can only be implemented in digital systems.

5.2. REGULATORS

In most plants, numerous regulators exist. They are simple, inexpensive, and require no external source of power. Regulators are proportional-only controllers of a mechanical nature, so understanding how they function can further one's understanding of the proportional mode. Also, terms such as "proportional band" and "droop" originated in the context of regulators.

Simple Level Regulator. We could analyze a commercial product. However, commercial regulators for industrial service must be designed so that they can be fabricated in a practical way. Although crucial to a successful commercial product, fabrication issues have nothing to do with control. We want to focus on the control issues. Consequently, a construction will be proposed for a simple level regulator that addresses all of the control issues but that is not encumbered by the fabrication issues required for practical industrial products.

Figure 5.1 illustrates a simple level regulator. The float senses the level in the tank. The float is attached to a bar, with its pivot point being at the right. One end of a cable is attached to the bar; the other end is attached to the stem of the valve on the discharge flow. A turnbuckle is inserted into the cable so that its length can easily be adjusted. The purpose of the turnbuckle will be explained shortly.

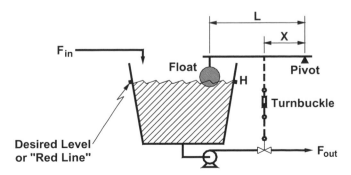

Figure 5.1. Simple level regulator.

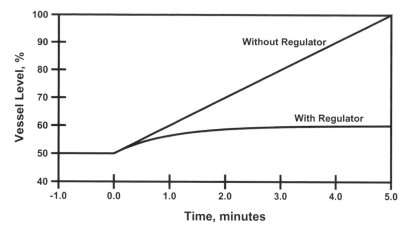

Figure 5.3. Response to an increase in feed flow.

Figure 5.3 illustrates the response to an increase in the feed flow for two cases:

With the level regulator. An increase in the inlet flow leads to an increase in the equilibrium value for the vessel level. The regulator is providing control action; it is just not able to maintain the vessel level at its original value.

Without the level regulator. An increase in the inlet flow would cause the vessel level to increase at a constant rate until the tank overflows. The level process is an integrating process.

The initial rate of change of vessel level (just after the increase in the inlet flow) is the same both with and without the regulator. But with the regulator, the increase in level leads to an increase in the outlet flow. This causes the rate of change in vessel level to decrease. This continues until a new equilibrium is established.

"Droop" is the general term used to describe this phenomenon. As the flow increases, the vessel level "rides up." As the flow decreases, the vessel level "droops down." The equilibrium value of the vessel level (the controlled variable) depends on the throughput (the flow through the vessel).

Effect of Sensitivity. For a given change in the flow, the amount that the level rides up or droops down depends on the sensitivity of the regulator. The higher the sensitivity, the less the level rides up or droops down. This suggests that the sensitivity of the regulator should be set as high as possible. However, if the sensitivity is too high, cycling will be observed in the level.

Issues pertaining to sensitivity have an impact on the general acceptability of regulators for industrial applications:

Pressure regulators. The sensitivity can normally be set sufficiently high that the change of pressure with throughput will be small and tolerable in industrial applications.

Level regulators. The sensitivity usually cannot be set so high. The change of level with throughput will be larger, but most applications can tolerate such changes.

Temperature regulators. The sensitivity cannot be set very large. Consequently, the effect of throughput changes on the temperature will be larger, usually too large to be acceptable.

Pressure Regulators. Pressure regulators come in two versions:

- Maintain the upstream pressure (sometimes referred to as a back-pressure regulator).
- Maintain the downstream pressure.

The flow through the regulator is the demand (in the same sense that the hot water flow is the demand for the hot water process). All regulators exhibit droop on a change in the demand, the behavior depending on the type of regulator:

Controlled Pressure	Action	Droop on Increase in Demand
Upstream	**Direct.** On an increase in the upstream pressure, regulator must open.	Upstream pressure "rides up" to increase regulator opening.
Downstream	**Reverse.** On an increase in the downstream pressure, regulator must close.	Downstream pressure "droops down" to increase regulator opening.

5.3. THE PROPORTIONAL BAND

One option for the tuning coefficient for the proportional mode is the proportional band. This term originated in the context of regulators, so we shall explain proportional band for the simple level regulator.

Limits on Control Action. All regulators, indeed all industrial controllers, have limits on the control action that they can take. Where the final control element is a valve, the limits are usually (but not always) imposed at the valve, with the limits being valve fully closed and valve fully open.

For most processes, it is possible to operate the process in such a manner as to require more flow through a valve than it can deliver. For the level process, there is a maximum flow capacity of the valve on the discharge flow.

If the inlet flow is greater than this value, the level will increase until the regulator fully opens the valve. But since the inlet flow still exceeds the outlet flow, the level continues to increase. The controller can do nothing about this; it has taken all of the available control action.

If the inlet flow is shut off, the level will drop to the point where the level regulator completely closes the discharge valve. Ideally, the level should drop no further. However, if the valve leaks or if some other loss of material from the vessel occurs, the level will continue to drop. But again, the controller can do nothing to stop this; it has taken all of the available control action.

The Proportional Band. The proportional band is the interval of the measured variable over which the controller is taking proportional control action. If the measured variable is within the proportional band, the output to the final control element is within the limits of the available control action. If the measured variable is outside the proportional band, the controller output is at one of the limits.

For the level regulator, the proportional band is established as follows:

- At what value of the level will the control valve be fully open? This establishes one end of the proportional band.
- At what value of the level will the control valve be fully closed? This establishes the other end of the proportional band.

The proportional band for the level regulator is illustrated in Figure 5.4. The following observations can be made:

- If the process attains equilibrium, the lineout value for the level will be within the proportional band.
- For low throughputs, the lineout value will be near the lower end of the proportional band.

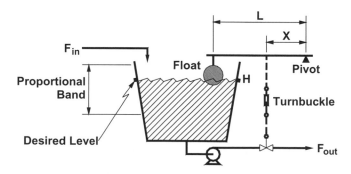

Figure 5.4. Proportional band.

• For high throughputs, the lineout value will be near the upper end of the proportional band.

Effect of Sensitivity on the Proportional Band. The proportional band is inversely proportional to the sensitivity of the controller.

For the level regulator, Figure 5.5 illustrates moving the attachment point to the left. This increases the sensitivity of the regulator. A smaller change in level is required to move the valve from fully closed to fully open. We say that the proportional band is "narrow."

Figure 5.6 illustrates moving the attachment point to the right. This decreases the sensitivity of the regulator. A larger change in level is required to move the valve from fully closed to fully open. We say that the proportional band is "wide."

For wide proportional bands (greater than the measurement range), one end of the proportional band will be beyond the achievable values of the measured variable. For the wide proportional band illustrated in Figure 5.6, the upper end is above the top of the tank. Since the level can never attain the upper end of the proportional band, it is not possible for the level regulator to drive the discharge valve fully open.

Effect of the Turnbuckle on the Proportional Band. Suppose the turnbuckle is adjusted so as to shorten the cable. As illustrated in Figure 5.7, the

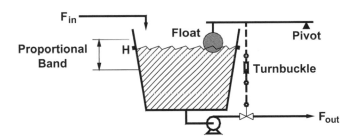

Figure 5.5. Narrow proportional band.

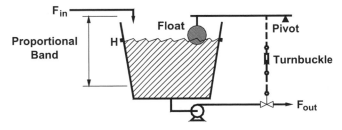

Figure 5.6. Wide proportional band.

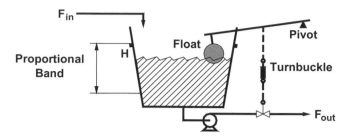

Figure 5.7. Shifting the proportional band down.

effect is to shift the proportional band down. The width of the proportional band is unchanged.

Adjusting the "Set" knob on a pressure regulator is equivalent to adjusting the turnbuckle on the level regulator. The effect is to shift the proportional band. This definitely affects the equilibrium or lineout value. However, droop will still occur should there be a change in throughput. Perhaps the label for the knob implies that you are adjusting the set point; but in reality, you are affecting the lineout value by shifting the proportional band.

Units for the Proportional Band. In most commercial controllers, the proportional band is expressed as a percent of the span of the measured variable.

The proportional band is occasionally expressed in the engineering units of the measured variable. For example, consider a temperature controller that accepts a direct thermocouple or RTD input. What is the span of the measured variable? One possibility is to use a span that reflects the measurement range of the basic sensor, such as $-200\,°C$ to $800\,°C$ for a type J thermocouple. For process applications, such a span is unreasonably large; its use would necessitate very small values of the proportional band. A more reasonable approach is to express the proportional band in the engineering units of the measured variable, specifically, in $°C$.

Set Point for a Regulator. For the simple level regulator, what is the set point? The turnbuckle has a large influence on the value at which the level lines out. However, the turnbuckle does this by providing a bias on the output of the control mechanism. For a given value of the level, the position of the bar is fixed. Provided the adjustment range of the turnbuckle is sufficient, the turnbuckle can be adjusted so that the valve stem position that corresponds to this level can be any value between 0% and 100%.

The term "set point" suggests a mechanism for directly specifying the desired value of the level. In this sense, the level regulator does not have a set point. And with no set point, the control error is also meaningless. Two consequences of this are as follows:

- Offset is the difference between the set point and the process variable at line out. With no set point, the term "offset" is not applicable to regulators. Instead, the term "droop" should be used.
- For the definition to be applicable to regulators, the controller action must be based on the process variable or measured variable. By defining a direct acting controller as one that increases its output in response to an increase in the measured variable, the ISA's definition of controller action is applicable to regulators.

5.4. BUMPLESS TRANSFER

Bumpless transfer describes the behavior of the controller output at the instant the controller is switched from manual to automatic. Figure 5.8 presents the three possible cases:

Top case. The switch from manual to automatic has no effect on the controller output. The value in automatic is the same as the value in manual. Furthermore, the rate of change of the controller output is zero in automatic as well as in manual. Clearly this is bumpless.

Middle case. There is an abrupt change in the controller output at the instant the controller is switched from manual to automatic. This type of behavior is not bumpless.

Bottom case. There is no abrupt change in the controller output at the instant the controller is switched from manual to automatic. However, the rate of change abruptly changes from zero to some finite value. This type of behavior is considered to be bumpless.

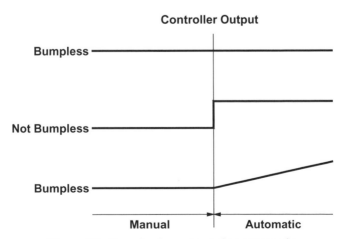

Figure 5.8. Transfer from manual to automatic.

Initialization Calculations. The term "balancing" is commonly used in lieu of initialization. Both terms refer to the actions required to achieve bumpless transfer. With conventional controls, the operators were usually responsible for balancing the controller so as to achieve bumpless transfer. With digital controls, the controllers perform the balancing calculations. The term "balanceless" is usually applied to such controls; however, "self-balancing" would be more descriptive.

The proportional mode computes the controller output M from the control error E, the controller gain K_C, and the controller output bias M_R:

$$M = K_C E + M_R$$

The initialization or balancing calculations compute the initial value $M_{R,0}$ for the controller output bias M_R from the initial controller output M_0, the initial error E_0, and the controller gain K_C:

$$M_{R,0} = M_0 - K_C E_0$$

This equation is simply the control equation solved for the controller output bias M_R.

The objective of the initialization or balancing calculations is to compute a value for the controller output bias M_R so that the value calculated for M just after the switch to automatic is the same as the value of the controller output just prior to the switch. This computation must be done at the time of the switch. However, most digital controllers perform this computation at every iteration that the controller is in manual.

Options Pertaining to Bumpless Transfer. Manual balancing procedures for conventional controls always began with "change the set point to the current value of the process variable." This is commonly referred to as "PV tracking." With PV tracking enabled, the initialization is as follows:

$SP_0 = PV_0$ PV tracking.

$E_0 = 0$ Calculate control error.

$M_{R,0} = M_0 - K_C E_0 = M_0$ Initialization calculations.

Since the value of the set point is updated to the current value of the process variable, the control error is zero. The value for the output bias M_R is the value of the controller output M_0. This simplicity is very appealing when manually balancing the controller.

But with digital systems performing the balancing calculations, such simplicity is not necessary. Consequently, most digital systems also provide the option to turn off the PV tracking and use the following calculations:

$E_0 = PV_0 - SP_0$ if controller is direct acting.

$E_0 = SP_0 - PV_0$ if controller is reverse acting.

$M_{R,0} = M_0 - K_C E_0$.

Basically, the initial value calculated for the controller output bias reflects the fact that the initial value of the control error may not be zero.

Operator Training Issues. Most digital systems provide the PV tracking option on an individual loop basis. For some loops, issues pertaining to advanced control applications require the PV tracking option to be configured in a specific way. But for other loops, the PV tracking option is entirely at the discretion of the plant. Ultimately, the choice reduces to how the plant prefers to train the operators.

Let us start with one premise:

> When switching a loop to automatic, the operators are responsible for making sure that the value of the set point is correct.

With PV tracking enabled, the set point cannot be changed when the loop is in manual. Thus, the operator can only proceed as follows:

1. Switch the controller to automatic.
2. Enter the desired value for the set point.

When PV tracking is disabled, the operator can change the set point with the loop in manual as well as in automatic. But it is essential that the operator specifies an appropriate value for the set point before switching to automatic. The operator should proceed as follows:

1. Enter the desired value for the set point. Before switching to automatic, the initialization calculations must compute the initial value $M_{R,0}$ for the controller output bias using an appropriate value for the set point.
2. Switch the controller to automatic.

Why proceed in this manner? Switching the loop to automatic with an unreasonable value of the set point means that the initial value $M_{R,0}$ for the controller output bias is equally unreasonable.

PV Tracking Enabled. Using the hot water process, let us illustrate switching the temperature controller to automatic when PV tracking is enabled. The controller is a PI controller with a controller gain of 5%/%, a reset time of 7.5 min, reverse acting, and PV tracking enabled.

The initial conditions are as follows:

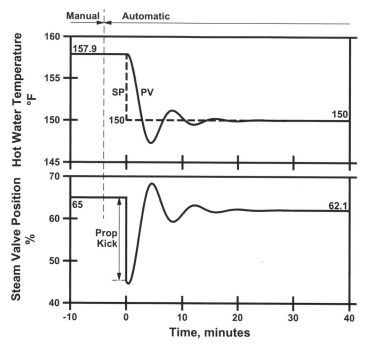

Figure 5.9. PV tracking enabled, PI control.

- The controller is on manual.
- The controller output is 65%.
- The process has lined out at a hot water temperature of 157.9 °F.
- Since PV tracking is enabled, the set point is also 157.9 °F, giving a control error of zero.
- The initial value $M_{R,0}$ for the controller output bias is 65% (the same as the controller output).

These initial conditions provide the starting point for the trends in Figure 5.9. With PV tracking enabled, the operator proceeds as follows:

Step 1—switch controller to automatic. Since the control error is zero, the controller maintains the output at 65%. The transition to automatic is indicated in Figure 5.9, and it is bumpless.

Step 2—enter value for set point. Change the controller set point to 150 °F. In Figure 5.9, this occurs at time zero. The change in set point causes an abrupt change in the controller output. This change is caused by the proportional control action, and it is often referred to as the "proportional kick." The change in set point is −7.9 °F. The control error also changes by −7.9 °F (the controller is reverse acting, so E = SP − PV) or −3.95% of span (measurement range is 50 °F to 250 °F). The change in controller output is −18.8%, which is the change in error (−3.95%) multiplied by the controller gain (5%/%).

PV Tracking Disabled. Again using the hot water process, let us illustrate switching the temperature controller to automatic when PV tracking is disabled. The controller is a PI controller with a controller gain of 5%/%, a reset time of 7.5 min, reverse action, and PV tracking disabled.

The initial conditions are as follows:

- The controller is on manual.
- The controller output is 65%.
- The process has lined out at a hot water temperature of 157.9 °F.
- The set point is 147.0 °F, giving a control error of 147.0 °F − 157.9 °F = −1 0.9 °F or 5.45% of span (measurement range is 50 °F to 250 °F).
- The initial value for the controller output bias is computed as follows:

$$M_{R,0} = M_0 - K_C E_0 = 65\% - (5\%/\%)(-5.45\%) = 92.2\%.$$

These initial conditions provide the starting point for the trends in Figure 5.10. With PV tracking disabled, the operator proceeds as follows:

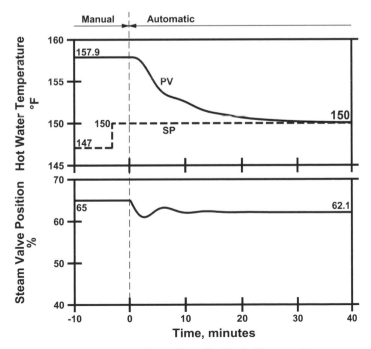

Figure 5.10. PV tracking disabled, PI control.

Step 1—enter value for set point. Since PV tracking is disabled, we can change the set point with the controller in manual. Let us change the set point to 150 °F, giving a control error of $150.0\,°F - 157.9\,°F = -7.9\,°F$ or 3.95% of span. This affects the initial value for the controller output bias, the new value being computed as follows:

$$M_{R,0} = M_0 - K_C\,E_0 = 65\% - (5\%/\%)(-3.95\%) = 84.7\%$$

Step 2—switch controller to automatic. This occurs at time zero on the trend in Figure 5.10. There is no abrupt change in the controller output upon the transition to automatic, so the transition is bumpless. However, the controller immediately begins to decrease its output, the objective being to decrease the hot water temperature from 157.9 °F to 150 °F. This is actually driven by the integral mode.

Proportional-Only Control with PV Tracking Disabled. Figure 5.10 illustrated switching the hot water temperature controller to automatic, with the controller being a PI controller. For Figure 5.11, everything is the same except that the controller is proportional-only.

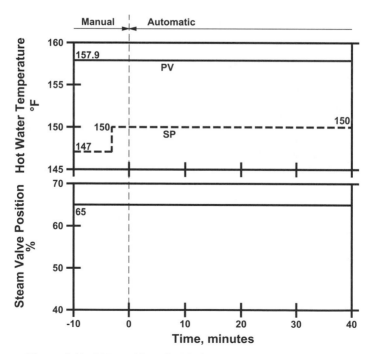

Figure 5.11. PV tracking disabled, proportional-only control.

The initial conditions are the same. Prior to switching the controller to automatic, the operator changes the set point from 147.0 °F to 150.0 °F. The values for the control error and the controller output bias are the same as before; neither depends on whether the controller is proportional-only or PI. For a set point of 150.0 °F, the control error is −7.9 °F (the controller is reverse acting, so E = SP − PV) and the initial value for the controller output bias is 84.7%.

At time zero in Figure 5.11, the controller is switched to automatic. But nothing changes! The controller maintains the controller output at 65%, despite the fact that the control error is 150.0 °F − 157.9 °F = −7.9 °F or −3.95%. The computation for the controller output multiplies the control error by the controller gain (5.0%/%) and then adds the controller output bias (84.7%) to obtain a value of 65% for the controller output. A valve position of 65% gives a hot water temperature of 157.9 °F. Both controller and process are at equilibrium, so nothing changes.

Although it might seem strange, the objective of the bumpless transfer or initialization calculations is to obtain the behavior illustrated in Figure 5.11. The objective of these calculations is for the value computed for the controller output just after the switch to automatic should be the same as the value of the controller output just prior to the switch to automatic. This is clearly achieved in Figure 5.11. As for the control error being nonzero, there is nothing in the proportional-only control equation $M = K_C\,E + M_R$ that forces the control error E to be zero at equilibrium. If E = 0 is desired (as it usually is), then the controller must contain integral or reset action.

5.5. SET-POINT CHANGES

We face a dilemma with regard to offset. With the exception of the field-mounted regulators, the integral or reset mode is used in virtually all controllers in process applications. As controllers with integral action do not exhibit offset, it seems that we could simply ignore offset. However, consider the following:

- In the trial-and-error tuning procedures for PI and PID controllers, step 1 is to obtain the proper value for the controller gain. The best approach is to first remove all integral and derivative action, and then to adjust the controller gain until the desired performance is attained. Under these conditions, the loop will exhibit offset.
- When troubleshooting, the recommended approach is to start with the simplest configuration that should at least work, although usually with some deficiencies that can be subsequently addressed. For controllers, this means proportional-only control. If you cannot get a proportional-only controller to provide some degree of control, there is no need to proceed

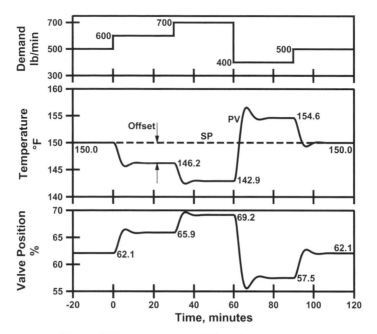

Figure 5.13. A sequence of disturbance changes.

The controller has opened the steam valve from its initial value of 62.1% to its final equilibrium value of 65.9%, a change of 65.9% − 62.1% = 3.8%. Since the controller gain is 2%/%, a control error of 3.8%/2%/% = 1.9% is required to change the output by 3.8%.

Multiple Load Changes. Figure 5.13 illustrates the response to a sequence of hot water demand changes, from 500 lb/min to 600 lb/min to 700 lb/min to 400 lb/min and finally to 500 lb/min. The values at each equilibrium are as follows:

Hot Water Demand	Process Variable	Offset	Controller Output
500 lb/min	150.0 °F	0 °F	62.1%
600 lb/min	146.2 °F	3.8 °F	65.9%
700 lb/min	142.9 °F	7.1 °F	69.2%
400 lb/min	154.6 °F	−4.6 °F	57.5%
500 lb/min	150.0 °F	0 °F	62.1%

The controller was balanced at a hot water demand of 500 lb/min, and it exhibits no offset at this point. At any other hot water demand, the loop exhibits offset. For hot water demand changes, the process exhibits nonlinear behavior. Note that the offset at 600 lb/min is 3.8 °F, but the offset at 400 lb/min is −4.6 °F. For a linear process, the values would be the same except for the sign.

5.7. PROPORTIONAL CONTROL OF SIMPLE MODELS

Our objective is to use simple models to illustrate the following characteristics of the proportional mode:

1. Increasing the controller gain leads to a faster response.
2. Increasing the controller gain reduces the stability margin.
3. For a PID controller, dead time degrades performance more than a time constant.

These characteristics are true for processes in general.

Time Constant Process. Let the process gain be K and the time constant be τ. For various values of the loop gain K K_C, Figure 5.14 presents the responses to a step change in the set point. Probably the most apparent effect of increasing K K_C is that the loop lines out with the PV closer to its set point; that is, the offset decreases. However, we want to focus on the speed of response.

When a proportional-only controller is applied to a time constant process, the closed-loop response to a set-point change is also a first-order lag, its time constant being the "closed-loop time constant" τ_{CL} (the process time constant τ is sometimes referred to as the "open-loop time constant"). The response of a first-order lag attains 63% of the final value in one time constant. The 63% point is indicated on each of the responses in Figure 5.14. The closed-loop time constant decreases as the loop gain increases.

The closed-loop time constant is a measure of the speed of response of the loop. The smaller the closed-loop time constant, the faster the response. For

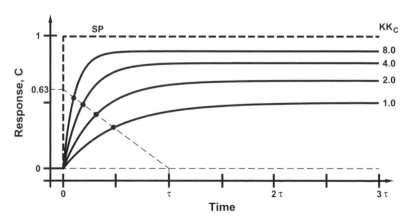

Figure 5.14. Effect of gain on response to a change in set point, proportional-only control of a time constant process.

proportional-only control of a time constant process, the closed-loop time constant is given by the following expression:

$$\tau_{CL} = \frac{\tau}{1 + K\,K_C}$$

The closed-loop time constant decreases as the loop gain increases, approaching zero for very large values of the loop gain. For a pure time constant process, there will never be any overshoot, even for extremely large values of the controller gain.

Process Consisting of Two Time Constants. Next consider using a proportional-only controller for a process consisting of two time constants. We designate the two time constants as follows:

τ_1 = Major process time constant.
τ_2 = Minor process time constant, that is, $0 \le \tau_2 \le \tau_1$.

For $\tau_2 = 0.5\tau_1$, Figure 5.15 presents the responses to a step change in the set point for various values of the loop gain $K\,K_C$. We make the following observations:

- For values of the loop gain of 2.0 and above, the responses clearly exhibit overshoot (actually the response for a loop gain of 1.0 exhibits a very small overshoot). Recall that the definition of overshoot is relative to the line-out value, not relative to the set point.

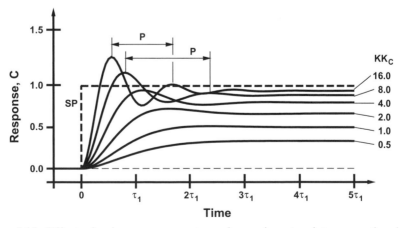

Figure 5.15. Effect of gain on response to a change in set point, proportional-only control of a two time constant process.

- For high values of the loop gain, the responses exhibit oscillations. However, the loop never becomes unstable. The limiting condition $(K\ K_C \to \infty)$ is a sinusoidal response with a constant amplitude.
- For responses that exhibit a cycle, a good measure of the speed or response is the period of the cycle. As indicated in Figure 5.15, the period of the response for $K\ K_C = 16$ is clearly less than the period of the response for $K\ K_C = 8$.

A process consisting of one time constant does not exhibit overshoot even for very large values of the loop gain. A process consisting of two time constants exhibits overshoot for very large values of the loop gain, but it is stable even for very large gains. But if either a third time constant or a dead time is added to the model, the loop will be unstable for large gains.

Process Consisting of a Time Constant and a Dead Time. Next consider using a proportional-only controller for a process consisting of a time constant τ and a dead time θ. For $\theta = 0.5\ \tau$, Figure 5.16 presents the responses to a step change in the set point for various values of the loop gain $K\ K_C$. For a loop gain of 4.0, the loop is unstable. In fact, the loop is unstable for loop gains above 3.8 (the product of the process gain and the ultimate gain K_U).

A dead time has a more detrimental effect on loop performance than a time constant of the same magnitude. Compare the responses in Figure 5.16 ($\theta = 0.5\ \tau$) with those in Figure 5.15 ($\tau_2 = 0.5\ \tau_1$). With dead time, the overshoot appears at a lower value of the loop gain. As the loop gain is increased, oscillations quickly appear, eventually leading to an unstable loop.

The responses in Figure 5.16 are more typical of actual processes than those in Figure 5.15. In practice, most processes exhibit some dead time, with a few

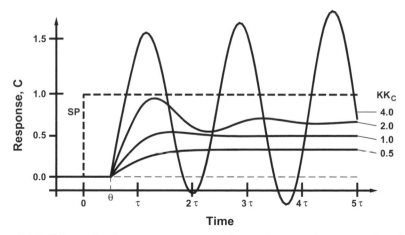

Figure 5.16. Effect of gain on response to a change in set point, proportional-only control of a time constant plus dead time process.

being dominated by dead time. Consequently, dead time must be included in a process model in order to be representative of true process behavior.

5.8. ADJUSTING THE CONTROLLER GAIN

Figure 5.17 presents the response of the temperature control loop for the hot water process for various values of the controller gain. All responses start from the same initial equilibrium conditions, namely, SP = PV = 150 °F and M = M_R = 62.1%. All responses in Figure 5.17 are to a set-point change from 150 °F to 160 °F. Responses are presented for values of the controller gain ranging from 0.5%/% to 8.0%/%.

The controller gain affects three aspects of loop performance:

Offset. Increasing the controller gain decreases the offset. But since the reset mode is added to almost all controllers to eliminate the offset, this is not of much significance in practice.

Stability margin. Increasing the controller gain decreases the stability margin, leading to more overshoot and oscillations. Normally this consideration places an upper limit on the acceptable values of the controller gain.

Speed of response. Increasing the controller gain makes the loop respond more rapidly to both set-point changes and load changes. This should be evident from the period of the cycle in the responses in Figure 5.17.

Overshoot and Decay Ratio. Suppose we start with a controller gain of 2.0%/%. The first response in Figure 5.18 is the response to a step change in the set point with this value of the controller gain. There is a first peak, but no

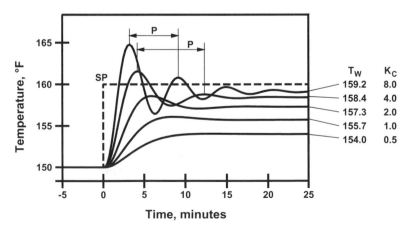

Figure 5.17. Effect of controller gain on the response to a change in set point.

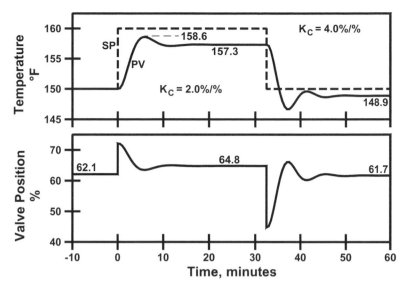

Figure 5.18. Adjusting controller gain when the control error is not zero.

second peak. Therefore, we can compute the percent overshoot for the response, but we cannot compute a decay ratio.

For responses that exhibit offset, the overshoot and decay ratio must be based to the initial and final values of the process variable, not the set point. Although the response never exceeds the set point of 160 °F, it does exceed the final value of 157.3 °F. Consequently, the response exhibits overshoot.

To compute the percent overshoot, the values of interest from the response in Figure 5.18 are as follows:

Initial value of PV: 150.0 °F.
Final value of PV: 157.3 °F.
Value of PV at peak: 158.6 °F.

Values such as these are normally very easy to retrieve from the trending facilities of modern control systems. The percent overshoot is then computed as follows:

Steady-state change in PV: 157.3 °F − 150.0 °F = 7.3 °F.
Overshoot: 158.6 °F − 157.3 °F = 1.3 °F.
Percent overshoot: $100 \times (1.3 \,°F / 7.3 \,°F) = 18\%$.

Although precise numbers are relatively easy to compute, the percent overshoot is often just estimated. For our purposes, a value of 20% for the overshoot of the response in Figure 5.18 is perfectly adequate.

For responses with a second peak, the decay ratio is computed in a similar fashion. That is, all peaks are relative to the final lineout value, not to the set point.

Controller Gain Adjustments. Let us select the quarter decay ratio as our performance objective for the temperature control loop for the hot water process. The first response in Figure 5.18 is for a controller gain of 2%/%. Changing the set point from 150 °F to 160 °F gives a response that has a modest overshoot, but no second peak is present.

To obtain a quarter decay ratio, we need to increase the controller gain. But by how much? The cautious approach would be to make small increases in the controller gain. The down side of this approach is that a large number of set-point changes will be required to tune the controller satisfactorily, and every set-point change is a disturbance to the process. A more aggressive approach is in order, but not so aggressive that we make the loop unstable. As a point of reference, the gain for a quarter decay ratio is approximately half the ultimate gain. Since we are well below a quarter decay ratio, let us go for it and change the controller gain to 4%/%.

Bumpless Tuning Parameter Adjustments. We are proposing to change the controller gain from 2%/% to 4%/% at a time when the control error is not zero. The set point is 160 °F, and the process variable is 157.3 °F, so the control error is 2.7 °F or 1.35% of span. The following calculations for the controller output illustrate the two options for handling a change in the controller gain:

Option	Calculation	K_C	Proportional Mode
	Current conditions	2%/%	$M = K_C E + M_R$ $= 2\%/\% \times 1.35\% + 62.1\%$ $= 64.8\%$
1	Calculate M from current M_R	4%/%	$M = K_C E + M_R$ $= 4\%/\% \times 1.35\% + 62.1\%$ $= 67.5\%$
2	Calculate M_R from current M	4%/%	$M_R = M - K_C E$ $= 64.8\% - 4\%/\% \times 1.35\%$ $= 59.4\%$

The effect of the two options is as follows:

Option 1. When the current value of the controller output bias M_R is retained, changing the controller gain would lead to an immediate change in the controller output M from 64.8% (for $K_C = 2\%/\%$) to 67.5% (for $K_C = 4\%/\%$). The result would be a bump to process.

Option 2. Changing the controller gain leads to a change in the controller output bias M_R, not to a change in the controller output M.

Option 2 is generally the preferred approach, and most (but not all) commercial control systems take this approach.

There is one consequence of option 2 that should be understood. Figure 5.18 illustrates the results of changing the set point back to 150 °F. Specifically, the sequence is as follows:

- The controller is initially balanced at $SP = PV = 150 °F$ and $M = M_R = 62.1\%$.
- The set point is changed to 160 °F. The process lines out with $PV = 157.3 °F$ ($E = 2.7 °F$ or 1.35% of span).
- The controller gain is changed from 2%/% to 4%/%. To achieve bumpless behavior (option 2), the controller output bias M_R is changed from 62.1% to 59.4%.
- The set point is changed to 150 °F. The process lines out with $PV = 148.9 °F$.

In a previous example (Figure 5.12), we started at 150 °F with no offset and then made several set-point changes. But when we returned the set point to 150 °F, the process lined out at 150 °F with no offset. This is because there was no change in the controller output bias M_R. But for the responses in Figure 5.18, the controller output bias M_R changed when the controller gain was changed from 2%/% to 4%/%. Consequently, when the set point is returned to 150 °F, the hot water temperature lines out at 148.9 °F, not at the set point of 150 °F.

5.9. TUNING

We have previously noted that the use of proportional only control is essentially limited to field-mounted regulators. Then why spend time on the subject? There are two reasons:

1. Tuning a proportional-only controller is step #1 in the trial-and-error tuning procedure in Table 1.1 for PI and PID controllers. It is essential that the first step be executed properly.
2. The speed of response of a loop is largely determined by the proportional mode setting. A common misconception is that the reset mode should be relied on to deliver fast response. But in reality, the integral mode has little impact on speed of response. We shall revisit the speed of response issue in the chapter on the integral mode.

Any controller without integral action will exhibit offset. Herein, we will not be concerned at all to the magnitude of the offset. We will assume that integral will eventually be added to the controller, which will eliminate the offset.

Choice of Performance Objective. The approach to tuning a proportional-only controller (in fact, any type of controller) is independent of the performance objective. For most loops, the performance objective will be one of the following or somewhere between:

Quarter decay ratio. If you ask an instrument technician to define "good control," the answer is likely to be a response with a quarter decay ratio. This response will exhibit a significant initial peak, a small second peak (one-quarter of the first), and inconsequential subsequent peaks. This is sometimes expressed as "big hump, little hump."

Minimal overshoot. Process engineers are by nature a bit on the conservative side. Overshoot and oscillations are unlikely to be consistent with their concept of good response. They prefer a response that approaches the set point as rapidly as possible, but with little overshoot if any. We shall refer to this as minimal overshoot, but "no overshoot" or "critically damped" would give the same result.

Tuning to Obtain a Quarter Decay Ratio. Let us illustrate tuning the temperature controller for the hot water process to give a response with a quarter decay ratio. We start from the initial equilibrium of SP = PV = 150 °F and with a controller gain of 2%/%. As illustrated in Figure 5.19, we proceed as follows:

Trial 1. Change the set point to 160 °F. The response does not have a second peak.

Trial 2. Increase the controller gain to 4%/%, and then change the set point to 150 °F. The second peak is now visible, but the decay ratio is closer to one eighth than one quarter.

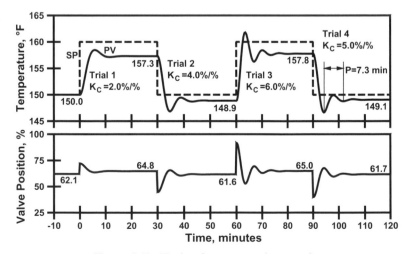

Figure 5.19. Tuning for quarter decay ratio.

Trial 3. Increase the controller gain to 6%/%, and then change the set point to 160 °F. The decay ratio is still not quite one quarter (probably around one sixth), which suggests a further increase in the controller gain. However, the overshoot of the first peak is substantial, the magnitude of the proportional kick is large, and the third peak is clearly visible. Increasing the gain even more is not advisable; in fact, a lower value for the gain should be considered.

Trial 4. Decrease the controller gain to 5%/%, and then change the set point to 150 °F. The overshoot of the first peak is considerably less, making this response preferable to the one for a gain of 6%/%.

Offset exists in Figure 5.19, so perhaps a few observations regarding offset are appropriate:

1. No attempt is made to minimize the offset. Instead, the offset was ignored.

2. Many instrument technicians get nervous when reset is removed from a controller. They correctly observe that offset is unacceptable. For the final product, this is certainly correct. But while tuning is being performed, offset can be tolerated.

3. All decisions are based on how the transition is made from one equilibrium state to the next, not on what the equilibrium states turn out to be. For trial #2 in Figure 5.19, the process begins at 157.3 °F and ends at 148.9 °F. These must be within the acceptable region of operation, but otherwise there is no concern as to the starting point and the ending point. All decisions are based on how the transition is made.

4. For a proportional-only controller, the closed-loop gain will be a number less than 1.0. Therefore, the change in the process variable from the preset equilibrium state to the next will be less than the change in the set point. For example, if the set point is increased by 10 °F, the next equilibrium value for the process variable will be no more than 10 °F higher than the current.

Tuning to Obtain Minimal Overshoot. We now illustrate tuning the hot water temperature controller to give a response with little or no overshoot. We start from the initial equilibrium of SP = PV = 150 °F and with a controller gain of 2%/%. As illustrated in Figure 5.20, we proceed as follows:

Trial 1. Change the set point to 160 °F. The response clearly exhibits overshoot.

Trial 2. Decrease the controller gain to 1%/%, and then change the set point to 150 °F. The overshoot is quite small, certainly less than 10%. One could probably consider this to meet the criteria of little or no overshoot. However, let us continue.

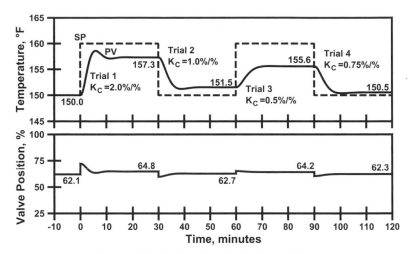

Figure 5.20. Tuning for minimal overshoot.

TABLE 5.1. Comparison of Controller Performance

		K_C	T_I	T_D	P	E_{MAX}
Controller	Objective	%/%	min	min	min	°F
P-Only	Quarter Decay	5.0	—	—	7.3	—
P-Only	No Overshoot	0.75	—	—	—	—

Trial 3. Decrease the controller gain to 0.5%/%, and then change the set point to 160 °F. This response clearly has no overshoot.

Trial 4. Increase the controller gain to 0.75%/%, and then change the set point to 150 °F. The overshoot in this response is just barely visible. This certainly meets our criteria.

Comparison of Tuning. This is just the start of our discussion of tuning the various types of controllers. Table 5.1 compares various aspects of controllers, their tuning, and the performance objectives used for tuning. We will eventually add PI and PID to this table.

One aspect that we want to include in the comparison is speed of response. One parameter for assessing speed of response is the period of the response, which is the time between the first two peaks. Table 5.1 provides a value for the period only for the controller tuned to the quarter decay ratio. A period cannot be calculated for the controller tuned to minimal overshoot.

Table 5.1 contains a column for E_{MAX}, which is the maximum departure from the set point. Because of the presence of offset, we cannot apply this parameter to proportional-only controllers, but we will be doing so for both PI and PID controllers.

5.10. SUMMARY

In any loop where aggressive action from the controller is desirable, the proportional mode must provide the primary component of the response from the controller. But in practice, the proportional mode is too often underused, with the controller gain being too low.

The best way to be sure that the proportional mode is properly adjusted is to remove all integral action and see how a proportional-only controller performs. But with no reset, the controller will not line out at its set point, exhibiting offset or droop. Hopefully you are now sufficiently comfortable with offset to ignore it.

CHAPTER 6

Integral Mode

With the exception of field-mounted regulators, virtually all controllers contain the integral mode. Our focus is on the proper use of the integral or reset mode:

Proper tuning. Situations where the reset time is too short (too much reset) are common. We present ways to detect these situations.

Speed of response. We show that it is proportional, not reset, that determines the speed of response of a loop.

Performance objective. We examine the minimal overshoot and quarter decay ratio performance objectives for both set-point and load changes.

6.1. CONTROL EQUATION

There are two possible single-mode controllers:

- Proportional-only control.
- Integral-only control.

Proportional-only was discussed in the previous chapter. We will briefly examine integral-only control, and then we proceed to PI control.

Integral-Only Control. The output of the controller is proportional to the integral of the control error, with the equation being

$$M = K_I \int_0^t E \, dt + M_0$$

where

Practical Process Control: Tuning and Troubleshooting, by Cecil L. Smith
Copyright © 2009 John Wiley & Sons, Inc.

E = Control error, %

   = SP – PV for reverse acting controller

   = PV – SP for direct acting controller

K_I = Integral gain, (%/%)/min

M = Controller output, %

M_0 = Initial controller output, %

The tuning coefficient is the integral gain K_I, whose units are (%/%)/min.

In practice, integral-only control is rarely used. Most flow loops could be, but the tradition is to use PI control with a very small controller gain. Integral-only control should be considered for loops with a very noisy measured variable. But where possible, the reasons why noise is present in the measurement should be understood, and the appropriate corrections should be made. In no case should filtering or smoothing be blindly applied.

Most plants will not have a single controller that is integral only. Therefore, it does not seem appropriate to devote any more time to integral-only control.

Proportional-Integral Control. The equation for PI control involves two terms:

Proportional—based on current value of the control error

Integral—based on the integral of the control error

The equation for a PI controller is often written as follows:

$$M = K_C \left[E + \frac{1}{T_I} \int_0^t E \, dt \right] + M_{R,0}$$

where

E = Control error, %

   = SP – PV for reverse acting controller

   = PV – SP for direct acting controller

K_C = Controller gain, %/%

T_I = Integral or reset time, min

M = Controller output, %

$M_{R,0}$ = Initial value for the controller output bias, %

   = $M_0 - K_C \, E_0$

E_0 = Initial value for control error, %

M_0 = Initial value for controller output, %

The integral mode completely eliminates offset. Let us assume the control loop lines out. At line out, both M and E must be constant. In the above equation, the left side M is constant. On the right side, the proportional term $K_C E$ is constant. The only remaining term is the integral term, which must also be constant. But in order for the output of the integrator to be constant, its integrand E must be zero. When the integral mode is present in a controller, the only possible equilibrium state is when the control error is zero. Thus, there can be no offset.

Implementation of the PI Control Equation. The following equations for a PI controller are equivalent to the equation presented previously:

$$M = K_C E + M_R$$

$$M_R = \int_0^t \frac{K_C}{T_I} E \, dt + M_{R,0}$$

where

  M_R = Controller output bias, %

In this formulation, the PI control equation is broken into a proportional part and an integral part. The proportional mode equation is the same as for a pure proportional controller, but with one exception. Being the output of the integrator, the controller output bias M_R is not constant, but changes with time. The initial condition $M_{R,0}$ for the integrator is determined by the initialization calculations to achieve bumpless transfer.

When the equations are expressed in this manner, the integral mode is adjusting the controller output bias for the proportional mode equation. Basically, the integral mode is shifting the operating line for the proportional mode. This is the basis for the observation that "integral works through proportional."

Bumpless Tuning Parameter Adjustments. Implementation of the control equations requires considerable attention to the details. Consider the following two expressions for the integral term:

$$M_R = \int_0^t \frac{K_C}{T_I} E \, dt + M_{R,0} \qquad M_R = \frac{K_C}{T_I} \int_0^t E \, dt + M_{R,0}$$

Under most situations, the two expressions are equivalent. However, there is one case where they are not: when the value of one of the tuning parameters K_C or T_I is changed. The tuning parameters are not constants; they are coefficients that are occasionally changed. In the expression on the left, changing K_C or T_I will have no effect on the current value of M_R; that is, tuning parameter adjustments will be bumpless. This is not the case for the expression on the right. The integral of the control error is unlikely to be zero. Therefore, any change in K_C or T_I will immediately affect the value of M_R, which means that tuning parameter adjustments will not be bumpless. If such details are missed, the controller can still get the job done; however, we say that the product has some "warts."

Lag on Controller Output. As the name implies, the integral mode is based on the integral of the control error. However, the integral mode can be implemented in a different way: as a first-order lag on the controller output. Starting with the equation for the integral mode expressed in its integrated form, the equivalent first-order lag equation for the integral mode is derived as follows:

$$M_R = \int_0^t \frac{K_C}{T_I} E \, dt + M_{R,0} \quad \text{Integral mode equation in integrated form.}$$

$$\frac{dM_R}{dt} = \frac{K_C}{T_I} E \qquad\qquad \text{Integral mode equation in differentiated form.}$$

$$T_I \frac{d M_R}{dt} = K_C E \qquad\qquad \text{Multiply both sides by the reset time } T_I.$$

$$T_I \frac{d M_R}{dt} = M - M_R \qquad \text{Replace } K_C E \text{ by } M - M_R \\ \text{(proportional mode equation).}$$

$$T_I \frac{d M_R}{dt} + M_R = M \qquad \text{All terms involving } M_R \text{ on left side of equation.}$$

The final form of the equation for the reset mode is the differential equation for a first-order lag. The input to this lag is the controller output M. The output of this lag is the controller output bias M_R. The time constant for the first-order lag is the reset time T_I.

Feedback Structure for PI Controller. The reset mode of a PI controller can be implemented via a first-order lag, giving the feedback configuration for the PI controller illustrated in Figure 6.1:

- The input to the first-order lag is the controller output M.
- The output of the first-order lag is the controller output bias M_R.
- The time constant of the first-order lag is the reset time.

The PI implementation in pneumatic and electronic controllers was based on this feedback structure. Some digital systems continue to provide this implementation.

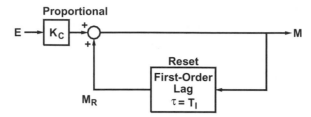

Figure 6.1. Feedback structure for PI controller.

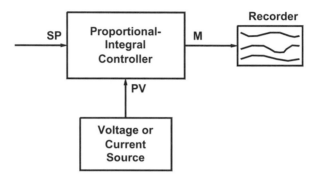

Figure 6.2. Configuration for open-loop test.

6.2. OPEN-LOOP BEHAVIOR

Most presentations of the integral mode begin by examining the open-loop response of a PI controller to a step change in the control error. The degree to which this enhances one's understanding of the integral mode is questionable. However, there are two aspects that definitely make this worth the effort:

- The term "repeats" is often incorporated into the engineering units for the reset mode tuning parameter. The basis for this reflects the characteristics of the response of a PI controller to a step change in the control error.
- A controller that contains integral action is "open-loop unstable." Understanding why this occurs is important because of another consequence of this behavior, namely, reset windup.

Configuration for Open-Loop Test. As illustrated in Figure 6.2, relatively simple equipment is required to perform the open-loop test. We will use a voltage or current source to provide the measured variable input. This permits us to specify any value we wish for the measured variable. We will connect the controller output to a recorder.

There is no connection from the output of the controller to the measured variable. Even with the controller in automatic, the configuration is still open loop. When controllers are installed in the process, changes in the controller output will lead to changes in the measured variable. In the configuration in Figure 6.2, this does not happen.

Open-Loop Test Execution. To execute the open-loop test, we have to introduce a step change in the control error. One way to do this is as follows:

1. Start with the controller in manual.
2. Set the output of the controller to 50%.
3. Adjust the voltage or current source so that the PV is 50% of span.
4. If PV tracking is not enabled, adjust the set point to equal the PV.
5. Switch the controller to automatic. There should be no change in the controller output.
6. Change the set point to 60% of span (an increase of 10% of span).
7. Record the response in the controller output, which is illustrated in Figure 6.3.

The choice of 50% for the initial values of the controller output and the PV is arbitrary. The test can be executed starting from any values for the controller output and the PV.

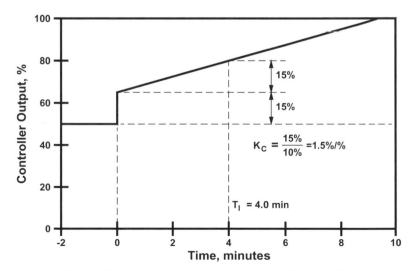

Figure 6.3. Results of open-loop test of a PI controller.

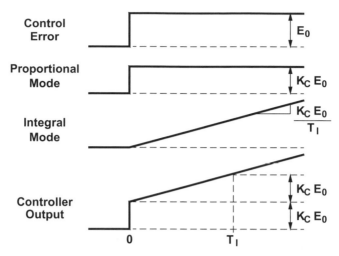

Figure 6.4. Open-loop response of PI controller to a step change in control error.

Origin of the Term "Repeats". The test procedure changes the control error from zero to some finite value E_0 at time zero. As illustrated in Figure 6.4, the combined output of the controller is a step and a ramp, which is the sum of the outputs of the individual modes:

Proportional mode—the output is a step of height $K_C E_0$.
Integral mode—the output is a ramp whose slope is $(K_C/T_I)E_0$.

A step change of height E_0 in the control error produces a step change of height $K_C E_0$ in the output from the proportional mode. There is no immediate reaction from the integral or reset mode. Instead, the output of the reset mode is a ramp of slope $(K_C/T_I)E_0$. But given enough time, the output of the reset mode will eventually change by amount $K_C E_0$. To be more specific, the time required for this to occur is the reset time T_I.

This behavior is often expressed as follows:

The reset mode output repeats the proportional mode output once each reset time.

Based on this point of view, it is logical to state the reset time as "minutes per repeat" and the reset rate as "repeats per minute." This is more commonly done for the reset rate than for the reset time.

Response to a Step Change in the Control Error. Figure 6.3 presents an example of the results of an open-loop test. The response is to the step change of 10% of span in the control error. When the set point is changed to introduce the control error, the controller output changes abruptly by 15% (the proportional kick). The reset mode then integrates the control error to produce the ramp in the controller output.

The values of the tuning parameters can be determined from the response in Figure 6.3:

1. A change of 10% in the control error led to an immediate change of 15% in the controller output. Therefore, the controller gain must be 1.5%/%.

2. It takes 4 minutes for the ramp to change the controller output by another 15%. Therefore, the reset time is 4.0 min.

These values should agree with the controller's tuning coefficients. For digital systems, they will. But for conventional pneumatic and electronic controllers, large discrepancies are sometimes observed, especially for controllers that have been in service for many years.

Open-Loop Unstable. The addition of the integral mode to a controller eliminates offset. This advantage is usually sufficient to justify the addition of the integral mode; however, there are some "side effects," one of which is very serious in a few loops.

Any controller that contains the integral mode is open-loop unstable. This is illustrated by the response in Figure 6.3. The introduction of the control error causes the output to begin to increase, and the output continues to increase until a limiting condition is attained, such as valve fully open or valve fully closed (the direction depends on the sign of the error and the action of the controller). With this behavior, the controller is open-loop unstable.

For a control loop to function properly, changes in the controller output must affect the measured variable. Such will certainly be the case under normal process operating conditions. But if for any reason the controller output ceases to affect the measured variable, the controller is in an open-loop configuration and is unstable. Usually this situation develops during some infrequent or unusual mode of process operations, and if so, the controller responds by driving its output to a limit. We shall revisit this subject later in this chapter when we examine a phenomenon called reset windup.

6.3. EFFECT OF RESET TIME

The objective of this section is to explain the effect of the reset time T_I on the behavior of the control loop. In doing so, we will use the temperature loop for hot water process as an example. The initial conditions are SP = PV = 150 °F, $M = M_R = 62.1\%$, and controller in automatic.

Effect of Reset Time on the Response. Since reset is almost always used in conjunction with proportional, we shall illustrate the effect of changing the reset time on the performance of a PI controller. We shall fix the value of the controller gain and then examine how the loop responds for various values of the reset time. We shall do this for changes in the set point and for changes in the disturbance.

As for values of the controller gain, we shall use values from Table 5.1 that were developed in the previous chapter on the proportional mode:

- For quarter decay ratio, $K_C = 5\%/\%$.
- For minimal overshoot, $K_C = 0.75\%/\%$.

We shall start with the latter of these values.

Effect of Reset Time on the Response to a Set-Point Change. The effect of reset action on offset is illustrated by the responses in Figure 6.5. For proportional-only control with a controller gain of $0.75\%/\%$, the loop lines out at $155\,°F$, which is $5\,°F$ below the set point of $160\,°F$. But whenever any reset action is present, the loop lines out at the set point. For very long reset times (such as 10 or 20 min), it takes an excessively long time to attain the set point, but it eventually does.

Short values of the reset time lead to overshoot and oscillations. For example, the response with a reset time of 1.25 min has a decay ratio of about one half. If we continue to shorten the reset time, the loop will become unstable.

What is the proper value for the reset time? The reset time should be adjusted based on the same performance criterion used to establish the controller gain. The value of $0.75\%/\%$ for the controller gain is based on minimal overshoot. We should use that same criterion for the reset time. This suggests a reset time of 5.0 min.

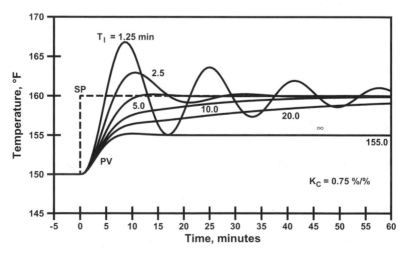

Figure 6.5. Effect of reset time on responses to a change in the set point.

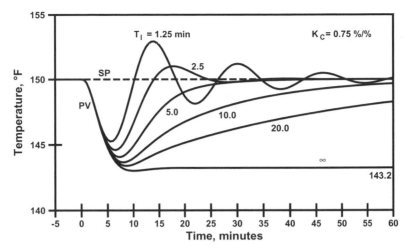

Figure 6.6. Effect of reset time on the responses to a change in hot water demand.

Effect of Reset Time on the Response to a Change in Hot Water Demand.

For the responses in Figure 6.6, the change in hot water demand is from 500 lb/min to 600 lb/min. For proportional-only control with a controller gain of 0.75%/%, the loop lines out at 143.2 °F, which is 6.8 °F below the set point of 150 °F. But whenever any reset action is present, the loop lines out at the set point. For very long reset times (such as 10 or 20 min), the hot water temperature approaches the set point very slowly.

Short values of the reset time lead to overshoot and oscillations. The response with a reset time of 1.25 min has a decay ratio of about one half. If we continue to shorten the reset time, the loop will become unstable.

The reset time should be adjusted based on the same performance criterion used to establish the controller gain, which was minimal overshoot. On this basis, the reset time should be less than 5.0 min but greater than 2.5 min. This result is slightly different from that based on the response to a set-point change.

Tuning to a Set-Point Change versus Tuning to a Load Change.

Tuning involves matching the characteristics of the controller to the characteristics of the process. The change used to elicit the process response is of secondary importance.

Of the three modes in the PID controller, only the reset mode is affected enough to be noticeable. The trend in Figure 6.7 presents the responses for reset times of 5.0 min and 3.5 min to both a change in set point and a change in hot water demand. For the set-point change, a reset time of 5.0 min gives a response with a very small overshoot. For the load change, a reset time of 5.0 min gives no overshoot at all. To obtain approximately the equivalent degree of overshoot in the load response, the reset time must be shortened to 3.5 min.

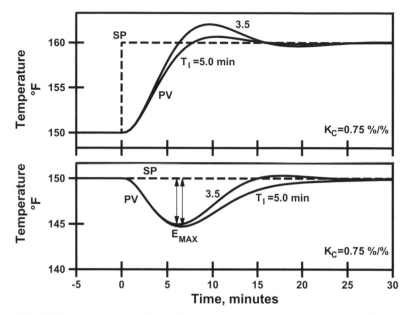

Figure 6.7. Difference in the effect of reset time on responses to a change in set point and responses to a change in load.

When tuned to a load change, the reset time will be approximately 25% shorter than when tuned to a set-point change. In the field, if you can tune to an accuracy of 25%, you are doing extremely well.

Analysis of the Response to a Change in Hot Water Demand. When there is an increase in the hot water demand, the temperature will drop below the set point by an amount that depends on two factors:

1. The magnitude of the increase in hot water demand.
2. The performance of the controller.

Suppose the minimum acceptable hot water temperature is 140 °F. We must operate with a set point above 140 °F; how much above depends on the above two factors. We have chosen to operate with a set point of 150 °F. But since it costs more to heat water to 150 °F than to 140 °F, there is a cost for this.

In response to an increase in hot water demand, the crucial point is at the maximum departure from the set point or, alternatively, the maximum value of the control error E_{MAX}. This value is indicated on the responses in Figure 6.7. If the corresponding hot water temperature is above 140 °F, the performance is acceptable. But if it is below 140 °F, we have a problem. In the responses in Figure 6.7, the reset time has little effect on this value. For a given upset, the maximum departure from the set point is largely determined by the controller gain.

6.4. PI CONTROL OF SIMPLE MODELS

Our objective is to illustrate the following points for the integral mode:

1. For nonintegrating processes, the appropriate value of the reset time is the time constant of the process.
2. For integrating processes, use of the integral mode in the controller leads to a "double-integrator" in the loop, which raises serious issues with regard to the stability of the loop.

PI Control of a Time Constant Process. Consider using a PI controller for a time constant process. For a loop gain of 2.0, the responses in Figure 6.8 illustrate the following aspects pertaining to the value of the reset time:

- $T_I > \tau$. The response approaches the set point from below. The longer the value of the reset time, the slower the approach.
- $T_I < \tau$. The response exhibits overshoot. The shorter the reset time, the greater the overshoot.
- $T_I = \tau$. The response lines out at the set point with no overshoot.

The above observations apply regardless of the value of the controller gain. The conclusion is that the appropriate value for the controller's reset time T_I is the value of the process time constant τ.

With the reset time T_I set equal to the process time constant τ, the responses in Figure 6.9 show the effect of the loop gain KK_C on the closed-loop behavior. The effect of the loop gain on the closed-loop time constant τ_{CL} is given by the following expression:

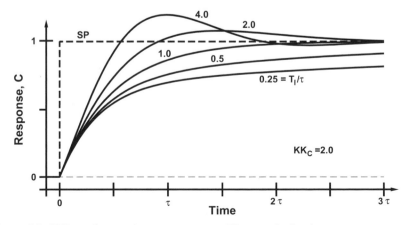

Figure 6.8. Effect of reset time on response, PI control of a time constant process.

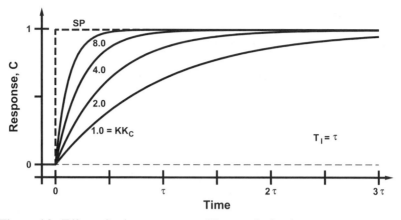

Figure 6.9. Effect of gain on response, PI control of a time constant process.

$$\tau_{CL} = \frac{\tau}{K\,K_C}$$

Increasing the controller gain decreases the closed-loop time constant, which leads to a faster response to changes in the set point.

Integral-Only Control of an Integrating Process. Now consider using an integral-only controller for an integrating process. The controller provides only one adjustable parameter, the integral gain K_I whose units are $(\%/\%)$/min.

For integrating processes, one has to be very careful when using a controller that contains integral or reset action. There is one integrator in the process. With another integrator in the controller, serious issues develop with regard to stability. This should become apparent when we examine the responses.

For integral control of an integrating process, the loop gain is $K\,K_I$. However, let us examine the engineering units for this product:

$$[(\%/\%)/\min]\times[(\%/\%)/\min] = 1/\min^2 = \min^{-2}$$

The $(\%/\%)$ component of K_I is the reciprocal of the $(\%/\%)$ component of K, so these cancel (just as in the product $K\,K_C$ for nonintegrating processes). The quantity $(K\,K_I)^{-\frac{1}{2}}$ has the units of time.

The responses in Figure 6.10 show the effect of the loop gain $K\,K_I$ on the closed-loop behavior. All responses are undamped sinusoids. Increasing the loop gain has the following effect:

1. The frequency of the sinusoid increases. The frequency (in radians per unit time) of the sinusoid is $(K\,K_I)^{\frac{1}{2}}$.
2. The amplitude of the sinusoid decreases.

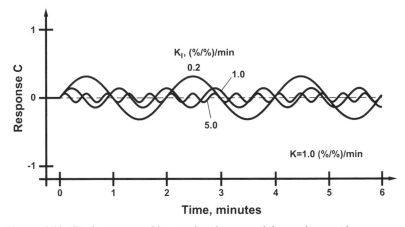

Figure 6.10. Performance of integral-only control for an integrating process.

Increasing the reset gain decreases the variance in the controlled variable. The reset gain can often be increased to the point that the resulting sinusoid in the controlled variable is totally within the acceptable range. If one makes all decisions solely on the behavior of the controlled variable, the logical approach would be to increase the reset gain until the variance is reduced to an acceptable value. Occasionally one encounters level loops where this procedure has been followed. However, the reduction in the variance in the controlled variable is often at the expense of a large increase in the variance in the manipulated variable, which usually has undesirable side effects.

Consider using a PI controller for an integrating process. If the controller has a low controller gain and a short reset time, the behavior approaches that of an integral-only controller. In practice, the process will have dynamic components in addition to the integrator, so the loop will be unstable if the reset time is too short.

In tuning level controllers, a common mistake is to make the reset time too short, resulting in oscillations. If overshoot or oscillations are excessive, the conventional wisdom is to decrease the controller gain. But this does not always work for level loops. If the process is integrating and the reset time is too short, reducing the controller gain will not compensate for this mistake; it will only make the period of the cycle longer and longer.

6.5. TUNING

The trial-and-error approach for tuning a PI controller involves steps 1 and 2 of the tuning procedure in Table 1.1. The approach is to first adjust the controller gain (with all reset action removed), and then to adjust the reset time. Basically we begin by tuning a proportional-only controller. Reset is then added to remove the offset.

This approach suggests that the controller gain for a PI controller is the same as the controller gain for a proportional-only controller. According to the Ziegler–Nichols tuning equations, the controller gain for a PI controller is 10% less than the controller gain for a proportional-only controller. This difference is insignificant in the context of trial-and-error tuning.

The addition of reset action to a controller will reduce the stability margin somewhat. If the procedure in Table 1.1 is followed, the reduction is small. But reset never increases the stability margin. Therefore, when adjusting the controller gain, one is advised to stay on the conservative side of the performance objective.

Tuning for Minimal Overshoot. For the hot water process, we will tune a PI controller to give a response with little or no overshoot. In Chapter 5 on the proportional mode, it was concluded that a proportional-only controller with a controller gain of 0.75%/% gives a response with little or no overshoot (Table 5.1). Using this as the starting point, let us add reset, as illustrated in Figure 6.11:

Trial 1. Starting with a reset time of 15 min, change the set point to 160 °F. The response approaches the set point far too slowly.

Trial 2. Shorten the reset time to 7.5 min, and then change the set point to 150 °F. The response still approaches the set point too slowly.

Trial 3. Shorten the reset time to 4.0 min, and then change the set point to 160 °F. This response has a bit too much overshoot.

Trial 4. Lengthen the reset time to 5.0 min, and then change the set point to 150 °F. The overshoot in this response is just barely visible. This certainly meets our criteria.

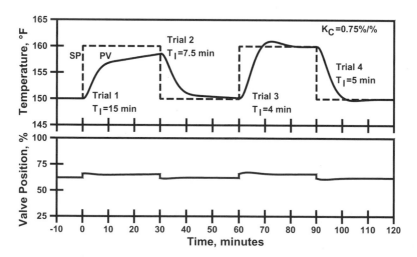

Figure 6.11. Tuning to minimal overshoot criterion.

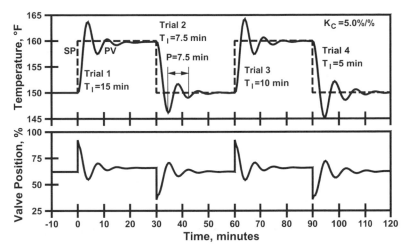

Figure 6.12. Tuning to quarter decay ratio criterion.

Tuning for Quarter Decay Ratio. For the hot water process, we will tune a PI controller to give a response with a quarter decay ratio. In Chapter 5 (Table 5.1), it was concluded that a proportional-only controller with a controller gain of 5.0%/% exhibited a quarter decay ratio. Using this as the starting point, let us add reset, as illustrated in Figure 6.12:

Trial 1. Starting with a reset time of 15 min, change the set point to 160 °F. The response approaches the set point too slowly.

Trial 2. Shorten the reset time to 7.5 min, and then change the set point to 150 °F. The response looks pretty good, but we shall try a couple more values of the reset time.

Trial 3. Lengthen the reset time to 10 min, and then change the set point to 160 °F. The slow approach to the set point can be detected in this response.

Trial 4. Shorten the reset time to 5.0 min, and then change the set point to 150 °F. This response exhibits more overshoot and oscillation than the response for a reset time of 7.5 min. It seems that a value of 7.5 min for the reset time is appropriate.

Alternative Quarter Decay Ratio. For the hot water process, a controller gain of 0.75%/% and a reset time of 5.0 min produce a response with a very small overshoot. But if you want a response with a quarter decay ratio, one way to obtain it is to shorten the reset time. As illustrated in Figure 6.13, we proceed as follows:

Trial 1. We know that a reset time of 5.0 min produces a response with a very small overshoot. Therefore, let us start with a reset time of 2.5 min.

and change the set point to 160 °F. The response exhibits overshoot, but the small second peak is much too small.

Trial 2. Shorten the reset time to 1.5 min (a reduction of not quite 50%), and then change the set point to 150 °F. The peaks are −5.4 °F and −1.8 °F, which results in a decay ratio of 1/3.

Trial 3. Lengthen the reset time to 1.8 min (an increase of 20%), and then change the set point to 160 °F. The peaks are 4.7 °F and 1.1 °F, which results in a decay ratio only slightly below one quarter.

We now have two different tuning adjustments that will produce a response with a quarter decay ratio. Many more adjustments exist, which creates a dilemma when tuning a PI controller.

Comparison of Controller Performance. Table 6.1 is Table 5.1 after adding the PI tuning results for the following three cases:

No Overshoot. The final results were a controller gain of 0.75%/% and a reset time of 5.0 min.

Quarter decay ratio. The final results were a controller gain of 5%/% and a reset time of 7.5 min. For this response, we can also obtain the period, which is 7.5 min (see Figure 6.12). For proportional-only control, the period was 7.3 min. The addition of reset to the controller has almost no effect on the period, which is a good indicator of the speed of response of the loop. Also note that the period (7.5 min) is exactly equal to the reset time (7.5 min). We shall have more to say about this shortly.

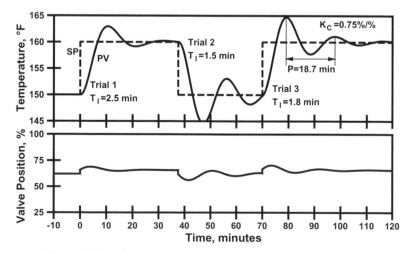

Figure 6.13. Alternate tuning to quarter decay ratio criterion.

TABLE 6.1. Comparison of Controller Performance

Controller	Objective	K_C %/%	T_I min	T_D min	P min	E_{MAX} °F
P-Only	Quarter Decay	5.0	—	—	7.3	—
P-Only	No Overshoot	0.75	—	—	—	—
PI	No Overshoot	0.75	5.0	—	—	—
PI	¼ Decay	5.0	7.5	—	7.5	—
PI	Alternative ¼ Decay	0.75	1.8	—	18.7	—

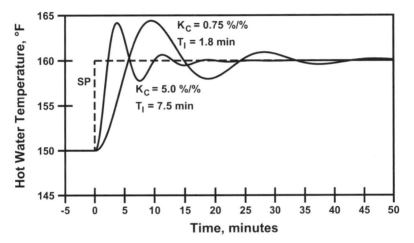

Figure 6.14. Two responses with a quarter decay ratio.

Alternative quarter decay ratio. The final results were a controller gain of 0.75%/% and a reset time of 1.8 min. For this response, we can also obtain the period, which is 18.7 min (see Figure 6.13). This response has a much shorter reset time (1.8 min versus 7.5 min); however, it has a much longer period (18.7 min versus 7.5 min). Does shortening the reset time (increasing the reset rate) lead to faster response? This example suggests that this is not the case.

Tuning Dilemma for a Proportional-Integral Controller. With proportional-only control, a controller gain of 5%/% gives a response with a quarter decay ratio. For any value of the controller gain less than 5%/%, the reset time can be adjusted to obtain a response with a quarter decay ratio.

Figure 6.14 presents two responses that have the same decay ratio. They are not equivalent. The tuning for the two cases is as follows:

- $K_C = 5\%/\%; T_I = 7.5$ min.
- $K_C = 0.75\%/\%; T_I = 1.8$ min.

The first has the higher controller gain; the second has the shorter reset time. Which produces the faster response? The one with the higher gain, not the one with the shorter reset time. The speed of response of a loop depends on the value of the controller gain; the value of the reset time has little effect on the speed of response.

6.6. SPEED OF RESPONSE

Although we routinely tune controllers based on the response to a change in the set point, the primary role of most loops, including the temperature loop for the hot water process, is to respond to changes in a disturbance or load. The responses in Figure 6.15 are for a change in hot water demand from 500 lb/min to 600 lb/min, and then back to 500 lb/min Because of the nonlinear behavior of the hot water process with regard to changes in throughput, the responses to an increase in hot water demand are not quite the same as the responses to a decrease.

From just an inspection of the responses, it should be apparent that the dynamics of the loop are far faster when the controller gain is 5%/%. This response was achieved by

- Selecting the quarter decay ratio criterion over the minimal overshoot criterion.
- Tuning the controller such that the controller gain is increased as much as possible.

We want to examine the importance of both of these.

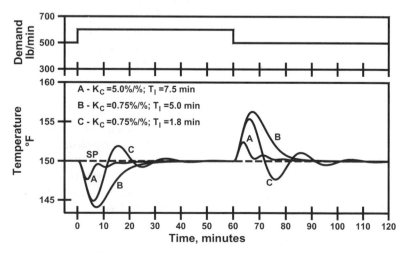

Figure 6.15. Responses to changes in hot water demand.

Maximum Departure from the Set Point. When analyzing the response to a disturbance or load, the most important aspect is usually the maximum departure from the set point (or the maximum value of the control error). For the above responses, the values are as follows:

Controller gain	5.0%/%	0.75%/%	0.75%/%
Reset time	7.5 min	5.0 min	1.8 min
Peak error, increase in demand	−2.3 °F	−5.9 °F	−5.1 °F
Peak error, decrease in demand	2.4 °F	6.3 °F	5.4 °F

The magnitude of the peak error reflects the speed of response of the loop (the smaller the peak error, the faster the loop). Increasing the controller gain has a dramatic effect on reducing the peak error. Shortening the reset time has only a small-to-modest effect. For a gain of 0.75%/%, shortening the reset time from 5.0 to 1.8 min (a factor of 2.8) decreases the peak error on an increase in hot water demand from 5.9 °F to 5.1 °F, which is a reduction of only 14%.

Comparison of Controller Tuning. Table 6.2 is an update of Table 6.1 to include the values for the maximum departures from the set point. These values are in the column for E_{MAX} and are for an increase in hot water demand, which causes the hot water temperature to drop. For the hot water process, decreases in hot water temperature are of more concern (if the decrease is too much, the water is no longer "hot"). The sign has also been dropped from the values for the maximum departure from the set point.

Selection of the Performance Objective. As we have stated many times, most process engineers are concerned about overshoot, usually preferring responses with little or no overshoot. However, this is invariably based on responses to a change in the set point. For the set-point change, the quarter decay ratio responses typically have an overshoot of 30% to 40%. Most process engineers will advance an argument that such overshoot is simply not acceptable.

TABLE 6.2. Comparison of Controller Performance

Controller	Objective	K_C %/%	T_I min	T_D min	P min	E_{MAX} °F
P-Only	Quarter Decay	5.0	—	—	7.3	—
P-Only	No Overshoot	0.75	—	—	—	—
PI	No Overshoot	0.75	5.0	—	—	5.9
PI	¼ Decay	5.0	7.5	—	7.5	2.3
PI	Alternative ¼ Decay	0.75	1.8	—	18.7	5.1

However, the tuning that gives the most overshoot for a change in the set point gives the smallest peak error for the responses to a change in hot water demand. Especially since changes in the hot water temperature set point are very infrequent, we should base our decisions on the response of the hot water temperature to a change in hot water demand. The smallest peak error is for the quarter decay ratio response obtained with the controller gain set to 5%/%.

Benefit of Tighter Control (faster response).   Why is the hot water process being operated at a set point of 150 °F? Suppose the minimum acceptable value for the hot water temperature is 140 °F. We cannot operate with a set point of 140 °F; any increase in hot water demand will cause the hot water temperature to fall below the minimum. By operating with a set point of 150 °F, the hot water temperature must drop 10 °F before we are in trouble.

Why operate only 10 °F above the minimum? Since it costs more to heat water to 150 °F than to 140 °F, a cost is associated with operating above the minimum. We have to operate somewhat above the minimum, but no more than is necessary. Normally the margin is established from operating experience. It depends on the magnitude of the disturbances and on the performance of the controller. If we can improve the performance of the control system so as to reduce the peak error for a given disturbance, then we should be able to operate with a set point closer to 140 °F. The incentive for improved control is to adjust operating targets to more efficient values. For the hot water process, this means moving the set point closer to 140 °F.

6.7. AVOIDING SLOPPY TUNING

There is a "quick-and-dirty" approach to tuning a PI controller:

- Specify an arbitrary value for the controller gain.
- Adjust the reset time to give a quarter decay ratio.

How one obtains a value for the controller gain is obscure. However, no effort is spent properly tuning the gain. Conservatively low values are the norm. The users of this approach feel good because the response has a quarter decay ratio; however, the controller does not deliver the performance of which it is capable.

This approach reduces the time required to tune the controller. It also avoids operating the controller for a period of time when offset is present, which is a situation that many find uncomfortable. But the result is a loop that responds more slowly than necessary. A penalty is always administered for this. If the temperature controller for the hot water process is tuned this way, the plant will operate with a set point farther above 140 °F than necessary. This consumes extra steam, which costs money.

Detecting Sloppy Tuning. The response to the third set-point change in Figure 6.13 is for a sloppily tuned controller, specifically with a controller gain of 0.75%/% and a reset time of 1.8 min. The decay ratio for this response is very close to one quarter. Suppose you are shown only this response. How can we tell that the controller is poorly tuned?

There are two ways to avoid sloppy tuning:

- Check the proportional-only response.
- Verify that the current value of the reset time is consistent with the period of the response.

Checking the proportional-only response is probably the most reliable approach. However, it is easier to compare the value of the reset time with the period of the response.

Compare Reset Time with Period. The appropriate value for the reset time is related to the period P of the response and to the current value of the reset time. The period can be determined from either the load response as in Figure 6.15 or the set-point response as in Figure 6.13. The values will be modestly, but not sufficiently, different to change the conclusions. There are two options:

- In our original example of tuning a PI controller to a quarter decay ratio, we noted that the reset time was equal to the period of the response. This is not always exact, but the reset time should satisfy the following inequality:

$$T_I > \frac{P}{2}$$

  As the period is 18.7 min, this suggests that the reset time should be longer than 9.4 min. The reset time of 1.8 min is substantially shorter.
- Compute the appropriate value of reset time T_I' from the period P and the current value of T_I (1):

$$T_I' = 0.5 \, P \, \tan^{-1}(2\,\pi\,T_I/P)$$
$$= 0.5 \times 18.7 \times \tan^{-1}(6.28 \times 1.8/18.7)$$
$$= 5.1 \, \text{min}$$

Our reset time is much shorter than this, which suggests that the controller is not properly tuned.

6.8. SUPPRESSING THE PROPORTIONAL KICK

For the hot water process, the response in Figure 6.16 for "Proportional on E" is to a change in the hot water temperature set point from 150°F to 160°F, with the controller tuned to a quarter decay ratio. There is an abrupt change in the controller output at the time the set-point change is introduced. This change, known as the "proportional kick," depends on two values:

- The change in the set point. For the example in Figure 6.16, the set-point change is 10°F or 5% of span (measurement range is 50°F to 250°F).
- The controller gain. The responses in Figure 6.16 are for a controller gain of 5%/%.

With the proportional mode based on the control error, the magnitude of the proportional kick is the product of these two values. For the example in Figure 6.16, the proportional kick is 5% × 5%/% = 25%. The set-point change of 10°F causes the controller to immediately change its output from 62.1% to 87.1%.

Consequence of Proportional Kick. In most loops, the proportional kick is advantageous because it enhances the speed of response to a set-point change. However, a couple of issues occasionally develop:

Final control element. In a few loops, the proportional kick causes problems for the final control element. It is rarely a problem for control valves. However, when the final control element is a relatively slow

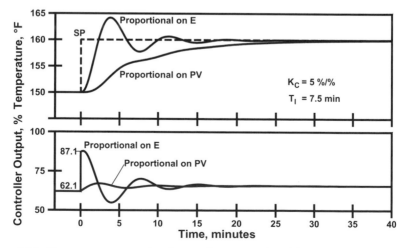

Figure 6.16. Response to set-point change for proportional based on E (proportional kick) and for proportional based on PV (no proportional kick).

mechanical device such as a screw feeder or a conveyor belt, abrupt changes in the speed are not desirable. We have to suppress the proportional kick somehow.

Process upset. Quickly opening a control valve from 62.1% to 87.1% causes a large increase in the flow through the valve. Sometimes there are undesirable consequences of this. Suppose the hot water process is the major consumer of steam at the plant. Increasing the flow of steam to the hot water heater would likely drop the pressure in the steam header, thus upsetting other parts of the plant.

Proportional Based on PV. One approach for eliminating the proportional kick is to base the proportional action on the process variable PV instead of the control error E. This modification to the control equation for a PI controller is as follows:

$$M = K_C \left[-PV + \frac{1}{T_I} \int_0^t E \, dt \right] + M_{R,0}$$

where

$M_{R,0}$ = Initial value for controller output bias, %
    = $M_0 + K_C \, PV_0$

When all modes are based on the control error, the action of the controller can be implemented by computing the control error E as SP – PV for a reverse-acting controller and PV – SP for a direct-acting controller. But when one or more modes are based on the process variable, the action must be incorporated via the controller gain K_C. If the control error E is always SP – PV, then the action of the controller is reverse if $K_C > 0$ and direct if $K_C < 0$.

When the proportional mode is based on the process variable, the controller must always contain some reset action. The set point only affects the controller output through the control error E. When the proportional mode is based on the PV as in the above control equation, the control error E only appears in the integral.

Performance for Set-Point Changes. For proportional based on the PV, Figure 6.16 presents the response of the temperature loop of the hot water process to a set-point change from 150 °F to 160 °F. Note the following:

- When the proportional action is based on the process variable, there is no proportional kick.
- As compared with the response with proportional based on the control error E, the response is significantly slower. Sometimes the control equation with proportional based on PV is referred to as the "slow set-point

equation." If fast response to a set-point change is required, the proportional mode should be based on the control error E.

For loops such as the hot water temperature loop, set-point changes are infrequent and usually small. The primary purpose of the controller is to respond
to changes in the hot water demand. To minimize the maximum departure
from the set point in such responses, the controller gain should be increased
as much as possible. But as the controller gain is increased, the magnitude of
the proportional kick for a set-point change also increases. When objections
are raised to such large changes in the controller output, one way to address
this concern is to base the proportional action on the PV.

Performance for Load Changes. Even with proportional based on the
control error E, there is no proportional kick associated with the response to
a load or disturbance change. With proportional based on the PV, the response
to a load change is identical to the response when the proportional is based
on the control error E. For the hot water process, Figure 6.17 presents the
response to an increase in hot water demand from 500 lb/min to 600 lb/min.

PV Tracking Disabled. When the proportional is based on the PV, the proportional kick is always suppressed. There is a way to make suppression of the
proportional kick at the discretion of the process operators:

1. Base the proportional action on the control error E.
2. Disable PV tracking.

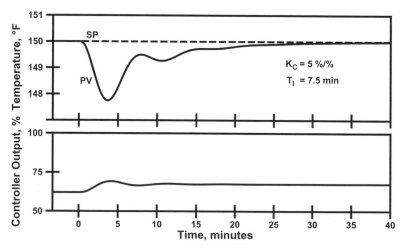

Figure 6.17. Response to demand change is the same for proportional based on PV as
for proportional based on E.

If the set point is changed with the controller in automatic, the proportional kick will occur. But when the proportional kick is not desired, the following procedure gives the same response as when the proportional is based on the PV:

1. Switch the temperature controller to manual.
2. Change the set point from 150 °F to 160 °F.
3. Switch the temperature controller to automatic.

Upon switching the loop to automatic, the response will be exactly the same as the response in Figure 6.16 for "Proportional on PV."

Ramping the Set Point. Most digital implementations of the PID control equation provide for ramping the change in the set point. However, they differ in the details.

Figure 6.18 presents an example of set-point ramping for the hot water process. The operator changes the temperature set point from 150 °F to 160 °F. But internally or "under the hood," the set-point change is implemented over 10 minutes, giving a set-point ramp rate of 1 °F/min. Although the proportional is based on the control error E, there is no proportional kick in the controller output.

With this approach, the following two factors are separated:

1. The controller is tuned so that its characteristics reflect the process characteristics.
2. The set-point ramp rate can be adjusted to address issues such as the proportional kick.

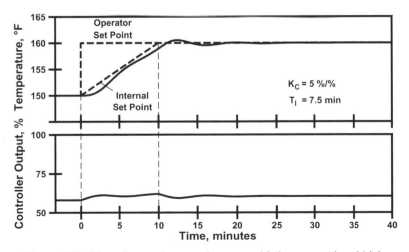

Figure 6.18. Use of set-point ramping to avoid the proportional kick.

Ramping the set point has no effect on the response to a load or disturbance change.

6.9. WINDUP PROTECTION

Any controller than contains the integral mode is susceptible to a phenomenon known as windup. Since this phenomenon is associated specifically with the integral mode, it is often referred to as "reset windup." Logic for windup protection is incorporated into the implementation of the PID control equation in all digital systems. There are several approaches to windup protection, four of which will be explained below. For most loops, any implementation of windup protection will be acceptable, but for an occasional loop, a subtle aspect of the windup protection logic proves to be significant.

We first begin with an example that illustrates windup. We then remove the reset mode to show that windup does not occur. We then explain different approaches to windup protection. We conclude by discussing some of the limitations of the windup protection mechanisms.

What Is Windup. We shall illustrate a situation where windup occurs in the temperature controller for the hot water process. For the process to produce hot water at a temperature of 150 °F, the maximum hot water demand is just over 1200 lb/min. Let us examine what happens when the hot water demand exceeds this value. We use the controller tuning for minimal overshoot. As per previous results, the controller gain is 0.75%/%. However, the process is dynamically faster at high water flows, so the reset time has been shortened to 3 min. For this example, the controller has no windup protection.

Figure 6.19 illustrates the following sequence of events:

1. Initially, the hot water demand is 1150 lb/min. The steam valve must be 95% open to maintain the hot water temperature at 150 °F.
2. The hot water demand increases to 1350 lb/min. The controller drives the valve fully open, and the hot water temperature drops to 142.7 °F. Actually, this is not a control problem. The process is unable to deliver 1350 lb/min of hot water at 150 °F.
3. Without windup protection, note what occurs while the process capacity is exceeded. The hot water temperature (142.7 °F) is below the set point (150 °F), so the controller continues to increase its output. With no windup protection, the controller is essentially unaware when the control valve is driven fully open. This will continue indefinitely.
4. The hot water demand is decreased to 1150 lb/min. At this time, the controller output has increased to 113.6%. However, this means that the control valve is fully open, and it will remain fully open for 30 minutes

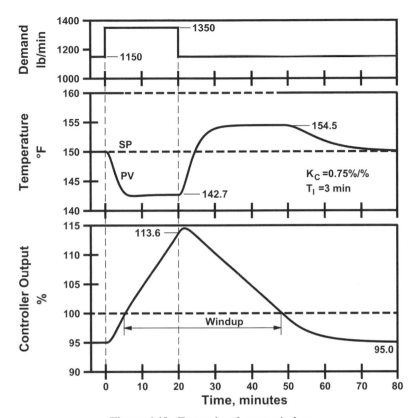

Figure 6.19. Example of reset windup.

or so. The hot water temperature increases to 154.5 °F, which is the equilibrium value for a hot water demand of 1150 lb/min and a fully open control valve. With the hot water temperature above its set point, the controller output begins to decrease slowly. However, the hot water valve remains fully open until the controller output decreases to 100%. Once the controller output drops below 100%, the loop recovers, eventually lining out at 150 °F.

The behavior in Figure 6.19 is referred to as windup. In this case, the controller is being asked to operate the process beyond its limit (it cannot produce 1350 lb/min of hot water at 150 °F). So the controller drives its output to a limit (valve fully open). But without windup protection, the controller drives its output beyond the limit. The controller is said to be "winding up." But if a controller "winds up," it subsequently has to "unwind." In the example in Figure 6.19, the controller "winds up" by driving its output to 113.6%. When the hot water demand is reduced to within the operating capacity of the process, the controller begins to "unwind," that is, reducing its output from

113.6% to 100%. But while the controller is "unwinding," the control valve remains fully open. For the hot water process, the hot water temperature is above its desired value throughout the time that the controller is "unwinding."

Facets of Windup. There are three statements that apply to windup:

- Windup occurs when a controller "unscrews its handle on the process." In the example in Figure 6.19, the controller increases its output above 100% (to 113.6%). This is analogous to unscrewing the handle on a hand valve. But before the valve can start to close, the handle has to be screwed back on. This is occurring as the controller reduces its output from 113.6% to 100%.
- A controller is "in windup" when changes in its output have no effect on the process variable. In the above example, this situation exists from the time the control valve is driven fully open to the time that the control valve begins to close. One way to determine whether windup can occur is to ask the question "are there any circumstances where changes in the controller output have no effect on the measured variable?"
- In the above example, the control valve is fully open at a time that the hot water temperature is 154.5 °F, which is above the set point of 150 °F. This makes no sense! If the water is too hot, then why is the control valve fully open? There are two aspects:
  1. The controller has driven its output to a limit (either fully open or fully closed).
  2. For the current value of the measured variable relative to the set point, this condition is unreasonable.

Reset Windup. Windup is commonly referred to as "reset windup." The potential for windup only exists when the integral or reset mode is being used. With the exception of field-mounted regulators, virtually all controllers are tuned with some integral or reset.

In the trend in Figure 6.20, the controller is proportional-only with a gain of 5%/%, which we previously concluded produces a response with a quarter decay ratio. The process is unable to heat 1350 lb/min of water at 150 °F. When the demand is increased to 1350 lb/min, the controller drives the valve fully open and the temperature drops to 142.7 °F. The controller output increases to 113.2% and lines out at this value. This type of behavior is exhibited by proportional action; in the trend in Figure 6.19, the reset action imparted a ramp to the controller output that continues until some limiting condition is reached.

When the hot water demand is reduced to 1150 lb/min, the hot water temperature begins to increase, which causes the proportional action to reduce the controller output promptly. The response in Figure 6.20 does not exhibit

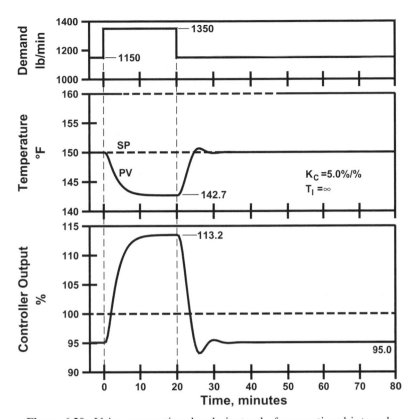

Figure 6.20. Using proportional-only instead of proportional-integral.

the situation that the control valve is fully open with the hot water temperature above its set point. The controller has driven its output beyond that required to open the valve fully, but there is no windup.

One way to prevent windup is to remove the integral or reset action from the controller. However, the controller no longer lines out at its set point and exhibits offset. Therefore, this solution to windup is rarely pursued.

Controller Output Limits. In the examples in Figures 6.19 and 6.20, no limits are imposed on the acceptable values for the controller output. An ideal control valve operates between 0% and 100%. But just because a controller drives its output beyond this range does not necessarily mean that windup is present. Windup results when the reset action drives the output above 100% or below 0%.

Most digital implementations provide configuration parameters for the controller output limits M_L and M_U, the values being 0% and 100% for an ideal control valve. But at the expense of some windup, the controller output limits are normally set slightly beyond the limits for an ideal control valve. To make sure that the controller can fully close the control valve, the lower output limit is typically set to a value slightly below 0%. If a control valve does not fully close, someone will probably notice. If a control valve does not fully open, no one is likely to notice, but an overrange is normally provided.

In commercial systems, the default for the overrange differs from one manufacturer to the next, but most are 5% or more. The user may specify lesser values, but not greater values. Herein we shall use a 2% overrange, making the lower output limit −2% and the upper output limit 102%.

Implementations of Windup Protection.   All digital implementations of the PI or PID control equation provide some form of windup protection. To prevent reset windup, the reset integrator must be somehow disabled when the output attains either output limit. We will describe four options for providing windup protection:

1. Integral limits.
2. Reset feedback with output limits.
3. Bias freeze.
4. Velocity algorithm windup protection.

For most loops, any of the above forms would be acceptable. However, there are a few exceptions:

1. Loops that function intermittently. Under normal process operations, the output of the controller is at one of the output limits and should remain there. But at times, the controller must quickly and smoothly bring the valve off the limit. In these applications, the velocity algorithm windup protection usually brings the valve off the limit prematurely.
2. Batch heatups. Material is charged to a vessel, and a large increase made in the temperature controller set point. The velocity algorithm windup protection is preferable for this application.

A few manufacturers provide options pertaining to the form of windup protection, but most provide only one form. Just finding out exactly what a given manufacturer has implemented can be challenging. You are often told "we have windup protection; don't worry about it." But there are a few loops where it does matter. When a digital controller drives its output to a limit, it should come off that limit smoothly and expeditiously. If not, something is wrong, and it is likely to involve some windup issue.

PI Control with Integral Limits.   The relationships for this form of windup protection are as follows:

Integral mode: $M_R = \int_0^t \dfrac{K_C}{T_I} E\,dt + M_{R,0}$

Windup protection: $M_R = \max(M_R, M_L)$
$M_R = \min(M_R, M_U)$

Proportional mode: $M = K_C E + M_R$

The equations must be evaluated in the above order; that is,

- Calculate a new value for the controller output bias M_R.
- Apply the controller output limits to the controller output bias M_R.
- Perform the proportional-plus-bias calculation to obtain the controller output M. Although not required for windup protection, the output limits are normally applied to M as well as M_R.

Figure 6.21 presents the response of the hot water process when this form of windup protection is used. The loop recovers from the limiting condition more expeditiously. However, this form of windup protection has one characteristic of note. Specifically, once the controller output bias M_R has attained an output limit (in this case, the upper output limit of 102%), it will remain at the upper

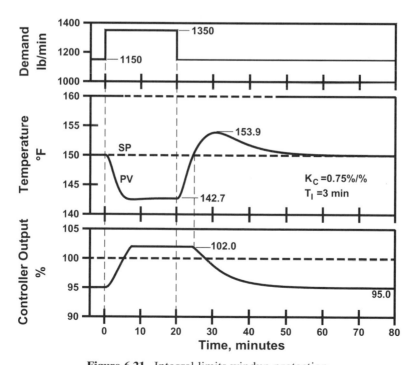

Figure 6.21. Integral limits windup protection.

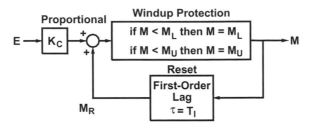

Figure 6.22. Feedback structure for PI controller with output limits for windup protection.

output limit until the process variable crosses the set point (the sign of the control error changes). As noted in Figure 6.21, the controller output remains at 102% until the hot water temperature exceeds the set point of 150 °F. In most cases, this assures some overshoot (3.9 °F in Figure 6.21), but an amount that is tolerable for most loops.

Reset Feedback PI with Output Limits. The relationships for this form of windup protection are as follows:

$$\text{Proportional mode:}\quad M = K_C\, E + M_R$$

$$\text{Windup protection:}\quad M = \max(M, M_L)$$
$$M = \min(M, M_U)$$

$$\text{Integral mode:}\quad T_I\,\frac{dM_R}{dt} + M_R = M$$

Incorporating the windup protection into Figure 6.1 gives the block diagram in Figure 6.22. The equations must be evaluated in the above order; that is,

- Perform the proportional-plus-bias calculation.
- Apply the controller output limits to the controller output M.
- With this value of M as the input, calculate the controller output bias M_R using the equation for a first-order lag.

In most cases, the performance of this form of windup protection is very similar to and sometimes (but not always) identical to the performance of the windup protection based on integral limits. This is the case for our example for the hot water process. The response when this form of windup protection is used is indistinguishable from the trend in Figure 6.21 for the integral limits form of windup protection.

PI Control with Bias Freeze. The relationships for this form of windup protection are as follows:

Integral mode: if $(M > M_L)$ and $(M < M_U)$

$$\text{then } M_R = \int_0^t \frac{K_C}{T_I} E \, dt + M_{R,0}$$

Proportional mode: $M = K_C E + M_R$

The equations must be evaluated in the above order; that is,

- If the current value of the control output M is within the output limits, then calculate a new value for the controller output bias M_R. If the current value of the control output M is at either output limit, the bias is "frozen."
- Perform the proportional-plus-bias calculation.

Figure 6.23 presents the response of the hot water process when this form of windup protection is used. Specifically note that the controller output comes off the upper output limit before the temperature crosses the set point. There is still some overshoot (3.5 °F for this trend, as compared with 3.9 °F for the previous trend). This overshoot is because a value of 102% for the upper output limit still results in some windup. If the upper output limit could be

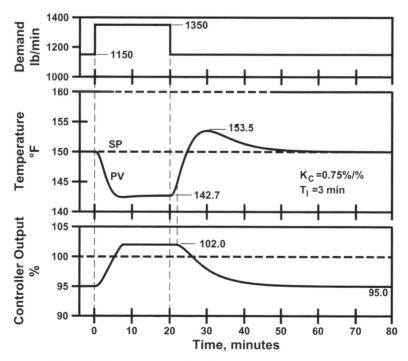

Figure 6.23. PI control with bias freeze windup protection.

perfectly matched to valve fully open, almost all of the overshoot would disappear. But in practice, perfect matching is not realistic, so some overshoot will result (unless the upper output limit is intentionally set so that the control valve does not quite fully open).

PI Control with Velocity Algorithm Windup Protection. This form of windup protection originates naturally when the velocity form of the control equation is used. The relationships are as follows:

Velocity algorithm: $\Delta M_n = K_C \left[(E_n - E_{n-1}) + \dfrac{1}{T_I} E_n \right]$

Windup protection: if $(M_n \le M_L)$ then $\Delta M_n = \max(\Delta M_n, 0)$
                              if $(M_n \ge M_U)$ then $\Delta M_n = \min(\Delta M_n, 0)$

Position calculation: $M_n = \Delta M_n + M_{n-1}$

The velocity algorithm calculates the change ΔM_n for the controller output. Windup protection consists of the following logic:

- If the output is at the lower output limit M_L, then negative values of ΔM_n are ignored. However, positive values of ΔM_n are accepted.
- If the output is at the upper output limit M_U, then positive values of ΔM_n are ignored. However, negative values of ΔM_n are accepted.

The final step is to add the change ΔM_n to the previous control output M_{n-1} to obtain the current control output M_n.

For the position form of the control equation, the following implementation is equivalent:

Integral mode: $M_R = \displaystyle\int_0^t \dfrac{K_C}{T_I} E\, dt + M_{R,0}$

Windup protection: $M_R = \max(M_R, M_L - K_C E)$
                              $M_R = \min(M_R, M_U - K_C E)$

Proportional mode: $M = K_C E + M_R$

The equations must be evaluated in the above order; that is,

- Calculate a new value for the controller output bias M_R.
- Limit the controller output bias M_R to the value required to drive the controller output M to the output limit.
- Perform the proportional-plus-bias calculation.

For the hot water process example, the performance of this form of windup protection is very similar to that of the bias freeze form. The trends are indistinguishable, so the responses in Figure 6.23 also apply to the velocity form of windup protection. The controller output normally comes off the limit before the temperature crosses the set point. But for some applications, it comes off the limit too soon.

Limitations of Standard Forms of Windup Protection. The condition for windup to occur can be stated very simply:

> Windup occurs whenever changes in the controller output have no effect on the measured variable.

The most common occurrence is when the controller drives its output beyond the control valve's operating range (ideally 0% to 100%). But to take into account the nonidealities present in all control valves and to provide some overrange (especially for valve closed), the configuration parameters for the PID controller permit values to be specified for the controller output limits. The standard forms of windup protection are invoked on a specific event: The controller output attains one of the controller output limits.

In a few loops, windup commences before one of the controller output limits has been attained. The liquid outlet temperature loop for the fired heater in Figure 6.24 is such a loop. For fired heaters, one limit that must not be violated is the minimum firing rate. For fuels with a constant heating value, this limit can be translated into a minimum fuel flow rate and, for gas fired heaters, to a minimum burner header pressure. One way to impose a minimum burner header pressure is to install a bypass pressure regulator in parallel with the main fuel valve. This is also illustrated in Figure 6.24.

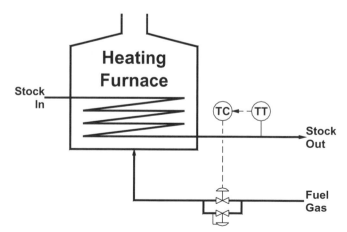

Figure 6.24. Furnace outlet temperature control with bypass pressure regulator to impose minimum firing rate.

When does windup commence in the temperature loop in Figure 6.24? As the temperature controller closes the main fuel valve, the firing rate decreases until the bypass regulator begins to open. This fixes the firing rate, so further closing the main fuel valve has no effect on the liquid outlet temperature. Windup commences in the temperature controller the instant the bypass regulator opens.

The windup protection mechanisms in most digital systems are capable of invoking the windup protection on any event, not just the controller output attaining one of the controller output limits. However, how do we know when the bypass regulator begins to open? Normally there is no indication to the controls. About all one can do is to try to match the lower controller output limit to the position of the main fuel valve at which the bypass regulator begins to open. This will be imperfect at best, especially if the fuel gas supply pressure varies.

To address the windup issues completely, a different control configuration is normally required. For the fired heater, one approach is to install a temperature-to-pressure cascade, with the inner loop being a burner header pressure loop.

6.10. SUMMARY

In the process industries, mode selection reduces to selecting PI or PID. That means every controller contains the reset mode. Without reset, a loop exhibits offset or droop; that is, the controller lines out at a value of the measured variable that is different from the set point.

In most loops, offset or droop is definitely undesirable. However, there are exceptions, two of them being as follows:

Level control. In most, the impact of vessel level on the plant profit and loss statement is obscure. High-level and low-level switches are installed in the vessel, and a shutdown of some type is initiated should either level switch react. But as long as the level is between the locations of these two switches, the process functions properly.

Inner loop of a cascade. Offset cannot be tolerated in the outer loop. But offset in the inner loop is a problem only if it impairs the performance of the outer loop, which it does not.

We have not and will not argue strongly that reset should be removed. From a process performance perspective, reset is not beneficial in either of the above cases. But when reset is removed, offset or droop will appear. This always seems to raise issues from the process operators, with all sorts of unrelated problems being attributed to the presence of the offset. A better approach is to use reset in the loops, but in the tuning of the loop, make sure that proportional is doing most of the work. Instead of getting the reset totally removed, let us try to get the loop properly tuned.

LITERATURE CITED

1. Smith, C. L., "Process Engineers: Take Control", Chemical Processing Progress, Vol. 96, No. 8 (August 2000), p. 19.

CHAPTER 7

Derivative Mode

Relatively few controllers (maybe 10%) are tuned with any derivative action. However, these controllers are invariably in loops that have a direct impact on plant performance, so the importance of the derivative is far greater than just the percentage of loops in which it is used. We explore the following issues:

Control equation. There are many variations in the implementation of a PID controller (parallel vs serial, smoothing the derivative, etc).

Tuning. After examining the effect of derivative action on loop performance, we then demonstrate the trial-and-error tuning procedure for PID.

7.1. CONTROL EQUATION

The derivative mode cannot be used alone. It must always be used in conjunction with the proportional mode. The minimum number of modes is proportional-derivative (PD), but such controllers are rare. In practice, if the derivative mode is used in a controller, the controller is almost always proportional-integral-derivative (PID).

Our first objective is to explain how the derivative is used to improve the performance of a controller. We need to do this in as simple a configuration as possible. For this reason, we shall introduce the derivative in the context of a PD controller. The equation for a PD controller is often written as follows:

$$M = K_C \left[E + T_D \frac{dE}{dt} \right] + M_R$$

where

E = Control error, %

    = SP − PV for reverse acting controller

    = PV − SP for direct acting controller

K_C = Controller gain, %/%

T_D = Derivative time, min

M = Controller output, %

M_R = Controller output bias, %

Another reason for discussing a PD controller is that it is sometimes an intermediate step in tuning a PID controller. In some trial-and-error tuning procedures for a PID controller, the approach is to tune proportional first, then proportional-derivative, and finally proportional-integral-derivative. If one wants to tune a PID controller optimally, this approach has merits. However, it is extremely time-consuming, especially for slow processes. In practice, the preferred approach is to tune proportional, then tune proportional-integral, and finally proportional-integral-derivative. The tuning procedure in Table 1.1 takes this approach.

Derivative of Set-Point Changes. Step changes in the controller set point are routinely made. These cause problems for the derivative mode. A step change in the set point translates into a step change in the control error. Theoretically, the derivative of a step change is an impulse function.

In digital control systems, this translates into a pulse with a very large amplitude and a very short duration (one sampling time). There are various ways of addressing this issue, a common one being to base the derivative on the process variable instead of the control error:

$$M = K_C \left[E - T_D \frac{dPV}{dt} \right] + M_R$$

With this approach, set-point changes are not reflected in the derivative. Most PID implementations provide the user the option of basing the derivative on either the process variable or the control error.

Why a Derivative Term. Figure 7.1 illustrates three cases that differ only by the value of the rate of change of the process variable. For each case, the value of the process variable is the same, which means that the control error E is the same.

A proportional-only controller would take the same control action for each of these cases. But is this really appropriate? Consider Case 2 relative to Case 1. In Case 2, the process variable is moving toward the set point. It seems reasonable to take a less aggressive control action in Case 2 as compared with Case 1. Now consider Case 3 relative to Case 1. In Case 3, the process variable

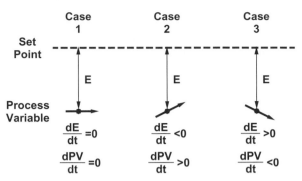

Figure 7.1. Cases with the same control error but a different rate of change of the PV.

is moving away from the set point. It seems reasonable to take a more aggressive control action in Case 3 as compared with Case 1.

By concluding that different control action is appropriate for the three cases in Figure 7.1, we are making a case for including a derivative term in the control equation.

Predicting Future Values of the Control Error. A controller with the derivative mode can be thought of as a type of predictive controller. Suppose the current value of the process variable is PV, and its rate of change is dPV/dt. The current control error is $E = SP - PV$ (assuming a reverse acting controller).

An elementary predictor can be formulated by simply assuming that the current rate of change in the PV will continue for some period of time. Specifically, let the period of time be the derivative time T_D. At one derivative time in the future, the predicted value for the process variable is as follows:

$$\widehat{PV} = PV + T_D \frac{dPV}{dt}$$

At one derivative time in the future, the predicted control error is (for a reverse-acting controller)

$$\hat{E} = SP - \widehat{PV}$$

A Predictive Controller. Instead of the current control error E, suppose we base our proportional control action on the predicted error $\hat{E}$. The proportional mode equation is as follows:

$$M - M_R = K_C \, \hat{E}$$
$$= K_C \left(SP - \widehat{PV} \right)$$

Now substitute the equation used to predict the future PV from its current value and rate of change:

$$M - M_R = K_C \left[SP - \left(PV + T_D \frac{dPV}{dt} \right) \right]$$

$$= K_C \left[(SP - PV) - T_D \frac{dPV}{dt} \right]$$

$$= K_C \left[E - T_D \frac{dPV}{dt} \right]$$

The final result is the equation previously presented for a PD controller with the derivative based on the process variable.

The Derivative Time. When the PD control equation is developed via the predictor formulation, we can attach a significance to the derivative time. For the process being controlled, how far in the future can we extrapolate the process variable based on its current rate of change? The answer to this question is the appropriate value for the derivative time.

Some processes will maintain the current rate of change for some time in the future; others will not. Flow processes will not maintain the current rate of change for any appreciable time in the future; thus, use of derivative in these loops is not appropriate. Most temperature processes will maintain the current rate of change for some period of time. If the temperature in a process vessel is increasing at a rate of 1 °F per minute, the process tends to maintain this rate of change. The traditional guidelines called for the derivative only in temperature loops, and in practice, temperature loops are the principal application for the derivative mode.

Pre-Act. When the derivative was initially introduced, it was called "pre-act." By basing the control action on the rate of change of the measured variable, the controller was endowed with the ability to anticipate based on this rate of change. This enabled the controller to take action in advance of the process attaining the conditions that, without the derivative, would be required for these actions to be taken. This notion of "acting in advance" was the basis for the term "pre-act."

A model predictive controller uses a process model to predict future values of the measured variable from the current and past control actions. Although one can think of a derivative mode controller as a predictive controller, it is not a model predictive controller. The derivative mode controller bases the prediction solely on the current rate of change of the process variable.

Which Loops Should Have the Derivative Mode. The derivative mode basically projects the process variable into the future based on the current rate of change. Derivative makes sense in those processes that will maintain their current rate of change for a significant time into the future. What does this tell us about loops?

Flow. A flow can change very quickly. Derivative is not recommended in flow loops.

Level. A change in an inlet or outlet flow is quickly reflected in the rate of change of vessel level. Derivative is not recommended.

Pressure. All liquid pressures and gas pressures associated with small volumes can change very quickly. However, gas pressures associated with large volumes change quite slowly. Their rates of change can usually be projected, thus making them a potential candidate for derivative.

Temperature. With very few exceptions, temperature loops respond very slowly. They also tend to maintain a rate of change. That is, they do not make "sharp right turns." Most applications of derivative are in temperature loops.

7.2. INCORPORATING DERIVATIVE INTO THE CONTROL EQUATION

Incorporating the derivative mode calculations into the control equation is the same for both a PD controller and a PID controller. As we shall see, proceeding from a PD to a PID controller is essentially the same as proceeding from a proportional-only controller to a PI controller. Specifically, the controller output bias becomes the output of an integrator. But for PID, other issues arise, which we will examine shortly.

But first, let us focus on how the derivative mode calculations are incorporated into the control equation. Just to keep the equation as simple as possible, we shall do this for the PD controller; however, the approach is exactly the same for the PID controller.

Derivative on E versus Derivative on PV. The previous formulation of the PD controller led to an equation with the derivative based on the measured variable or PV. A minor change in the formulation leads to an equation with the derivative based on the control error E. The two formulations are as follows:

Derivative based on E:

$$\text{Control error:} \quad E = SP - PV \text{ for reverse acting controller}$$
$$= PV - SP \text{ for direct acting controller}$$

$$\text{Derivative:} \quad \hat{E} = E + T_D \frac{dE}{dt}$$

Proportional: $M = K_C \hat{E} + M_R$

Derivative based on PV:

Derivative: $\widehat{PV} = PV + T_D \dfrac{dPV}{dt}$

Control error: $\hat{E} = SP - \widehat{PV}$ for reverse acting controller

$= \widehat{PV} - SP$ for direct acting controller

Proportional: $M = K_C \hat{E} + M_R$

In practice, smoothing or filtering must be incorporated into the derivative calculation. When this is done, the derivative can be based on either E or PV. The difference in performance is minor.

Smoothing the Derivative. In a practical implementation of the derivative mode, smoothing must be applied to whatever variable is being differentiated. Smoothing is provided by the usual first-order lag smoothing equation. For the derivative mode, the filter time constant is normally αT_D, where α is the derivative mode smoothing coefficient. We shall discuss this shortly, but first let us present the control calculations.

If the derivative mode is based on E, the sequence of calculations is as follows:

1. Compute control error E.

$$E = SP - PV \text{ for reverse-acting controller}$$
$$= PV - SP \text{ for direct-acting controller}$$

2. Apply smoothing to the control error to obtain the smoothed control error E'.

$$\alpha \, T_D \, \frac{dE'}{dt} + E' = E$$

3. Compute projected control error $\hat{E}$ by projecting the rate of change of the smoothed control error E'.

$$\hat{E} = E' + T_D \frac{dE'}{dt}$$

4. Perform the proportional mode calculations using $\hat{E}$.

$$M = K_C \, \hat{E} + M_R$$

If the derivative mode is based on PV, the sequence of calculations is as follows:

1. Apply smoothing to the PV to obtain the smoothed process variable PV′.

$$\alpha\, T_D \frac{dPV'}{dt} + PV' = PV$$

2. Compute projected process variable $\widehat{PV}$ by projecting the rate of change of the smoothed process variable PV′.

$$\widehat{PV} = PV' + T_D \frac{dPV'}{dt}$$

3. Compute projected control error $\hat{E}$ as the difference between the set point and the projected process variable $\widehat{PV}$.

$$\hat{E} = SP - \widehat{PV} \text{ for reverse-acting controller}$$
$$= \widehat{PV} - SP \text{ for direct-acting controller}$$

4. Perform the proportional mode calculations using $\hat{E}$.

$$M = K_C\, \hat{E} + M_R$$

Lead-Lag Implementation of the Derivative. In conventional (analog) controller implementations, the proportional-derivative controller (and the PD part of a PID controller) was implemented via a lead-lag element:

$$\tau_{LG} \frac{dY}{dt} + Y = \tau_{LD} \frac{dX}{dt} + X$$

where

  X = Input to the lead-lag
  Y = Output of the lead-lag
  τ_{LD} = Lead time, min = T_D for a PD or PID controller
  τ_{LG} = Lag time, min = αT_D for a PD or PID controller

In the above formulations of the PD controller, separate equations are provided for smoothing and for projecting one derivative time into the future. If these equations are combined, the result is the equation for a lead-lag element. With this approach, a single equation provides both smoothing of the input to the derivative calculations and projecting the current rate of change one derivative time into the future.

If the derivative mode is based on E, the sequence of calculations is as follows:

1. Compute control error E.

$$E = SP - PV \text{ for reverse-acting controller}$$
$$= PV - SP \text{ for direct-acting controller}$$

2. Compute projected control error $\hat{E}$ using the lead-lag equation:

$$\alpha\, T_D\, \frac{d\hat{E}}{dt} + \hat{E} = T_D\, \frac{dE}{dt} + E$$

3. Perform the proportional mode calculations using $\hat{E}$.

$$M = K_C\, \hat{E} + M_R$$

If the derivative mode is based on PV, the sequence of calculations is as follows:

1. Compute the projected process variable $\widehat{PV}$ using the lead-lag equation.

$$\alpha\, T_D\, \frac{d\widehat{PV}}{dt} + \widehat{PV} = T_D\, \frac{dPV}{dt} + PV$$

2. Compute the projected control error $\hat{E}$ as the difference between the set point and the projected process variable $\widehat{PV}$.

$$\hat{E} = SP - \widehat{PV} \text{ for reverse-acting controller}$$
$$= \widehat{PV} - SP \text{ for direct-acting controller}$$

3. Perform the proportional mode calculations using $\hat{E}$.

$$M = K_C\, \hat{E} + M_R$$

Derivative Gain Limit. The concept of the derivative gain limit originated when the lead-lag element was used to implement the derivative mode. Suppose there is a one-unit change in the PV. With no derivative, this would cause a one-unit change in the control error. For a controller gain of 1%/%, the resulting change in the controller output would also be one unit.

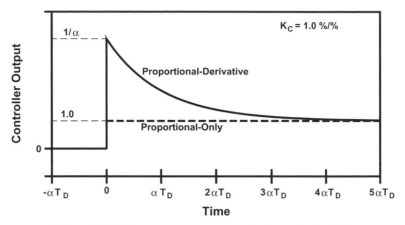

Figure 7.2. PD response to a one-unit change in the PV.

For a PD or PID controller, the one-unit change in the PV would result in a one-unit change in the input to the lead-lag element. As illustrated in Figure 7.2, the output of the lead-lag element changes by $1/\alpha$ units. If the derivative mode smoothing coefficient is 0.1, a one-unit change in the input to the lead-lag gives a 10-unit change in the output of the lead-lag. The input is amplified by a factor of 10, so the derivative gain limit is said to be 10.

Some commercial systems permit the derivative gain limit (or the derivative mode smoothing factor α) to be adjusted; most do not. Although sometimes included in the menu of tuning parameters, the derivative gain limit is not truly a tuning parameter. The appropriate value for the derivative mode smoothing depends on algorithm implementation issues, not process characteristics. In digital systems, the smoothing is typically a derivative gain limit of 10 or a derivative mode smoothing coefficient of 0.1. In conventional analog controls, considerably more smoothing was provided.

7.3. PID CONTROL EQUATIONS

Proceeding from PD to PID merely requires adding an integrator whose output is the controller output bias. But depending on how this integrator is implemented, two variations result for the PID control equation.

To keep the equations simple, we shall omit the smoothing for the derivative mode. However, smoothing is required in the final implementation.

Two Formulations of the PID. The two formulations are the result of how the integral mode is implemented, not how the derivative mode is implemented. In this section, we shall always base the derivative mode on the control error E and then consider adding the integral mode. Analogous results are obtained when the derivative is based on PV, but we will not present these equations.

In all previous formulations for the PD controller, the final calculation was the proportional mode calculation using $\hat{E}$:

$$M = K_C\,\hat{E} + M_R$$

To add integral action, an equation must be inserted ahead of the proportional mode equation to calculate the controller output bias M_R by integrating the control error. However, there are two options:

1. Integrate the current control error E:

$$M_R = \int_0^t \frac{K_C}{T_I}\, E\, dt + M_{R,0}$$

2. Integrate the projected control error $\hat{E}$:

$$M_R = \int_0^t \frac{K_C}{T_I}\, \hat{E}\, dt + M_{R,0}$$

Whether you integrate the current control error E or the projected control error $\hat{E}$ does not seem like a major issue, and indeed, it really is not. However, the resulting formulations for the PID control equation are not exactly the same and have been given different designations:

Parallel formulation. In the equation for the integral action, the input to the integrator is the control error E.

Series formulation. In the equation for the integral action, the input to the integrator is the projected control error $\hat{E}$.

The basis for these terms will hopefully become apparent from the following discussion.

The Parallel Formulation. The formulation of the PID control equation is as follows:

1. Compute control error E.

$$E = SP - PV \text{ for reverse-acting controller}$$
$$= PV - SP \text{ for direct-acting controller}$$

2. Compute projected control error $\hat{E}$ by projecting the rate of change of the control error E.

$$\hat{E} = E + T_D \frac{dE}{dt}$$

3. Compute a value for the controller output bias from the control error E.

$$M_R = \int_0^t \frac{K_C}{T_I} E \, dt + M_{R,0}$$

4. Perform the proportional mode calculations using $\hat{E}$.

$$M = K_C \hat{E} + M_R$$

The above expressions can be combined into a single equation that contains individual terms for each mode:

$$M = K_C \left[E + \frac{1}{T_I} \int_0^t E \, dt + T_D \frac{dE}{dt} \right] + M_{R,0}$$

This equation is expressed in terms of the traditional tuning coefficients, namely, the controller gain K_C, the reset time T_I, and the derivative time T_D. Alternatively, this equation can be expressed in terms of individual gains for each mode (the proportional gain K_P, the reset gain K_I, and the derivative gain K_D):

$$M = K_P E + K_I \int_0^t E \, dt + K_D \frac{dE}{dt} + M_{R,0}$$

The term "parallel" is based on the observation that the PID equation divides very cleanly into three terms that can be evaluated individually and then summed to obtain the controller output. Some argue that the true parallel equation is the one expressed in terms of the individual gains K_P, K_I, and K_D. However, simple algebraic expressions relate K_C, T_I, and T_D to K_P, K_I, and K_D.

The "parallel" nature of this formulation of the PID control equation should be apparent from Figure 7.3. Each mode can be evaluated individually (in "parallel") and the results summed to obtain the controller output.

This form of the equation is sometimes referred to as "ideal" or "noninteracting." The origin of these terms will be more apparent after the discussion of the series formulation.

The Series Formulation. The formulation of the PID control equation is as follows:

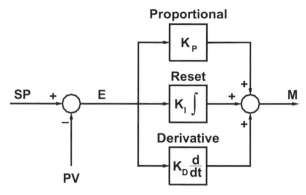

Figure 7.3. Parallel form of the PID control equation.

1. Compute control error E.

$$E = SP - PV \text{ for reverse-acting controller}$$
$$= PV - SP \text{ for direct-acting controller}$$

2. Compute projected control error $\hat{E}$ by projecting the rate of change of the control error E.

$$\hat{E} = E + T_D \frac{dE}{dt}$$

3. Compute a value for the controller output bias from the projected control error $\hat{E}$.

$$M_R = \int_0^t \frac{K_C}{T_I} \hat{E} \, dt + M_{R,0}$$

4. Perform the proportional mode calculations using $\hat{E}$.

$$M = K_C \hat{E} + M_R$$

The above expressions can also be combined into a single equation that contains individual terms for each mode:

$$M = K_C \left[\left(1 + \frac{T_D}{T_I} \right) E + \frac{1}{T_I} \int_0^t E \, dt + T_D \frac{dE}{dt} \right] + M_{R,0}$$

The proportional term is the coefficient of E in the final expression. In addition to the controller gain K_C, this coefficient also contains the ratio T_D/T_I. This

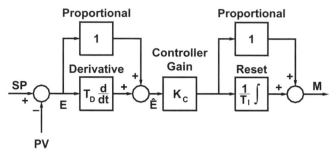

Figure 7.4. Series form of the PID control equation.

suggests that adjusting the reset time T_I or the derivative time T_D also affects the proportional action. This is the basis for applying the terms "nonideal" or "interacting" to the series formulation.

The evaluation of this equation is most logically performed in a sequential or "series" fashion as illustrated by the block diagram in Figure 7.4. The derivative is first applied to the control error E to obtain the projected control error $\hat{E}$. The projected control error $\hat{E}$ is the input to the proportional-integral equation.

The controller gain K_C is inherently applied to all terms in the control equation. Computing individual gains for each mode in this form of the control equation is possible, but the algebraic relationship of these gains to K_C, T_I, and T_D is messy. Consequently, individual mode gains are not applied to the series formulation of the PID control equation.

Reset Feedback Formulation for PID. A PI controller can be implemented by either integrating the control error or lagging the output of the controller. An analogous formulation for the series form of the PID is presented in Figure 7.5.

The sequence of the calculations for the reset feedback form of the PID are as follows:

1. Compute control error E.

$$E = SP - PV \text{ for reverse acting controller}$$
$$= PV - SP \text{ for direct acting controller}$$

2. Compute projected control error $\hat{E}$ by projecting the rate of change of the control error E.

$$\hat{E} = E + T_D \frac{dE}{dt}$$

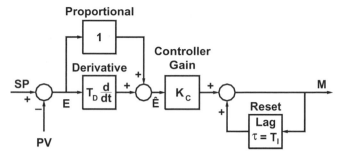

Figure 7.5. Reset feedback implementation of the PID control equation.

3. Perform the proportional mode calculations using $\hat{E}$.

$$M = K_C \, \hat{E} + M_R$$

4. Compute a value for the controller output bias using a first-order lag on the controller output M.

$$T_I \frac{dM_R}{dt} + M_R = M \quad \text{Initial condition: } M_R = M_{R,0}$$

Parallel versus Series. To most, the parallel formulation of the PID would seem to be preferred. However, all pneumatic and electronic analog controls used the series implementation. We manufactured a lot of chemicals using these controllers, so the series form must not be entirely bad. The option of series versus parallel appeared with digital systems. Some manufacturers of digital systems implement the series PID equation; some implement the parallel PID equation; and a few permit the user to choose.

The series PID equations and the parallel PID equations are not identical. However, the differences are small. One can observe differences on the results of simulations. But given the tolerance to which we can reasonably tune controllers in the field, the differences will not be noticeable. We make the following observations:

- The distinction between series and parallel is only significant for PID control. For PI, the two formulations are identical.
- The ratio T_D/T_I is a minor term. The expression in the proportional term of the series formulation is $1 + (T_D/T_I)$. The Ziegler–Nichols tuning equations suggest a value of 0.25 for the ratio T_D/T_I for the series form of the PID control equation, and in practice, the ratio is often less. At a value of 0.25 for the ratio, the effective proportional sensitivity increases from 1.0 to 1.25, which is a difference of 25%.

- Suppose you have tuned a series PID controller to deliver the performance required for a loop. Switching the PID control equation to parallel would give a slightly different response, but not a major difference. This statement also applies for tuning a parallel PID controller and then switching to a series PID.
- Suppose you tune a series PID controller and then tune a parallel PID controller. Which would deliver the better performance? The difference, if any, would be minor, and within the precision to which one can realistically tune a PID controller. If you can get the job done with one formulation, then you can also get the job done with the other.

Relating the Tuning Coefficients. The tuning coefficients in the parallel form of the PID control equation can be algebraically related to the tuning coefficients in the series form of the PID. Let us distinguish the tuning coefficients as follows:

- Tuning coefficients for the parallel form: K_C, T_I, and T_D.
- Tuning coefficients for the series form: K_C', T_I', and T_D'.

By equating the coefficients in the two formulations of the PID equation, we obtain the following relationships:

$$K_C = K_C'\left(1 + \frac{T_D'}{T_I'}\right)$$

$$\frac{K_C}{T_I} = \frac{K_C'}{T_I'}$$

$$K_C T_D = K_C' T_D'$$

These equations can be very easily rearranged to express the coefficients K_C, T_I, and T_D in the parallel form to the coefficients K_C', T_I', and T_D' in the series form, and vice versa. Table 7.1 presents the results.

The equations for converting the parallel coefficients to the series coefficients have an interesting aspect. If $T_D > \frac{1}{4}T_I$, the tuning coefficients K_C', T_I', and T_D' have an imaginary part. However, the Ziegler–Nichols tuning equations suggest that the ratio T_D/T_I should be 0.16 for the parallel form of the PID control equation. Although it is possible that the ratio T_D/T_I could exceed 0.25, it is unlikely and should it be encountered, the tuning should definitely be examined.

The relationships in Table 7.1 assume that all modes are based on the control error. If either the derivative mode or the proportional mode is based on the PV, there is no exact equivalent. But nevertheless, the relationships in Table 7.1 are still used to convert the coefficients.

TABLE 7.1. Tuning Coefficient Conversion Equations

Coefficients for parallel form: K_C, T_I, and T_D
Coefficients for series form: K'_C, T'_I, and T'_D

Mode	Series to Parallel	Parallel to Series
Proportional	$K_C = K'_C \dfrac{T'_I + T'_D}{T'_I}$	$K'_C = \dfrac{1}{2} K_C \left[1 + (1 - 4 T_D/T_I)^{1/2} \right]$
Integral	$T_I = T'_I + T'_D$	$T'_I = \dfrac{1}{2} T_I \left[1 + (1 - 4 T_D/T_I)^{1/2} \right]$
Derivative	$T_D = \dfrac{T'_I \, T'_D}{T'_I + T'_D}$	$T'_D = 2 T_D \Big/ \left[1 + (1 - 4 T_D/T_D)^{1/2} \right]$

7.4. EFFECT OF DERIVATIVE TIME

To illustrate the effect of derivative time on loop performance, we will use the hot water process. We will also use the results of previous tuning efforts, specifically, those from Table 6.2. The initial state of the hot water process is SP = PV = 150 °F, the controller output M = M_R = 62.1%, and the hot water demand is 500 lb/min.

Unless otherwise noted, the control equation will be the parallel formulation with the derivative based on the PV: The derivative mode smoothing factor α is 0.1.

Effect of Derivative Time on the Response to a Set Point Change. To illustrate the effect of derivative, we will add the derivative to a proportional-only controller. We previously determined that a controller gain of 5%/% gives a response with approximately a quarter decay ratio.

In Figure 7.6, the set point is switched between 150 °F and 160 °F (recall that the derivative is based on the PV). The effect of increasing the derivative time is as follows:

$T_D = 0$. A controller gain of 5%/% gives a response with a decay ratio somewhat less than one quarter.

$T_D = 0.5$. The first peak overshoot is much less; the second peak is insignificant.

$T_D = 1.0$. The first peak overshoot is only barely visible. The stability margin is clearly enhanced.

$T_D = 2.0$. No overshoot is present, but the response approaches the lineout value more slowly.

$T_D = 4.0$. The response is even slower. But also note the high-frequency oscillations in the controller output.

$T_D = 8.0$. The oscillations in the controller output are very pronounced. Small cycles are also present in the PV.

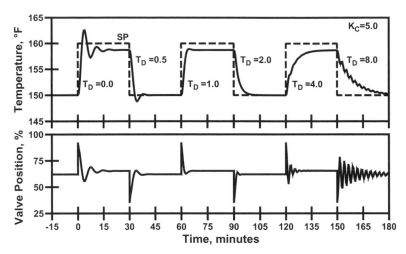

Figure 7.6. Effect of derivative time on response.

Up to a derivative time of 1.0 min, the derivative is enhancing the stability margin. However, derivative times beyond 1.0 min lead first to a slower response and then to high-frequency oscillations.

Impact of Process Dead Time. In the previous example, the derivative was very effective at removing overshoot and oscillations. However, the process had very little dead time (transportation lag). The effectiveness of derivative decreases as the dead time within the loop increases.

Dead time can be added to the hot water process by locating the temperature transmitter at some distance from the tank. Let us locate the temperature transmitter 200 ft from the tank, which gives a dead time of about 2 minutes at a hot water demand of 500 lb/min. As the trend in Figure 7.7 shows, the derivative is much less effective:

$T_D = 0$. Because of the dead time, the controller gain must be reduced to 2%/% to obtain a quarter decay ratio.

$T_D = 1.0$. The derivative has removed some of the overshoot and oscillations, but not as much as before.

$T_D = 2.0$. The oscillation in the response has a small amplitude but a decay ratio of about one half.

$T_D = 4.0$. The slowly decaying oscillation is caused by the derivative.

PI to PID. To most, the logical approach to tuning a PID controller is as follows:

1. Get the loop working with PI control. This is usually a necessary part of plant startup.

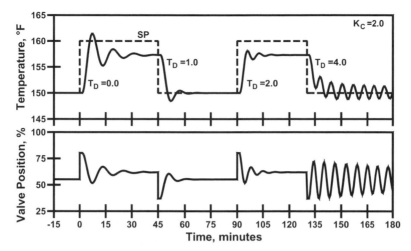

Figure 7.7. Impact of dead time on derivative mode performance.

2. Add the derivative to enhance loop performance. This may be done either during or subsequent to startup.

Although not always stated explicitly, the assumption is made that the reset time appropriate for PI control is also appropriate for PID control. We want to examine this assumption, beginning with the three responses in Figure 7.8:

A This is a PI controller with the tuning coefficients from our previous tuning efforts ($K_C = 5\%/\%$; $T_I = 7.5$ min).

B For the PD controller, a derivative time of 1.0 min gave a response with negligible overshoot. Response B uses PI tuning for the gain and reset, but with a derivative time of 1.0 min. The response has an overshoot of about 10%, which is caused by the reset action. This is more than observed for PD, but the difference would be acceptable.

C Extending the reset time to 12 min eliminates virtually all overshoot. Although a reset time of 7.5 min is likely to be acceptable, a reset time of 12 min is better. This is an increase of 60%.

Ziegler–Nichols Equations for PI and PID. Ziegler and Nichols presented equations that related the tuning coefficients (K_C, T_I, and T_D) to the ultimate gain K_U and ultimate period P_U. Their equations for the reset time $T_{I,PI}$ for PI control and the reset time $T_{I,PID}$ for PID control are as follows:

PI Control: $\qquad\qquad\qquad\qquad\qquad$ $T_{I,PI} = P_U/1.2$

PID Control (parallel formulation): $\quad$ $T_{I,PID} = 0.625\,P_U$

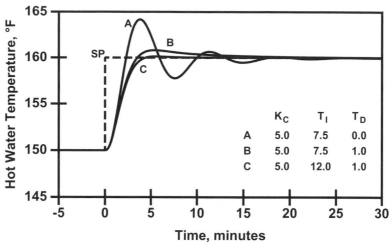

Figure 7.8. PI to PID.

Eliminating the ultimate period P_U suggests that $T_{I,PID}$ should be 25% shorter than $T_{I,PI}$:

$$T_{I,PID} = 0.625 \times (1.2\ T_{I,PI}) = 0.75\ T_{I,PI}$$

In most cases, the reset time for PID can be the same as the reset time for PI. However, the resulting PID controller is not optimally tuned, and occasionally some adjustment of the reset time will be necessary to obtain acceptable performance.

Derivative on PV versus Derivative on E. The two PID responses in Figure 7.9 differ in only one aspect. In one, the derivative is based on the measured variable or PV. In the other, the derivative is based on the control error. Both are parallel formulations of the PID, and both have the same tuning coefficients.

The two responses in Figure 7.9 are not identical, but the difference is modest. The response for the derivative based on E is slightly preferable to the other. However, this cannot be generalized. Whether such a change improves the response or makes it worse depends on how the controller was originally tuned. In practice, switching from derivative on PV to derivative on E (or vice versa) has such a small effect that it is not worth pursuing. Adjustments in the tuning coefficients have a much larger effect on the response.

Parallel versus Series Formulations. The two PID responses presented in Figure 7.10 differ in only one aspect. In the one, the PID equation is the parallel formulation. In the other, the PID equation is the series formulation. Both base the derivative on the PV, and both have the same tuning coefficients.

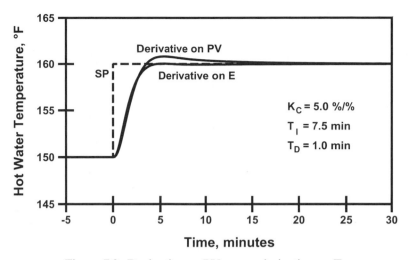

Figure 7.9. Derivative on PV versus derivative on E.

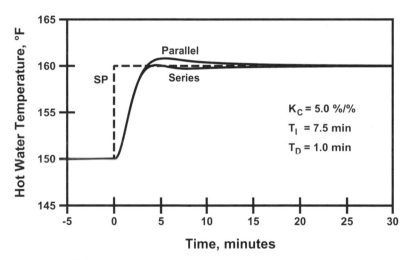

Figure 7.10. Parallel versus series PID.

The performance of the parallel formulation of the PID is not identical to the series formulation of the PID. Furthermore, any difference can be removed by adjusting the tuning coefficients using the equations in Table 7.1.

7.5. GETTING THE MOST FROM THE DERIVATIVE

In the examples that follow, the performance of the hot water temperature controller will be assessed on the maximum departure from the set point for

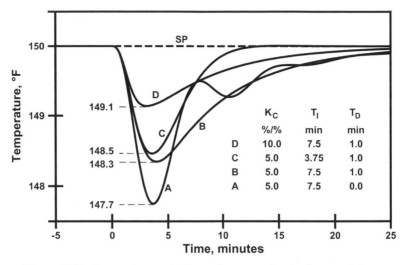

Figure 7.11. Increasing controller gain versus shortening reset time.

an increase in hot water demand from 500 lb/min to 600 lb/min. Derivative action will be added to the controller using the following procedure:

1. Tune a PI controller to provide a response with a quarter decay ratio. This is response A in Figure 7.11. On the increase of the hot water demand, the hot water temperature drops to 147.7 °F for a maximum departure from a set point of 2.3 °F.
2. Add derivative action, adjusting the derivative time to enhance the stability margin as much as possible. The response will now exhibit less than a quarter decay ratio. This is response B in Figure 7.11. The response clearly exhibits an enhanced stability margin. On the increase of the hot water demand, the hot water temperature drops to 148.3 °F for a maximum departure from a set point of 1.7 °F.
3. Since the addition of derivative action enhances the stability margin, this permits either of the following to be done:
   a. Shorten the reset time. The result is response C in Figure 7.11.
   b. Increase the controller gain. The result is response D in Figure 7.11.

We want to show that increasing the controller gain provides far more benefit than shortening the reset time.

Increasing the Controller Gain. To obtain response D in Figure 7.11, we arbitrarily doubled the controller gain. On the increase of the hot water demand, the hot water temperature drops to 149.1 °F for a maximum departure from a set point of 0.9 °F. The addition of derivative action alone reduces the maximum departure from 2.3 °F to 1.7 °F, which is a reduction of about 25%. But also increasing the controller gain reduces the maximum departure from 2.3 °F to 0.9 °F, which is a total reduction of over 50%.

Response D in Figure 7.11 exhibits neither overshoot nor oscillation. Is a further increase in the controller gain possible? The Ziegler–Nichols equations for the controller gain are as follows (K_U is the ultimate sensitivity for proportional-only control):

PI Control: $K_C = 0.45\ K_U$

PID Control, series form: $K_C = 0.6\ K_U$, an increase of 33% over PI

PID Control, parallel form: $K_C = 0.75\ K_U$, an increase of 67% over PI

In practice, increasing the controller gain by 50% is a good starting point. Doubling the controller gain is possible only when the process contains very little dead time.

Shortening the Reset Time. To obtain response C in Figure 7.11, the reset time was reduced from 7.5 min to 3.75 min. Reducing the reset time by a factor of 2 has only a minor effect on the maximum departure from the set point. The hot water temperature drops to 148.5 °F for a maximum departure from a set point of 1.5 °F. The response returns to the set point more expeditiously, but it is the peak error that does the damage.

As before, the addition of derivative action reduces the maximum departure from 2.3 °F to 1.7 °F, which is a reduction of about 25%. However, shorting the reset time does not contribute very much, with the total reduction being from 2.3 °F to 1.5 °F. By comparison, increasing the controller gain gave a total reduction of 2.3 °F to 0.9 °F.

Implications for Tuning a PID Controller. The usual approach to tuning a PID controller is to tune PI first and then to add the derivative. For controllers whose primary purpose is to respond to disturbances or loads (such as the temperature controller for the hot water process), the focus must be on reducing the maximum departure from the set point. The issues are as follows:

1. Just adding derivative action to a PI controller reduces the maximum departure from the set point, but more is possible.
2. Adding derivative action enhances the stability margin, which permits either the controller gain to be increased or the reset time to be shortened. Increasing the controller gain will significantly reduce the maximum departure from the set point; shortening the reset time has only a minor effect.
3. In most loops, the reset time for PI control can also be used for PID control. This is not the optimal value for the reset time, and in a few loops, an adjustment in the reset time will be necessary.

The benefit provided by the derivative decreases as the process dead time increases. The hot water process used in our examples has very little process dead time, so the derivative is effective. But as we demonstrated in a previous example by relocating the temperature transmitter to 200 ft from the tank, the derivative is much less effective for loops with significant dead time.

Excessive Proportional Kick. Increasing the controller gain increases the magnitude of the proportional kick. As we have illustrated, one benefit of the derivative is that it permits the controller gain to be increased, thereby giving a faster responding loop. When responding to a load or disturbance change, a large controller gain reduces the maximum departure from the set point, which is very desirable in loops such as the temperature controller for the hot water process. But for set-point changes, the large controller gain also increases the magnitude of the proportional kick, possibly to the point that the control valve is driven fully open or fully closed.

In applications such as the hot water process, this really is not a problem. However, it is often perceived to be a problem, and there are applications where it is a problem. One way to avoid this problem is to base the proportional action on the PV instead of the control error E. This completely eliminates the proportional kick, but at the expense of a slower response to set-point changes. Basing proportional on PV versus basing proportional on E has no effect on responses to a change in a disturbance or load. Another alternative is to ramp all set-point changes.

7.6. PID CONTROL OF SIMPLE MODELS

The derivative is most effective for processes whose characteristics include a major time constant, a significant minor time constant, and little dead time (meaning a dead time smaller than the minor time constant). In this section, our objective is to illustrate the following characteristics of derivative:

1. The derivative provides no benefit when the process is described by a single time constant.
2. For processes described by two time constants, starting with no derivative and increasing the derivative time is initially beneficial.
3. If the derivative time is increased to values exceeding the minor process time constant, the performance of the loop suffers. This characteristic is the basis for some tuning techniques to recommend that the derivative time should be equal to the minor process time constant.

For the examples, we will use a series PID with a derivative mode smoothing factor of 0.1 (derivative gain limit of 10). Why a series PID? We previously presented equations to relate the tuning coefficients in the parallel PID to the series PID, and vice versa. The coefficients in the series PID can always be converted to equivalent coefficients for the parallel PID. However, the coefficients for the parallel PID can be converted to equivalent coefficients for the series PID only if $T_D < \frac{1}{4}T_I$. In the examples, we will be presenting responses for a variety of derivative times, some of which will be greater than $\frac{1}{4}T_I$.

A Time Constant Process. Consider using a PID controller for a time constant process. As concluded in the previous chapter, the appropriate value for the reset time is $T_I = \tau$. For a loop gain of 4.0, the responses in Figure 7.12 illustrate the responses for various values of the derivative time. The most rapid response is for $T_D = 0$. Increasing the derivative time makes the responses progressively slower.

A Process with Two Time Constants. Consider using a PID controller for a process consisting of two time constants. In keeping with the conclusions in the previous chapter, the reset time T_I is equal to the major time constant τ_1. Let the minor time constant τ_2 equal $0.3\tau_1$.

For a loop gain $K K_C = 4.0$, Figure 7.13 presents responses for various values of the derivative time. The observations are as follows:

$T_D = 0$. This is PI control. The response exhibits an overshoot of almost 20%.

$T_D = 0.1\tau_1$. The overshoot is now less than 10%.

$T_D = 0.2\tau_1$. The overshoot is now trivial.

$T_D = 0.3\tau_1 = \tau_2$. There is definitely no overshoot, but the response is slower.

$T_D = 0.5\tau_1$. The response is even slower.

Adding the derivative is beneficial up to at least $T_D = 0.2\tau_1$, and possibly up to $T_D = 0.3\tau_1$. However, derivative times in excess of $0.3\tau_1$ are clearly not

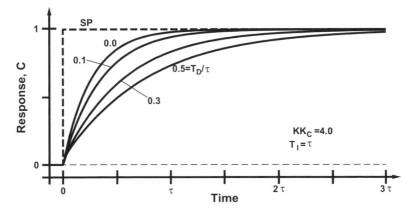

Figure 7.12. Effect of derivative time on response, process with one time constant.

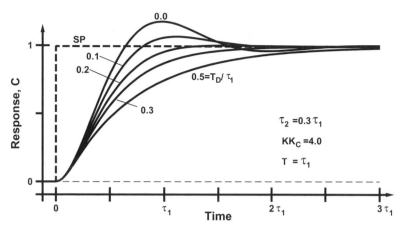

Figure 7.13. Effect of derivative time on response, process with two time constants.

beneficial. Note that the minor time constant τ_2 also equals $0.3\tau_1$. At least based on this example, the appropriate value for the derivative time T_D is the minor time constant τ_2, or perhaps slightly less. This is consistent with the recommendations of several tuning techniques.

A Process with Two Time Constants and a Dead Time. Consider using a PID controller for a process consisting of two time constants and a dead time. In keeping with the conclusions in the previous chapter, the reset time T_I is equal to the major time constant τ_1. Let the minor time constant τ_2 equal $0.3\ \tau_1$ and the dead time equal $0.2\ \tau_1$.

Figure 7.14 presents responses for various values of the derivative time. In our previous example, the time constants had the same values, but there was no dead time. The impact of the dead time is as follows:

- The loop gain KK_C must be 2.0 as compared with 4.0 without the dead time.
- The responses in Figure 7.14 are slower than those in Figure 7.13. Dead time influences response speed in two ways:

    Directly. The direct effect of dead time is to cause the response to be slower.

    Indirectly. Because of dead time, the loop gain KK_C must be lower, which further slows the response. In most loops, this is the major impact.

The observations on the responses in Figure 7.14 are as follows:

   $T_D = 0.$ This is PI control. The response exhibits an overshoot of almost 30%.

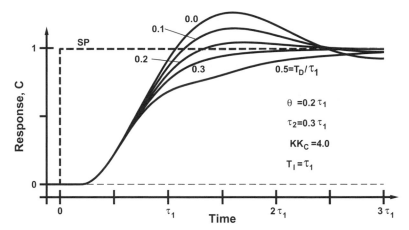

Figure 7.14. Effect of derivative time on response, process with two time constants and dead time.

$T_D = 0.1\tau_1$. The overshoot is now just over 10%.

$T_D = 0.2\tau_1$. The overshoot is now about 5%.

$T_D = 0.3\tau_1 = \tau_2$. There is no overshoot.

$T_D = 0.5\tau_1$. The response is even slower.

The conclusions pertaining to the appropriate value for the derivative time are the same as for the previous example. Adding the derivative is beneficial up to at least $T_D = 0.2\tau_1$, and possibly up to $T_D = 0.3\tau_1 = \tau_2$. Derivative times in excess of $0.3\tau_1 = \tau_2$ are clearly not beneficial.

7.7. TUNING

The following issues make trial-and-error PID tuning a daunting endeavor:

1. Tuning a PI controller involves adjusting the controller gain first and then adjusting the reset time. Each step involves adjusting a single coefficient. But when tuning derivative, the controller gain and the derivative time must be adjusted together.
2. The derivative mode is used primarily in temperature loops, most of which are quite slow. We put the most difficult mode to tune in the slowest loops in the plant.
3. Derivative is only used in the important loops in the plant. Trial-and-error tuning involves making a sequence of changes in either set point or load. Every change upsets the plant.

Given these obstacles, can we realistically expect the trial-and-error tuning approach to produce results that are close to the optimal values? We invariably have to compromise between the quality of the results and the time (and process disturbances) entailed in the tuning effort. If you want tuning even close to the optimum, you should be applying a tuning method. Trial-and-error tuning can get some of the benefits of the derivative but rarely gets all of it.

Approaches for Trial-and-Error Tuning of a PID Controller. There are two general approaches:

Tune PD and then PID. This approach provides the best assurance that the controller gain is properly tuned. By adjusting the controller gain in the absence of reset action, the enhanced stability margin resulting from derivative action will definitely be used to increase the controller gain.

Tune PI and then PID. This approach is most compatible with plant startup practices. Usually the first goal is to get the plant to produce salable product. This can normally be achieved with only PI control. The next goal is to get the plant to run as efficiently as possible. This entails, among other things, operating as close to limiting conditions as possible. We now need to pursue the use of PID control in the important loops.

Most prefer to pursue the second approach. However, this approach is effective only if the PI controller is properly tuned (that is, not tuned with excessive reset action).

Tuning Procedure for PID Controller. In the trial-and-error tuning procedure in Table 1.1, the first two steps are the same as those for tuning a PI controller. With this procedure, one first tunes PI, and then tunes PID.

The method in Table 1.1 is not claimed to tune a PID controller optimally. One defect is that it uses the same reset time for PID as for PI. The method will get some benefit from the derivative mode but rarely all of it. The method is a compromise between the benefits realized from the derivative mode and the time (including process upsets) expended on the tuning effort. A major weakness of this method is that it depends on the PI controller being properly tuned. Unfortunately, many are tuned with too much reset action. When the derivative is added, it still enhances the stability margin. However, it is largely compensating for the excess reset.

For PI, we have claimed that the trial-and-error approach is competitive with the tuning methods, both in regard to the quality of the results and the time required to tune the controller. Such claims are not made for PID.

Adjusting the Derivative Time. For the hot water process, our previous tuning efforts for a PI controller concluded that a controller gain of 5%/% and a reset time of 7.5 min gives a response with a quarter decay ratio. Using this as the starting point for the responses in Figure 7.15, we increase the derivative time and observe its effect on the response to a step change in the hot water temperature set point:

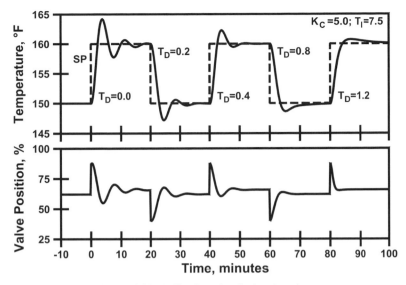

Figure 7.15. Adjusting the derivative time.

Trial 1. This is the performance of the PI controller. It is crucial that the PI tuning be done properly.

Trial 2. We begin with a small value for the derivative time, specifically 0.2 min. The effect on the response is very small.

Trial 3. We increase the derivative time to 0.4 min. As compared with the PI response, the overshoot of the first peak is less and the cycles decay more rapidly.

Trial 4. We increase the derivative time to 0.8 min. A peak is still visible, but there are no oscillations.

Trial 5. We have been doubling the derivative time, but such a large increase is probably not advisable. We increase the derivative time by 50%, to 1.2 min. The overshoot in this response is probably due to the reset.

Let us use 1.2 min for the derivative and add this to our tuning table. In Table 7.2, this case is the first entry for PID. The next step is to increase the controller gain.

Adjusting the Controller Gain. For the hot water process, we have concluded that a derivative time of 1.2 minutes is appropriate. The next step is to increase the controller gain. Using $K_C = 5\%/\%$, $T_I = 7.5$ min and $T_D = 1.2$ min as the starting point for the responses in Figure 7.16, we proceed as follows:

Trial 1. This is a repeat of the last response obtained when adjusting the derivative time.

TABLE 7.2. Comparison of Controller Performance

Controller	Objective	K_C %/%	T_I min	T_D min	P min	E_{MAX} °F
P-Only	Quarter Decay	5.0	—	—	7.3	—
P-Only	No Overshoot	0.75	—	—	—	—
PI	No Overshoot	0.75	5.0	—	—	5.9
PI	¼ Decay	5.0	7.5	—	7.5	2.3
PI	Alt. ¼ Decay	0.75	1.8	—	18.7	5.1
PID	¼ Decay	5.0	7.5	1.2	—	1.6
PID	¼ Decay	7.5	7.5	1.2	—	1.1
PID	¼ Decay	10.0	7.5	1.2	—	0.8

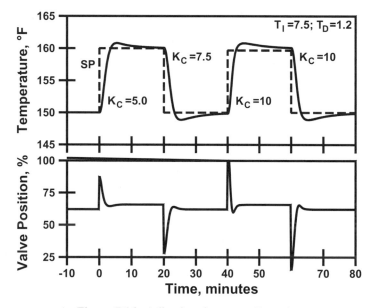

Figure 7.16. Adjusting the controller gain.

Trial 2. Ziegler–Nichols suggests a 33% increase, but let us increase the controller gain by 50% (to 7.5%/%). This has only a small effect on the response to a response to a set-point change, and not necessarily in the positive direction. We will subsequently examine the response to a hot water demand change, so this case is the second entry for PID in Table 7.2.

Trial 3. Let us increase the gain to 10%/% (twice its original value). This also does not have a significant effect on the response. However, the large gain causes the control valve to be driven fully open. When tuning, one should avoid such situations, usually by making smaller set-point changes (5 °F instead of 10 °F).

Trial 4. For a set-point change of 10 °F (5% of span), a gain of 10%/% gives a proportional kick of 50%. The controller output for a hot water temperature of 160° is 65.8%. Therefore, we can decrease the set point by 10 °F without driving the valve fully closed. In this case, the response is only slightly different from that for the 10 °F increase. This case is the third entry for PID in Table 7.2.

Response to a Change in Load. The tuning table in Table 7.2 contains three PID entries. All have the same reset time and derivative time; the only difference is the controller gain. Figure 7.17 presents the responses to an increase in hot water demand from 500 lb/min to 600 lb/min and then back to 500 lb/min. The maximum departures from the set point for both an increase in hot water demand and a decrease in hot water demand are as follows:

K_C	T_I	T_D	E_{MAX} (increase)	E_{MAX} (decrease)
5.0%/%	7.5 min	1.2 min	−1.6 °F	+1.6 °F
7.5%/%	7.5 min	1.2 min	−1.1 °F	+1.1 °F
10.0%/%	7.5 min	1.2 min	−0.8 °F	+0.8 °F

The responses to a change in the set point showed no benefit from increasing the controller gain. The responses in Figure 7.17 clearly show a benefit. Increasing the controller gain by 50% (from 5.0 to 7.5%/%) reduces the maximum departure from the set point by a third (from 1.6 °F to 1.1 °F); doubling the controller gain (from 5.0 to 10.0%/%) reduces the maximum departure from

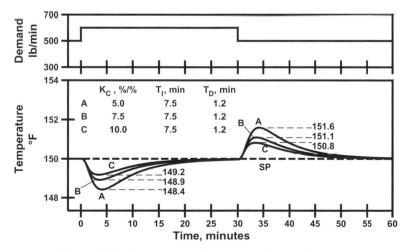

Figure 7.17. Response to change in hot water demand.

the set point by a half (from 1.6 °F to 0.8 °F). These values are included in Table 7.2.

7.8. SUMMARY

The traditional guideline for selecting PI versus PID goes as follows:

Install PID in temperature loops; PI in all others.

Some would modify statement to read "important temperature loops." The effort required to tune a PID controller must be justified by the returns from improved performance. This can only happen in loops that have a significant impact on process operations. Temperature loops most often fall into this category.

The two criteria for derivative to be really effective in process applications are as follows:

• The loop must contain little dead time.
• The measured variable must be free of noise.

Dead time degrades the performance of the derivative. Applying the derivative to a heavily smoothed process variable is not productive.

In practice, the derivative provides significant improvement in a few applications. In others, there is some improvement with the derivative, but the effort to obtain this improvement cannot be justified by the benefits realized from the derivative. How does one find the loops where the derivative is very effective? The two criteria stated above are the key.

CHAPTER 8

Tuning Methods

Herein the term "tuning methods" will encompass both of the following:

Traditional tuning techniques. These are designed to be usable by people with limited computational aids. The Ziegler–Nichols method appeared in 1942, and most others predate the personal computer.

Automated tuning. The simpler methods of this type are executed by the controller itself. For the more sophisticated methods, the data collected from the process is transferred to a personal computer for processing.

But before presenting any tuning method, we need to be upfront about the state of the technology: Too often a tuning method is used as a measure of last resort on an untunable loop. Basically, this is merely throwing technology at a problem. Rarely is this successful, and tuning is no exception. Always remember the following:

If you cannot tune the loop, neither will any tuning method.

To be useful, a tuning method must either tune the loop faster (and with less disruption to the process) or must yield better tuning coefficients.

The traditional tuning techniques are certainly not perfect. But especially for PI control, their predicted values for the tuning coefficients are usually close. However, their general reputation within the industry does not reflect this.

A large part of this is due to how the traditional tuning techniques are often applied. Usually the first attempt is to tune the controller using trial-and-error procedures. Often there is a second attempt by someone experienced in tuning. Then one of the traditional tuning techniques is applied. The results are almost always unsatisfactory. If a controller cannot be tuned by trial-and-error procedures, it is very unlikely that any tuning method will be successful. Such loops have a problem that will have to be identified and corrected before the loop can be successfully tuned.

Automated tuning has suffered likewise. In more than one plant, an action item from the "Monday morning goat rope" is to identify all loops that cannot be tuned and to purchase controllers with automatic tuning capability for those loops. This is merely throwing technology at the problem.

Some loops cannot be tuned by any approach, be it trial-and-error, traditional tuning techniques, or automated tuning. Some of the more common problems that render a loop untunable are as follows:

1. Process nonlinearities.
2. Deficiencies in the process and instrumentation (P&I) diagram.
3. Problems with the final control element (especially valves).
4. Noise.
5. Large dead times.
6. Loop interaction.
7. Improper nesting of cascade loops.

When one is faced with a loop with a history of tuning difficulties, one has to focus on the root of the problem. The tuning difficulty is merely a symptom that unfortunately too often receives all of the attention. Applying either traditional tuning techniques or automated tuning is treating a symptom, not the root cause of the problem.

Example Process for Tuning. We will illustrate the various tuning methods on the hot water temperature control loop for the hot water process. The starting point will always be PV = SP = 150 °F and M = M_R = 62.1%.

All tuning methods have one characteristic in common. When the process dead time is extremely small, the predicted value for the controller gain is very high, usually too high to be usable. In practice, the dead time present in most loops is such that this does not occur very often. However, the hot water process exhibits very little dead time. So that this process behaves in a manner more typical of industrial processes, a modest amount of dead time has been added by locating the hot water temperature transmitter 100 ft from the tank.

We do not want to give the wrong impression. We are inserting dead time into the hot water temperature loop so that it is more typical of an industrial control loop, not that this is good control practice. The intentional insertion of dead time into a control loop is never a recommended practice.

8.1. WHAT IS A TUNING METHOD

Applying a tuning method is a three-step procedure:

Step 1: Perform a process test. Every tuning method must somehow characterize the process. All rely on a process test to obtain response data for the process.

Step 2: Quantify the behavior of the process. The tuning techniques that appeared prior to the widespread availability of computers relied largely on graphical constructions to obtain values for the parameters from the response data. But relying on computers to analyze the response data permits the use of tests that are less disruptive to the process.

Step 3: Calculate the tuning coefficients. For each type of controller (PI, PID, etc), most tuning methods provide equations to compute the tuning coefficients that achieve a stated performance objective. But with computers, calculating the tuning coefficients by other means is possible.

Process Characterization for Controller Tuning. As used in controller tuning, the term "process" encompasses the characteristics of the final control element and the measurement device. The controller has to contend with whatever is between the controller output and the process variable input. The "process" is everything between these two signals.

Most tuning techniques characterize the process by evaluating the parameters in simple models consisting of some combination of a process gain, a process dead time, one or more time constants, and possibly an integrator. For the purposes of controller tuning, the following combinations are used:

- Integrator plus dead time (an integrating process).
- Time constant plus dead time (a nonintegrating process).
- Two time constant plus dead time (a nonintegrating process).

There are two approaches for analyzing process test data:

Graphical constructions. Prior to computers, all tuning techniques used graphical approaches to obtain the model coefficients. This effectively restricted the models to first order, either integrator plus dead time or one time constant plus dead time. These are adequate for tuning PI controllers, but perhaps not PID.

Regression techniques. With the availability of computers to collect and analyze data, regression fits can be used instead of graphical methods. The data are captured by the control system, but they must be transferred to a PC for further analysis. Commercial packages are available that both analyze the test data and compute values for the tuning coefficients.

Performance Objectives. Most tuning techniques adopt a stated performance objective. However, the options for the performance objective are not as broad as one might anticipate:

Simple performance measures. Of these, most tuning methods adopt the quarter decay ratio as the performance objective. And very little else is available.

Closed loop time constant. We shall describe two tuning techniques whose objective is to provide closed-loop behavior with a specified time constant.

Integral criteria. Tuning techniques are available for the common integral criteria, namely, ISE (integral of square error), IAE (integral of absolute error), and ITAE (integral of time and absolute error).

8.2. PROCESS CHARACTERIZATIONS

All traditional tuning techniques are based on a first-order representation of the process characteristics, with the choices being

Integrator-plus-dead-time model. An example is the Ziegler–Nichols (1) method. Although this model is integrating, the Ziegler–Nichols method is applied to both integrating and nonintegrating processes.

Time-constant-plus-dead-time model. Examples include the Lambda method (2), the IMC method (3), and the integral criteria method (4).

Our objective is to illustrate the impact of the choice for the process model on the form of the tuning equations. If two tuning methods rely on the same process characterization, the forms of their equations will be the same. The coefficients may be different but primarily because of the performance objective. For this comparison, we will compare the Ziegler–Nichols equations (based on integrator-plus-dead-time) and the integral criteria equations (based on time-constant-plus-dead-time).

Proportional Mode Equation. The equations for both methods suggest that the controller gain must be decreased as the process dead time increases:

Ziegler–Nichols: The proportional mode tuning equation can be expressed as follows:

$$K_C = \frac{1}{L_R\,R_R} = \frac{1}{K\,\theta} \quad \text{or} \quad K\,K_C = \frac{1}{\theta}$$

The loop gain $K\,K_C$ equals the reciprocal of the process dead time. Note that the process is an integrating process, so the units on K are (%/%)/min.

Integral criteria: The ITAE equation for the loop gain $K\,K_C$ is as follows:

$$K\,K_C = 0.586\left[\frac{\theta}{\tau}\right]^{-0.916} \cong 0.586\,\frac{\tau}{\theta}$$

The value of the exponent of θ/τ is −0.916, which is close to −1.0. With this approximation, the loop gain $K\,K_C$ is inversely proportional to the process dead time. Note that the process is a nonintegrating process, so the units on K are %/%.

The value of the coefficient differs largely because of the respective performance objectives. Ziegler–Nichols is based on the quarter decay ratio, and the coefficient is 1.0. The ITAE criterion is more conservative, which results in a smaller value (0.586) for the coefficient.

Integral Mode Tuning. The equation for the reset time equation is fundamentally different:

Ziegler–Nichols: The equation for the reset time is

$$T_I = 3.33\,L_R = 3.33\,\theta$$

The reset time is 3.33 times the process dead time θ.

Integral criteria: The ITAE equation for the reset time is

$$\frac{\tau}{T_I} = 1.03 - 0.165\frac{\theta}{\tau} \cong 1.0 \quad \text{or} \quad T_I \cong \tau$$

The term $0.165\,(\theta/\tau)$ is small. By dropping this term and adjusting the coefficient 1.03 to 1.0, the equation for the reset mode reduces to $T_I = \tau$.

For an integrating process, Ziegler–Nichols is correct to relate the reset time T_I to the process dead time θ. But for a nonintegrating process, the reset time should be related to the process time constant τ as the ITAE equation does. All tuning methods that are based on a time-constant-plus-dead-time characterization do likewise.

Derivative Mode Tuning. One of the most glaring deficiencies from using first-order characterizations of the process is the tuning equations for the derivative mode:

Ziegler–Nichols: For the parallel PID, the tuning equation for the derivative mode is

$$T_D = 0.4\,L_R = 0.4\,\theta$$

The equation for the series PID is the same except that the coefficient is 0.5. In either case, the derivative time is related to the process dead time θ.

Integral criteria: For the parallel PID, the ITAE equation for the derivative time is

$$\frac{T_D}{\tau} = 0.308\left[\frac{\theta}{\tau}\right]^{0.929} \cong 0.308\frac{\theta}{\tau} \quad \text{or} \quad T_D \cong 0.308\,\theta$$

The value of the exponent of θ/τ is 0.929, which is close to 1.0. Making this exponent 1.0 simplifies the equation to the derivative time being 0.308 times the process dead time θ.

The issue is not whether the coefficient should be 0.3, 0.4, or 0.5. The tuning equations for the derivative mode are theoretically unsound. The equations suggest that a process with a large dead time would require a large derivative time. However, the derivative mode is ineffective for processes with a large dead time.

The difficulty is with the process characterizations. To tune the derivative mode consistently and effectively, the process characterization must include dynamics beyond first order.

8.3. ZIEGLER–NICHOLS CLOSED-LOOP METHOD

The Ziegler–Nichols closed-loop method (1) calculates the controller tuning parameters from the ultimate gain K_U and the ultimate period P_U for proportional-only control of the process. Ziegler–Nichols recommended the direct testing approach:

1. With all reset and derivative action removed from the controller, adjust the controller gain until the loop cycles continuously. Note the value of the controller gain (the ultimate gain K_U) and the period of the cycle (the ultimate period P_U).
2. Substitute the values of the ultimate gain and the ultimate period into the tuning equations in Table 8.1 to compute values for the controller tuning coefficients that give a response with a quarter decay ratio.

For many loops, the severity of such a test is unacceptable. We shall present an alternate approach shortly.

The test is conducted with the controller in automatic, but with all reset and derivative action removed. We need a starting value for the controller gain. If the loop has previously been tuned, the current value of the controller gain

TABLE 8.1. Ziegler–Nichols Open-Loop Tuning Equations

P-Only	PI	PID—Series	PID—Parallel
$K_C = 0.5\,K_U$	$K_C = 0.45\,K_U$	$K_C = 0.6\,K_U$	$K_C = 0.75\,K_U$
	$T_I = P_U/1.2$	$T_I = 0.5\,P_U$	$T_I = 0.625\,P_U$
		$T_D = P_U/8$	$T_D = P_U/10$

could be doubled (a gain of half the ultimate gain gives a response with a quarter decay). Otherwise, a starting value of 2%/% or 4%/% could be used.

The next step is to upset the loop. A convenient way to do this is to move the set point up (or down), wait for the loop to begin to respond, and then return the set point to its original value. Provided the controller gain is sufficiently large, this should lead to a cycle in the controlled variable.

Figure 8.1 illustrates the test for the hot water process. With the proportional-only controller in automatic, the set point is increased from 150 °F to 160 °F, but returned to 150 °F as soon as the hot water temperature begins to change. For the test in Figure 8.1, the successive values for the controller gain are as follows:

$K_C = 4.0\%/\%.$ The cycle is rapidly decaying. We can safely double the controller gain.

$K_C = 8.0\%/\%.$ The cycle is increasing. We need to reduce the controller gain.

$K_C = 6.0\%/\%.$ The cycle is now slowly decaying.

$K_C = 6.5\%/\%.$ The cycle is slowly increasing.

$K_C = 6.2\%/\%.$ The amplitude of the cycle is essentially constant ($K_U = 6.2\%/\%$); the period is approximately 7.5 minutes ($P_U = 7.5\,\text{min}$).

A constant amplitude cycle with the same period will also appear in the controller output. We are not interested in the amplitude of either cycle, provided no component of the loop is being driven to a limit (we want a natural harmonic, not a limit cycle).

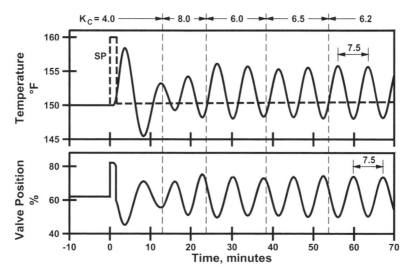

Figure 8.1. Process test as proposed by Ziegler and Nichols.

The severity of the test is a major issue. For slow temperature loops, the test interval would be measured in hours. For loops such as reactor temperature, cyclical conditions lasting for hours are simply not acceptable. This restricts the applicability of the method to fast loops, notably level and pressure.

Tuning Equations. Ziegler–Nichols provided the tuning equations in Table 8.1. For a performance objective, Ziegler–Nichols chose the quarter decay ratio, arguing that it is a good compromise between a rapid response and a short line out.

The original Ziegler–Nichols equations applied to the series form of the PID (the parallel form could not be implemented in the pneumatic hardware of 1942). The coefficients in Table 8.1 for the parallel form of the PID are obtained using the equations from Table 7.1 for converting from series to parallel. Otherwise, the coefficients in the Table 8.1 are those originally proposed by Ziegler–Nichols. Other values have been suggested for the coefficients in the equations; normally these are more conservative than those recommended by Ziegler–Nichols.

The usual approach for adjusting the sensitivity of a pressure regulator is to increase the sensitivity until the pressure cycles, and then to reduce the sensitivity by half—exactly what is suggested by the proportional-only equation in Table 8.1.

Observations for Controller Gain. The Ziegler–Nichols equations suggest that the controller gain for PI control is only 10% less than the controller gain for proportional-only control ($0.45\ K_U$ versus $0.5\ K_U$). This supports the procedure suggested for tuning a PI controller, namely, first to adjust the controller gain to obtain the desired performance and then to adjust the reset time.

When proceeding from PI to PID, how much can the controller gain be increased? Let us state the increase as a percentage of the controller gain for PI. The conclusions depend on the form of the PID:

Series. The equations suggest an increase of 33% ($0.45\ K_U$ to $0.6\ K_U$).
Parallel. The equations suggest an increase of 67% ($0.45\ K_U$ to $0.75\ K_U$).

In trial-and-error tuning endeavors for PID controllers, starting with an increase of 50% for the controller gain is consistent with the Ziegler–Nichols equations.

Observations for Derivative Time. The Ziegler–Nichols equations for the PID controller suggest that the derivative time should be set at a fixed fraction of the reset time:

Series: $T_D = ¼\ T_I$
Parallel: $T_D = 0.16\ T_I$

One's initial impulse is to conclude that the derivative time for a PID controller should be 1/4 (series PID) or 0.16 (parallel PID) of the reset time for a PI controller. But this interpretation assumes that the reset time for a PID controller is the same as the reset time for a PI controller. As we shall discuss shortly, this is not consistent with the Ziegler–Nichols equations.

Unfortunately, the relationship between reset time and derivative time is not so simple, and in fact, there may not even be one. One of the tuning methods that will be presented subsequently suggests that the reset time should equal the major process time constant and the derivative time should equal the minor process time constant. All that can be concluded regarding the relationship of derivative time to reset time is that the derivative time should be shorter than the reset time.

Observations for Reset Time. In the trial-and-error approach for tuning a PI controller, the first step is to tune a proportional-only controller to give a quarter decay ratio. For a starting value for the reset time, we recommended the period of the response with proportional-only. According to Ziegler–Nichols, the reset time should be the ultimate period divided by 1.2. The difference of about 20% is insignificant in the context of trial-and-error tuning.

For PI, Ziegler and Nichols suggest a reset time of $P_U/1.2$. For PID, their suggestion depends on the type of PID:

Series: $T_I = 0.5 P_U$, which is 0.6 times the reset time for PI.

Parallel: $T_I = 0.625 P_U$, which is 0.75 times the reset time for PI.

If one relies on the Ziegler–Nichols equations when proceeding from PI to PID, the approach should be as follows:

	Series PID	Parallel PID
Controller gain K_C	Increase by 33%	Increase by 67%
Reset time T_I	Shorten by 40%	Shorten by 25%
Derivative time T_D	¼ of T_I for PID	0.16 of T_I for PID

Tuning Coefficients for the Hot Water Process. From the process test illustrated in Figure 8.1, the ultimate gain K_U is 6.2%/% and the ultimate period P_U is 7.5 min. The tuning coefficients for a PI controller are computed as follows:

$$K_C = 0.45 K_U = (0.45)(6.2\%/\%) = 2.8\%/\%$$

$$T_I = P_U/1.2 = (7.5 \text{ min})/1.2 = 6.3 \text{ min}$$

Figure 8.2 illustrates the performance of the controller to an increase in the hot water temperature set point from 150 °F to 160 °F. The response has a decay ratio of approximately ¹/₃, which is reasonably close to ¼.

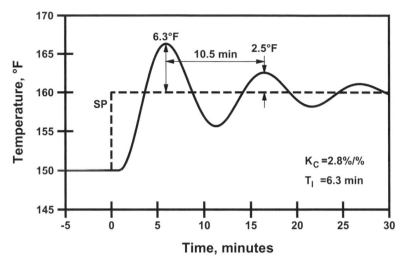

Figure 8.2. Tuning coefficients from the Ziegler–Nichols closed-loop method.

Advocates of applying the traditional tuning techniques readily acknowledge that the methods are not perfect. However, they argue that the tuning coefficients are usually (but not always) close to the appropriate values, in fact, close enough that the desired performance can be obtained by making "small refinements" in the values of the tuning coefficients. To obtain a response with a quarter decay ratio, there are two possibilities:

> **Reduce the controller gain.** For a nonintegrating process, this will be successful even if the reset time is too short. For the response in Figure 8.2, reducing K_C from 2.8%/% to 2.2%/% gives a response with a decay ratio of approximately ¼.
>
> **Lengthen the reset time.** If the controller gain is too high, this will not be successful.

Adjusting the controller gain is normally advocated. However, the following two checks can be applied to determine whether the reset time is excessively short:

1. The reset time should be at least half the period of the response. The response in Figure 8.2 has a period of 10.5 min; the reset time is 6.3 min. The reset time exceeds half the period, but not by much.
2. The following equation has been suggested for determining the appropriate value of the reset time T_I' from the current value of the reset time T_I and the period P of the cycle in the response:

$$T_I' = 0.5 \, P \tan^{-1}(2 \, \pi \, T_I / P)$$

For the response in Figure 8.2, the period of the cycle is 10.5 min and the value of the reset time is 6.3 min. This equation suggests that the reset time should be 6.9 min, an increase of about 10%. This change is insignificant.

8.4. THE RELAY METHOD

As compared with the direct testing approach suggested by Ziegler–Nichols, the relay method is a much less severe test for determining values for the ultimate gain K_U and the ultimate period P_U:

1. The amplitude of the cycle in the controlled variable is smaller.
2. The test period is shorter.

With the values of K_U and P_U from the relay test, the controller tuning coefficients are computed using the equations proposed by Ziegler–Nichols.

The Relay Test. The objective is to generate a cycle in the controlled variable by switching the controller output between two values, designated as M_U and M_L. The relay test is conducted by alternating the controller output between M_U and M_L based on the following switching logic:

$$M = \begin{cases} M_U \text{ if } PV \le SP \\ M_L \text{ if } PV > SP \end{cases}$$

where

M = Controller output
M_L = Value for M that will drive PV below target
M_U = Value for M that will drive PV above target
PV = Process variable for the controller
SP = Target for the switch (normally the set point of the controller)

For a direct-acting process (such as the hot water process), the most conservative approach would be to let M_L be 0% and M_U be 100% (for a reverse-acting process, M_U would be 0% and M_L would be 100%). However, the amplitude of the resulting cycle would be unnecessarily large.

Figure 8.3 illustrates the execution of the relay test on the hot water process. The cycle is generated by switching M between M_U and M_L each time the process variable crosses the target. For this example, the peak above the target is slightly larger than the peak below the target. Consequently, we will measure the amplitude A from peak-to-peak. Although major distortions should be avoided, minor distortions seem to have little effect on the results.

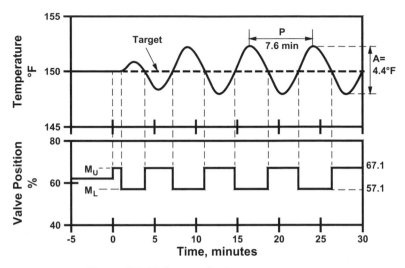

Figure 8.3. Relay test for hot water process.

The ultimate gain and ultimate period are related to the peak-to-peak amplitude A and period P of the cycle:

$$K_U = \frac{\pi \, |M_U - M_L|}{2 \, A}$$

$$P_U = P$$

Deadband for the Switch. Especially in level and pressure loops, noise on the process variable can lead to "chatter" (rapid switching between M_U and M_L) in the switching logic. To obtain a clean switch, a deadband DB must be incorporated into the switching logic:

$$M = \begin{cases} M_U \text{ if } PV \le SP - DB \\ M_L \text{ if } PV \ge SP + DB \\ \text{Unchanged otherwise} \end{cases}$$

The ultimate gain and ultimate period are related to the peak-to-peak amplitude A, the period P, and the deadband DB as follows:

$$K_U = \frac{\pi \, |M_U - M_L|}{2 \, (A^2 + 4 \, DB^2)^{1/2}}$$

$$P_U = P$$

The deadband should be set slightly larger than the noise band on the measured variable. The deadband must be large enough to eliminate the "chatter"; however, excessively large deadbands should be avoided.

Terminating the Test. In situations where no process disturbances occur (such as for the test in Figure 8.3), the amplitude and period become essentially constant after two or three cycles. However, should disturbances occur, this will not be the case. There are several possibilities for conditions for terminating the test:

1. Arbitrarily terminate the test after a fixed number (2 or 3) of cycles.
2. Terminate the test when the amplitude and period change by less than a specified amount.
3. Terminate the test when the integral of the cycle changes by less than a specified amount.

Although the second or third option might seem the best, how does one determine the tolerance on the change in amplitude/period or the integral of the cycle for terminating the test? Most users accept the default value from the supplier.

Utilization of the Relay Method in Automated Tuning. At least conceptually, the relay test could be executed manually, but this is unusual. Instead, the relay test is coupled with tuning equations such as Ziegler–Nichols to create an automated tuning procedure that we shall refer to as relay tune.

Relay tune is the most commonly available method for automated tuning, probably because it is very simple to implement. Versions of the relay tune method have been implemented in single-loop controllers as well as in distributed control systems and programmable logic controllers. The implementations differ mainly in the details, such as the condition for terminating the test.

The relay tune method is even used in such simple controllers as laboratory pot heaters. To initiate self-tuning via the relay method, the user enters the target for the pot temperature. The controller then switches power between full ON and full OFF to generate the cycle in the pot temperature. The tuning parameters for the controller are computed from the amplitude and period of the cycle.

Example. For the relay test illustrated in Figure 8.3, the test parameters are as follows:

Target $= 150\,°F$
M_U $= 67.1\%$
M_L $= 57.1\%$
DB $= 0\,°F$

After three cycles, the amplitude and period of the cycle are as follows:

A = 4.4 °F = 2.2% of span (measurement range is 50 °F to 250 °F)
P = 7.6 min

The values computed for the ultimate gain and ultimate period are as follows:

$$K_U = \frac{\pi |M_U - M_L|}{2 A} = \frac{\pi (10\%)}{2 (2.2\%)} = 7.1 \%/\%$$

$$P_U = P = 7.6 \text{ min}$$

From the direct test, the ultimate gain K_U was 6.2%/% and the ultimate period P_U was 7.5 min.

Using the equations originally proposed by Ziegler–Nichols (Table 8.1), the tuning coefficients for a PI controller are computed as follows:

$$K_C = 0.45 \, K_U = (0.45)(7.1 \%/\%) = 3.2 \%/\%$$

$$T_I = P_U/1.2 = (7.6 \text{ min})/1.2 = 6.3 \text{ min}$$

The direct testing approach originally suggested by Ziegler–Nichols gave 2.8%/% for the controller gain and 6.3 min for the reset time. The reset time is the same, but the controller gain is slightly larger. Consequently, the response will exhibit a somewhat larger second peak than for the response in Figure 8.2 for the original tuning. To obtain a quarter decay ratio, the controller gain will have to be reduced to 2.2%/%, which is a reduction of 31% from the tuning coefficients computed from the relay test.

8.5. OPEN-LOOP METHODS

As implied by the designation "open loop," the open-loop tuning methods execute the process test with the controller on manual. The test data consist of the response in the process variable to a known change in the controller output. The steps involved in an open-loop tuning method are as follows:

Step 1. With the controller on manual, obtain the response of the process to a known change in the controller output.
Step 2. Analyze the response data to obtain parameters that are indicative of the behavior of the process.
Step 3. Substitute the values of these parameters into the tuning equations to compute values for the controller tuning coefficients.

Never take process testing lightly. Probably the most appropriate description of process testing is as follows:

> Conducting a test on a production process can be done, but with great difficulty.

The most common problem in applying an open-loop tuning method is that the process test is not executed properly. It is important to recognize that there are two aspects of process testing:

Test execution. Generate the desired change in the controller output and collect the data. Computers are very good at this.

Test supervision. Make sure that the response is due solely to the change in the controller output. For this, computers are of no value.

Disturbance changes, load changes, and operator actions can easily influence the process variable, thus compromising the test data. Supervision of an open-loop test is a challenge.

Test Execution. In their 1942 article, Ziegler–Nichols suggested that the 2-inch strip chart recorder on pneumatic recording controllers was adequate. This is hardly the case, but in the pneumatic world, nothing else was available. Electronic recorders were definitely superior, but very few plants purchased one.

The advent of digital controls greatly improved our ability to collect the test data. We can collect data on multiple variables (at least the process variable and the controller output) on whatever time interval is desired. The data are collected in engineering units, so they can be subsequently plotted or processed in whatever manner is appropriate.

Disturbances and Loads. The analysis of the test data assumes that the response in the process variable is due solely to the change made in the controller output. If this assumption is not valid, the values for the parameters obtained from the process response data will not reflect the true nature of the process. "Garbage in, garbage out" definitely applies to open-loop test endeavors.

From a process perspective, the objective is to prevent all changes in disturbances and loads throughout the test period. Preventing such changes is rarely an easy endeavor, especially when the test period is measured in hours. Sources of changes include feed streams, ambient conditions, equipment malfunctions, and others. Some disturbances can be anticipated (for example, switching from one feed tank to another). However, others (such as equipment malfunctions) occur despite our best efforts to prevent them.

If one does not diligently supervise the test, disturbances are very likely to occur. But even with our best efforts at test supervision, disturbances can still invalidate a test.

Actions Taken by the Process Operator. A potential source of disturbances throughout the test is actions taken by the process operators. We are
not talking about the operator making unwanted changes in the controller
output; these are easy to prevent. But during the test period, the operator will
want to change the set points (and possibly outputs) of other controllers. Some
of these changes could affect the process variable for the loop for which the
test is being conducted.

Especially during a lengthy test, the operator will need to make changes to
address problems that originate in other parts of the process. But for each
change that the operator wishes to make, the test supervisor must ascertain
whether the change will affect the variable being recorded. If the change will
have an effect, making the change will invalidate the test data. The test supervisor must convince the operator that one of the following is acceptable:

- Delay the change until after the test is completed.
- Address the problem by making some other change that has no effect on
  the variable being recorded.

Test Supervision. Testing slow processes is far more difficult than testing
fast processes. A paper machine test requires about 5 minutes—1 minute or
so to make sure the process is lined out, a couple of minutes for the machine
to respond to the change in stock flow, and another minute or so to make sure
the machine again lines out. For a 5 minute test, the probability of a disturbance occurring is fairly low. Even if one occurs, repeating the test might be
less painful than trying to prevent the disturbances during the test. But if the
test interval is 5 hours, the probability of a disturbance occurring is quite high.
Repeating such tests is not very appealing either.

Most industrial processes are slow. Can we realistically execute an open-
loop test where the test interval is 5 hours? The answer is a qualified "yes."
Such tests can be executed, but the commitment required is not trivial. For
applications such as controller tuning, such a commitment is often difficult
to obtain. The alternative is trial-and-error tuning, which requires no such
commitment.

Change in Controller Output. Step 1 in the open-loop test procedure calls
for introducing one or more changes in the controller output. The impact on
process operations is minimized by using a small amplitude test, but factors
such as measurement noise and valve stiction/hysteresis will be more significant in a small amplitude test. We want the test data primarily to reflect the
process behavior. Consequently, we want to use as large a change in the controller output as process operations permit.

For any test, there is an acceptable range of values for the process variable.
The influence of the measurement device and final control element is minimized when the change in the controller output drives the process variable

over this entire range. We know what change in the process variable is acceptable, but translating these into corresponding values for the controller output can be a challenge.

Process Reaction Curve. Prior to the advent of computers, the process test data was analyzed by applying graphical constructions to the response of the process to a step change in the output of the controller. The test is as follows:

1. Allow the process to attain equilibrium (line out).
2. Switch the controller to manual (if not already on manual).
3. Change the controller output (increase or decrease).
4. Record the response in the process variable (the process reaction curve).

Ziegler–Nichols used the term "process reaction curve" for such a response. This term is indeed appropriate; with the controller on manual, the response data depend solely on the process (including measurement device and control valve).

Figure 8.4 presents the process reaction curve for the hot water process. The process is initially lined out at 150 °F; the final line-out value is approximately 164 °F. Two issues with this occur:

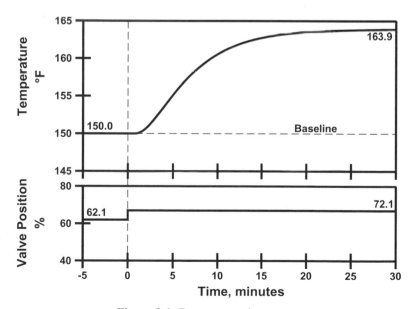

Figure 8.4. Process reaction curve.

1. We return the process to the process operator with the hot water temperature at 164 °F, but we got the process at 150 °F.
2. The hot water process is tolerant of increases in hot water temperature. For many processes, the acceptable range of operating conditions is very narrow.

In most respects, the use of a step test is dictated by the desire to use graphical constructions. With the availability of computers, other tools are available to analyze the test data. This also frees us to use tests other than the step test.

Pulse or Bump Test. In most cases, a pulse or bump test would be less disrupting to process operations. The procedure for a pulse or bump test is as follows:

1. Allow the process to attain equilibrium (line out).
2. Switch the controller to manual (if not already on manual).
3. Change the controller output (increase or decrease).
4. After some period of time, return the controller output to its original value.

Figure 8.5 presents the results of a pulse or bump test on the hot water process. The process is initially lined out at 150 °F with the controller output of 62.1%. The controller output is increased to 72.1%, held for 3 minutes, and then

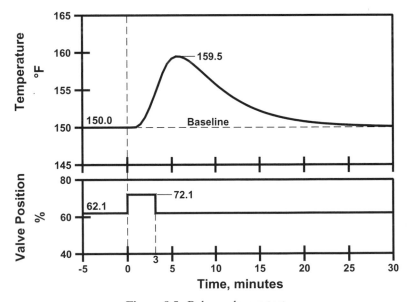

Figure 8.5. Pulse or bump test.

returned to 62.1%. The hot water temperature increases to almost 160 °F and then returns to 150 °F.

As compared with the step test, the advantages of the bump test are as follows:

1. We return the process to the process operator with the hot water temperature at 150 °F, which is the same temperature as when the operator turned the process over to us.
2. A larger change can be made in the controller output. This minimizes the effect of valve nonidealities on the test results.
3. By reducing the pulse amplitude or shortening the pulse duration, the maximum departure from the initial value can be reduced. This makes the pulse or bump test more acceptable when the allowable operating range is narrow. However, this increases the influence of factors such as measurement noise.

A few graphical constructions have been proposed for analyzing the results of a pulse or bump test, but none are practical. The analysis has to be by nonlinear regression techniques.

Ideally, the response to the pulse or bump returns exactly to the original base line. If a significant disturbance occurs during the test, the difference between the final equilibrium and the initial equilibrium will be substantial, and the test data must be discarded. However, factors such as hysteresis and/or stiction in the control valve will cause minor differences. The data can be retained, but the regression procedures must somehow take this into account. How does one distinguish between significant differences and minor differences? Usually this is at the discretion of the user.

8.6. GRAPHICAL CONSTRUCTIONS AND NONLINEAR REGRESSION

In presenting the graphical constructions, we will consider two categories of processes:

- Integrating.
- Nonintegrating.

Most of the attention will be directed to nonintegrating processes. Level is the most common integrating process. But in most cases, precise control of level is not required, and it can even be counterproductive. Such controllers are tuned very conservatively, and normally without the aid of a tuning method.

Logically, tuning methods should be applied to the important loops within a process, where any improvement in performance yields benefits to process operations. This makes temperature loops the most likely candidate, and these are nonintegrating processes.

Integrating Processes. The response of an integrating process to a step change is a ramp that continues until some limiting condition is attained. As undesirable consequences are usually associated with the limiting condition, the ramp must be reversed before limiting conditions are attained. To reverse the ramp, change the controller output in the opposite direction by twice the first change.

Figure 8.6 illustrates testing an integrating process in this manner. The controller output is initially changed from 70% to 75% (the change $\Delta M = 5\%$). After a constant rate of change is established for the ramp, the controller output is then changed to 65%. Once the response has approximately returned to its original value, the controller output is restored to its original value of 70%.

Such a test provides a ramp in both directions. For linear processes, the slope of the two ramps should be the same except for the sign. If the slopes are only modestly different, their values are usually just averaged. But if the slopes are significantly different, the process is directional in its behavior. The tuning coefficients should be computed using the smaller slope (slow processes are more difficult to control than fast processes).

Graphical Construction. For an integrating process, the simplest model is an integrator-plus-dead-time model with two parameters:

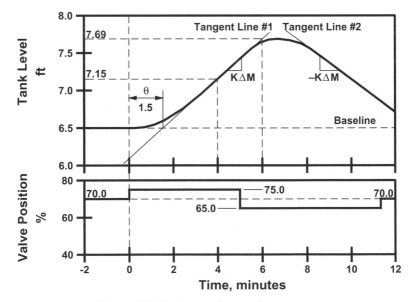

Figure 8.6. Testing an integrating process.

$$K = \text{Process gain}, (\%/\%)/\text{min}.$$

$$\theta = \text{Process dead time}, \text{min}.$$

The simple graphical construction illustrated in Figure 8.6 can be used to determine values for the gain K and the dead time θ. The following data are from the initial ramp:

- As time increases, the rate of change of the response should approach a constant value. Construct a line tangent to the response in the region of constant rate of change.
- Construct the base line for the response, and locate the intersection of the tangent line and the base line. Considering time zero to be the time that the step change was introduced, the dead time θ is the time at which the tangent line intersects the base line. For the response in Figure 8.6, the dead time θ is 1.5 min. But often this is a major weakness of the graphical construction—the value of a small dead time is difficult to determine accurately.
- The slope of the tangent line is the product of the gain K and the magnitude ΔM of the step change in the input. The slope of the tangent line is most easily determined from two points on the tangent line. The following two points are indicated in Figure 8.6 (span of tank level transmitter is 0 to 12 ft):

	Time	Tank Level
Point 1	4.0 min	7.15 ft or 59.6%
Point 2	6.0 min	7.69 ft or 64.1%

The process gain K is computed as follows:

$$K = \frac{\text{Slope}}{\Delta M} = \frac{(64.1\% - 59.6\%)/(6\,\text{min} - 4\,\text{min})}{5.0\%} = 0.45\,(\%/\%)/\text{min}$$

Similar constructions can be applied to the second ramp in Figure 8.6.

Nonintegrating Process. The response of a nonintegrating process to a step change in the controller output will be an S-shaped curve as previously illustrated in Figure 8.4. The response starts at the original base line and ends at the new steady state. The simplest model that can potentially represent the process is the time-constant-plus-dead-time model, which has three parameters:

- K = Process gain, %/%.
- τ = Process time constant, min.
- θ = Process dead time, min.

The easiest of these parameters to determine from the process reaction curve is the process gain K. From Figure 8.4, the following values are available:

	Controller Output	Hot Water Temperature
Initial	62.1%	150.0 °F or 50.00% of span
Final	67.1%	163.9 °F or 56.95% of span

The process gain is the ratio of the change in the PV to the change in the controller output M. The value of the process gain is computed as follows:

$$K = \frac{\Delta PV}{\Delta M} = \frac{56.95\% - 50.00\%}{67.1\% - 62.1\%} = 1.39\%/\%$$

We shall present three different ways to determine the dynamic parameters τ and θ; however, the process gain K is the same for all three.

Tangent Line. This approach determines values for both the process time constant τ and the process dead time θ from a tangent line. Figure 8.7 illustrates the following construction:

- Locate the point on the response where the rate of change is greatest (the steepest ascent point). Construct a line tangent to the response at this point.

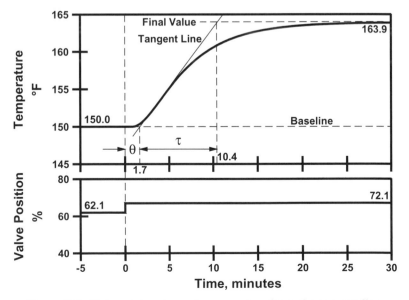

Figure 8.7. Determining dynamic parameters from the tangent line.

- Construct the base line for the response, and locate the intersection of the tangent line and the base line. Considering time zero to be the time that the step change was introduced, the dead time θ is the time at which the tangent line intersects the base line. For the example in Figure 8.7, the dead time θ is 1.7 min. Especially when the dead time is small, accurately determining a value for the dead time is difficult.
- Construct the final value line for the response, and locate the intersection of the tangent line and the final value line. The time constant τ is the time past the dead time that the tangent line intersects the final value line. For the example in Figure 8.7, the time constant τ is 8.7 min.

Constructing the tangent line is much more difficult than one initially suspects. Even with good resolution on the trend, the steepest ascent point is difficult to locate precisely. Even a minor amount of noise on the process variable makes drawing an accurate tangent line far more difficult. Unfortunately, the approach illustrated in Figure 8.7 obtains both dynamic parameters from a tangent line that is difficult to construct.

Tangent Line Plus 63% Point. An alternative approach is to determine the process dead time θ from the tangent line, but to determine the process time constant τ from the 63% point. Figure 8.8 illustrates the construction. The tangent line is constructed and the dead time θ determined to be 1.7 min as described above.

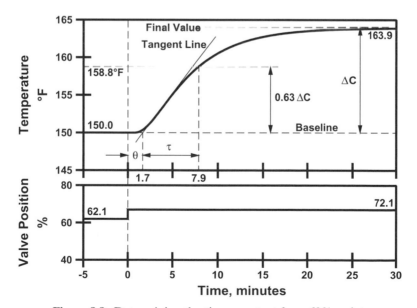

Figure 8.8. Determining the time constant from 63% point.

To locate the 63% point, first determine the total change ΔPV. This is 13.9 °F, 63% of which is 8.8 °F. The 63% point is when the response changes by 8.8 °F to attain 158.8 °F, the time being 7.9 min for the response in Figure 8.8. Subtracting the dead time θ of 1.7 min from this value gives 6.2 min for the time constant τ.

Two Points. Both of the previous approaches require constructing a tangent line, which is not easy. This approach replaces this by another issue that is also not easy to answer. Basically, this approach forces the model response to match the process reaction curve exactly at two points. This raises another issue: which two points? There is no way to answer this question, but before dismissing this approach as silly, try constructing a tangent line. Given the difficulty of doing this, the usual conclusion is that constructing the tangent line is equally silly.

There is another advantage of using two points. Most digital systems retain trend history data. From this data, one can retrieve the value of the PV at selected points in time. Consequently, it is very easy to determine two points on the response. To construct a tangent line, you have to retrieve the data and construct a trend plot that can be printed. Most systems permit this to be done, but it is far easier simply to retrieve two points. In the end, the ease of extracting two points from the trend history data is what makes this approach most appealing.

The proposed approach determines the process dead time θ and the process time constant τ by forcing the model response to match the process response at two points. Let us designate these two points as follows:

c_1 = Fraction of the total change ΔPV attained at time t_1.

c_2 = Fraction of the total change ΔPV attained at time t_2.

The general equations for computing the time constant τ and the dead time θ from these two points are as follows:

$$\tau = \frac{t_2 - t_1}{\ln\left[\dfrac{1 - c_1}{1 - c_2}\right]}$$

$$\theta = t_1 + \ln(1 - c_1) = t_2 + \ln(1 - c_2)$$

Using the 20% point ($c_1 = 0.20$; $t_1 = t_{20}$) and the 63% point ($c_2 = 0.63$; $t_2 = t_{63}$), the equations are as follows:

$$\tau = 1.29\,(t_{63} - t_{20}).$$

$$\theta = t_{63} - \tau.$$

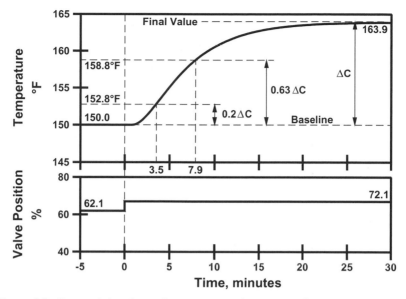

Figure 8.9. Determining dynamic parameters from two points on the response.

For the response in Figure 8.9, the values are as follows:

$$t_{20} = 3.5\,min$$
$$t_{63} = 7.9\,min$$
$$\tau = 1.29\,(7.9 - 3.5) = 5.7\,min$$
$$\theta = 7.9 - 5.7 = 2.2\,min$$

Why the 20% point and the 63% point? Using the 63% point yields a simple relationship—the 63% point is the sum of the time constant and the dead time. However, when retrieving two points from the trend history file, there is no need to retrieve the 63% point precisely—any point in the vicinity will do, as long as we use the time and the actual percentage change of the PV for this point.

Quality of Fit. Figure 8.10 compares the response of the model with the process response for each of the three fits. The observations are as follows:

Tangent line. The quality of this fit is poor.
63% point. The fit is reasonable. However, the model response is merely forced to match the process response at the 63% point.
Two points. The fit is reasonable. However, the model response is merely forced to match the process response at the 20% and the 63% points.

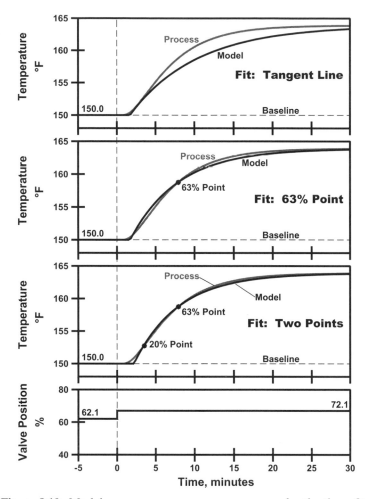

Figure 8.10. Model response versus process response for the three fits.

Would the use of regression significantly improve the quality of the fit? For the fits relying on one or two points on the response, the simplicity of the model is undoubtedly the limiting factor on the quality of the fit. Using regression would give slightly different values for θ and τ, but the overall quality of the fit would not improve significantly.

Nonlinear Regression. The first-order models are surprisingly effective at tuning a PI controller. But to tune a PID controller effectively and consistently requires a more sophisticated model, such as the two-time-constants-plus-dead-time model. Graphical constructions have been proposed to obtain the coefficients for such a model. But if you think constructing a tangent line is difficult, just try one of these constructions! It is even possible to obtain values for the two time constants and the dead time from three points on the response, but even this is impractical—very small errors in the data points have a large effect on the values of the model coefficients.

Prior to the availability of PCs, the regression approach was certainly inconvenient. But today's PCs are easily capable of quickly executing a nonlinear regression analysis. Most digital systems can collect the trend data from a process test, from which it can be retrieved for off-line analysis on the PC. Given that pulse or bump tests can be used in lieu of step tests, this approach deserves serious consideration.

To obtain values for the model parameters, the nonlinear regression approach requires the following information:

1. Process test data (data points for controller output and process variable over some interval of time).
2. The structure of the process model. To tune PID, the model should be the two-time-constants-plus-dead-time model. However, nonlinear regression can be applied to other models.
3. Starting values for the parameters in the process model.

The nonlinear regression consists of iterations. On each iteration, the following occurs:

1. Compute the response of the model from the data points for the controller output and the current values of the model parameters.
2. Compare the model response with the process variable data points from the process test data. Compute some measure of the fit, usually the sum of squares of the model errors (the difference between the model response and the data points from the process test).
3. Adjust the current values of the model parameters with the objective of reducing the value of the measure of fit. A variety of mathematical routines generally referred to as search techniques enable computers to do this.

The iterations continue until the measure of fit is minimized.

The test data are from the pulse or bump test illustrated in Figure 8.5. We previously used graphical constructions on the step response to obtain the parameters for a time—constant-plus-dead-time model. Let us begin by fitting a time-constant-plus-dead-time model to the pulse or bump test data. The following table presents the previous results plus the model parameters from a regression fit to minimize the sum of squares of the model errors:

Deficiencies. Two major criticisms have been directed to the tuning equations:

- The reset time is related to the process dead time (for PI, $T_I = 3.33\ L_R = 3.33\ \theta$). This may be appropriate for integrating processes, but for a nonintegrating processes, the reset time should be related to the process time constant.
- For the PID controller, the derivative time is related to the process dead time ($T_D = 0.5\ L_R = 0.5\ \theta$ for the series form of PID or $T_D = 0.4\ L_R = 0.4\ \theta$ for the parallel form of PID). For a large dead time process, this suggests a large derivative time. However, the derivative mode is not effective for a large dead time process.

The last criticism can be directed to any tuning technique that relies on a first-order characterization of the process (integrator-plus-dead-time model or time-constant-plus-dead-time model).

Example. We shall apply the Ziegler–Nichols open-loop method to the hot water process. The process reaction curve in Figure 8.13 is the result of a change ΔM of +5% in the output of the controller. After constructing a tangent line at the point of steepest ascent, the following values can be determined:

$$L_R = 1.67\ \text{min}$$

$$L_R\ R_R = 150.0\,°F - 147.4\,°F = 2.6\,°F$$

$$= 1.3\%\ \text{of span (measurement range is } 50\,°F \text{ to } 250\,°F).$$

$$R_R = 1.3\%/1.67\,\text{min} = 0.78\%/\text{min}.$$

Alternatively, a value of 0.79%/min for the reaction rate can be computed from the two points on the tangent line as indicated in Figure 8.13.

The tuning coefficients for a PI controller are computed as follows:

$$K_C = \frac{0.9\ \Delta M}{L_R\ R_R} = \frac{(0.9)(5\%)}{(1.67\ \text{min})(0.78\%/\text{min})} = 3.5\%/\%$$

$$T_I = 3.33\ L_R = (3.33)(1.67\ \text{min}) = 5.6\ \text{min}$$

Figure 8.14 presents the performance of the controller to an increase in the hot water temperature set point from 150°F to 160°F. The response has a decay ratio of approximately two thirds.

As noted, the customary approach is to adjust the controller gain until the desired performance is attained. Reducing the controller gain from 3.5%/% to 2.1%/% gives a response with a decay ratio of approximately one quarter.

As before, we can apply the following two checks on the value of the reset time:

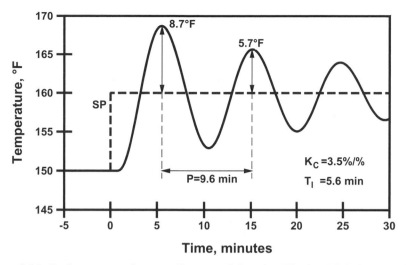

Figure 8.14. Performance of controller tuned by the Ziegler–Nichols open-loop method.

1. The reset time should be at least half the period of the response. The response in Figure 8.14 has a period of 9.6 min; the reset time is 5.6 min. The reset time exceeds half the period, but not by much.

2. The following equation has been suggested for determining the appropriate value of the reset time T_I' from the current value of the reset time T_I and the period P of the cycle in the response:

$$T_I' = 0.5\ P\ \tan^{-1}(2\ \pi\ T_I/P).$$

Substituting 9.6 min for P and 5.6 min for T_I gives a value of 6.3 min for the reset time. This is an increase of only 10%, which suggests that the reset time is not grossly in error.

8.8. THE LAMBDA METHOD

The Lambda method (2) originated in the paper industry and continues to enjoy popularity in that industry. It is essentially a synthesis method; that is, the controller is designed specifically for the process.

The block diagram in Figure 8.15 is a general representation of a control loop. The control loop establishes a relationship between the set point SP and the controlled variable C. To design a custom control equation, two items must be known:

1. A process model, including values for all model parameters (gain, time constant, etc.).

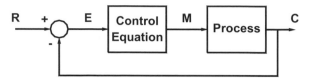

Figure 8.15. General representation of a control loop.

2. The performance specifications for the control loop. One could specify C for a stated SP. For example, one could describe how the loop is to respond to a unit step change in the set point.

The process model tells us how the process behaves. The performance specifications for the loop tell us how the control loop is to behave. The only unknown is the control equation (and any coefficients contained therein).

Suppose we specify that the closed-loop behavior is to be described by a first-order lag coupled with a dead time. There are three coefficients in this relationship:

Closed-loop gain K_{CL}. To force the controller to contain reset action, the closed-loop gain K_{CL} must be 1.0. A unit change in the set point SP must cause the controlled variable C to change one unit.

Closed-loop dead time θ_{CL}. The loop dead time θ_{CL} will be the same as the process dead time θ. It is possible to add more dead time to the loop ($\theta_{CL} > \theta$), but there is no incentive to do so.

Closed-loop τ_{CL}. A short closed time constant gives a fast response and a more aggressive controller.

There is only one value for the user to specify, specifically, the closed-loop time constant τ_{CL}. When originally presented, λ was used instead of τ_{CL} for the closed-loop time constant, hence, the name "Lambda method."

Tuning Equations. The Lambda method is not constrained to yield a PI or PID equation for the controller. But for the simple models typically used for controller tuning, the control equation from the design procedure turns out to be

- PI when the model is time-constant-plus-dead-time.
- PID when the model is two-time-constants-plus-dead-time.

For these models, the design procedure yields the tuning equations in Table 8.3. The performance specification τ_{CL} and the process dead time θ only affect the controller gain K_C. The reset time T_I and the derivative time T_D depend only on the process time constant(s).

TABLE 8.3. Lambda Method Tuning Equations

	Proportional	Integral	Derivative
PI	$KK_C = \dfrac{\tau}{\tau_{CL} + \theta}$	$T_I = \tau$	
PID—Series	$KK_C = \dfrac{\tau_1}{\tau_{CL} + \theta}$	$T_I = \tau_1$	$T_D = \tau_2$
PID—Parallel	$KK_C = \dfrac{\tau_1 + \tau_2}{\tau_{CL} + \theta}$	$T_I = \tau_1 + \tau_2$	$T_D = \dfrac{\tau_1\,\tau_2}{\tau_1 + \tau_2}$

The value for τ_{CL} is usually within the following range:

$$\theta \le \tau_{CL} \le \tau$$

The minimum possible value for τ_{CL} is zero. If $\tau_{CL} = 0$, then for PI control, $K\,K_C = \tau/\theta$. Setting $\tau_{CL} = \theta$ gives a loop gain of half this value. Tuning a PI controller can be approached as follows:

1. Set the reset time to τ.
2. Set the controller gain to the starting value suggested by either $\tau_{CL} = \theta$ (to begin with a large value for the controller gain and then reduce it) or $\tau_{CL} = \tau$ (to begin with a small value for the controller gain and then increase it).
3. Adjust the controller gain until the desired performance is attained.

Example. We shall apply the Lambda method to the hot water process. Let us use the model parameters from Figure 8.8:

Process gain $K = 1.39\%/\%$.
Process dead time $\theta = 1.7$ min.
Process time constant $\tau = 6.2$ min.

Figure 8.16 illustrates the performance of a PI controller tuned via the Lambda method. Responses are provided for the following values of τ_{CL}:

τ_{CL}, min	$K\,K_C$	K_C, %/%	T_I, min
0	3.6	2.6	6.2
θ	1.8	1.3	6.2
$2\,\theta$	1.2	0.86	6.2
τ	0.78	0.56	6.2

For this example, one could start with $K_C = 1.3\%/\%$ ($\tau_{CL} = \theta$) and then decrease the controller gain until the desired performance is attained. Alternatively, one

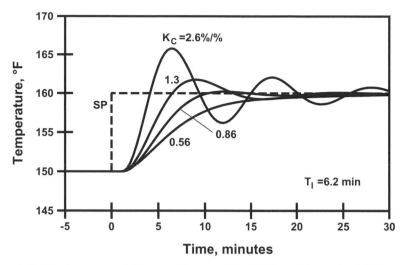

Figure 8.16. Performance of PI controller tuned by the Lambda method (also applies to the IMC method).

could start with $K_C = 0.56\%/\%$ ($\tau_{CL} = \tau$) and then increase the controller gain until the desired performance is attained.

8.9. IMC METHOD

Internal model control (IMC) is a controller synthesis method that derives a control equation from the process model and a performance specification. When the process contains dead time, the IMC control equation provides for dead time compensation. But when the dead time is small relative to the process time constant, an approximation can be substituted for the dead time to give tuning equations for the following controllers (3):

- PI control for a time-constant-plus-dead-time model.
- PID control for a time-constant-plus-dead-time model (a different approximation is used for the dead time).
- PID control for a two-time-constants-plus-dead-time model.

Table 8.4 presents the tuning equations. The following observations apply:

1. The closed-loop time constant τ_{CL} is exactly the same as for the Lambda method.
2. The following relationships are identical to those provided by the Lambda method:

TABLE 8.4. IMC Tuning Equations for the Nonintegrating Process

	Model	Proportional	Integral	Derivative
PI	K, τ, θ	$KK_C = \dfrac{\tau}{\tau_{CL} + \theta}$	$T_I = \tau$	
PID—Series	K, τ, θ	$KK_C = \dfrac{2\tau}{2\tau_{CL} + \theta}$	$T_I = \tau$	$T_D = \dfrac{\theta}{2}$
PID—Parallel	K, τ, θ	$KK_C = \dfrac{2\tau + \theta}{2\tau_{CL} + \theta}$	$T_I = \tau + \dfrac{\theta}{2}$	$T_D = \dfrac{\tau\theta}{2\tau + \theta}$
PID—Series	$K, \tau_1, \tau_2, \theta$	$KK_C = \dfrac{\tau_1}{\tau_{CL} + \theta}$	$T_I = \tau_1$	$T_D = \tau$
PID—Parallel	$K, \tau_1, \tau_2, \theta$	$KK_C = \dfrac{\tau_1 + \tau_2}{\tau_{CL} + \theta}$	$T_I = \tau_1 + \tau_2$	$T_D = \dfrac{\tau_1 \tau_2}{\tau_1 + \tau_2}$

     a. The relationships for the PI controller.

     b. The relationships for the PID controller that are based on the two time-constant-plus-dead-time model.

3. In the relationships for the PID controller that are based on the time-constant-plus-dead-time model, the derivative time is determined primarily if not exclusively by the process dead time. As observed, this is theoretically unsound.

Since the tuning equations for PI control are identical, the responses in Figure 8.16 apply to the IMC method as well as to the Lambda method.

Integrating Process. The IMC equations can be used to obtain the tuning equations in Table 8.5 for an integrating process. For an integrating process, the closed-loop time constant τ_{CL} affects the controller gain, the reset time, and the derivative time (except for the series PID).

    To tune a loop, one can still start with $\tau_{CL} = \theta$, and then increase τ_{CL} until the desired performance is obtained. But as τ_{CL} is changed, all tuning coefficients must be recomputed.

8.10. INTEGRAL CRITERIA METHOD

The integral criteria method (4) is an open-loop method that calculates the controller tuning parameters from the coefficients in a time-constant-plus-dead-time model. The relationships provide the values of the tuning coefficients that minimize an integral criterion. Although others are available, we shall present relationships for only the following two integral criteria:

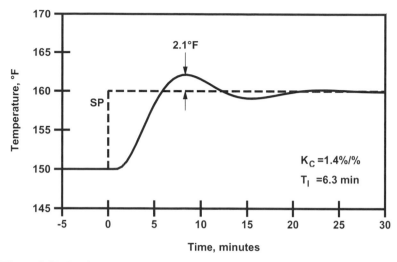

Figure 8.17. Performance of PI controller tuned for ITAE integral criterion.

8.11. SUMMARY

In the 1960s, advocates of using computers for process control promised automatic tuning. Considerable effort was directed to this technology but with little concrete results. It was not until the 1990s that automatic tuning became a common feature in commercial control products.

But despite all of this, most controllers are tuned by the traditional trial-and-error or "knob-twiddling" approach, the reasons being:

- Automatic tuners only work in those loops that you can tune. There are untunable loops, and in those loops, tuning is treating a symptom, not the problem.
- As compared with an automatic tuner, anyone reasonably proficient in tuning can tune a PI controller in a comparable time and obtain comparable results. Then why bother? As the saying goes, "If it ain't cheaper, it had better be better."
- The simple (not computer based) tuning methods will not consistently and effectively tune PID controllers. Any method that relates the derivative time to the process dead time has to be viewed with suspicion.

As for automatic tuning, it is certainly good to have this technology available, but in reality, its effect on the practice of process control has been minimal.

What does the future hold? Tuning PID controllers in slow-temperature loops is where tuning assistance would be of great value. The regression methods are capable of tuning such loops, provided a quality test can be performed on the process. But once the decision is made to invest the time and effort to conduct a process test, two options are now possible:

1. Apply regression techniques to the data to obtain the coefficients for a two time-constant-plus-dead-time model, and then from these coefficients, compute the tuning parameters for a PID controller.
2. Use the test data as the basis for developing a model predictive controller for the loop.

If option 2 is selected, a test other than a pulse or bump may be conducted, but the overall effort is about the same.

LITERATURE CITED

1. Ziegler, J. G., and N. B. Nichols, "Optimum Settings for Automatic Controllers", *ASME Transaction Journals*, Vol. 64, No. 11 (November 1942), p. 759.
2. Dahlin, E. B., "Designing and Tuning Digital Controllers", *Instruments and Control Systems*, Vol. 41, No. 6 (June 1968), p. 77.
3. Chien, I., and P. S. Fruehauf, "Consider IMC Tuning to Improve Controller Performance", *Chemical Engineering Progress*, Vol. 86, No. 10 (October 1990), p. 33.
4. Rovira, A. A., P. W. Murrill, and C. L. Smith, "Tuning Controllers for Set Point Changes", *Instruments and Control Systems*, Vol. 42, No. 12 (December 1969), p. 67.

CHAPTER 9

Measurement Devices

Being the eyes of the controller, the measurement device is a critical component of every control loop, although not the most probable source of problems. We examine three issues:

Steady-state characteristics. We specifically examine the consequences of a span that is much wider than required for control.

Dynamic characteristics. We use a temperature probe to illustrate the impact of measurement device dynamics.

Smoothing or filtering. We point out the perils of excessive smoothing or filtering.

9.1. STEADY-STATE BEHAVIOR

The focus here is on three aspects of the measurement device:

- Resolution.
- Linearity.
- Turndown ratio.

Resolution. A digital value is an approximation of an analog value to a finite number of digits. The resolution is a change of one in the least significant digit of the digital representation. The difference between the digital approximation and the analog value is the quantization error. In the context of controls, its effect is similar to that of noise, and it is sometimes referred to as quantization noise.

The upper part of Figure 9.1 illustrates the effect of resolution on converting an analog value to a digital value. The range of the analog signal is 1 to 5 volts. This signal is converted to a digital value as a percent of span with a resolution of 0.1%. The resulting three-digit digital representation provides

---

Practical Process Control: Tuning and Troubleshooting, by Cecil L. Smith
Copyright © 2009 John Wiley & Sons, Inc.

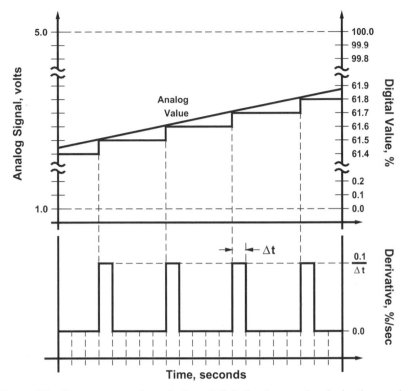

Figure 9.1. Consequence of resolution of digital value on the derivative mode.

1000 values, ranging from 0.0% to 99.9% in increments of 0.1%. The resolution is usually stated as 1 part in 1000.

In Figure 9.1, the analog signal is slowly increasing with a constant rate of change. However, the digital value increments by 0.1, resulting in a function that consists of a succession of step changes of magnitude 0.1. The impact of this on the individual modes is as follows:

Integral. The impact is minimal.

Proportional. Suppose the controller gain is 5%/%. Each time the digital value changes by 0.1%, the controller output changes by 0.5%. The larger the controller gain, the more serious the impact of the resolution.

Derivative. The lower part of Figure 9.1 illustrates computing the derivative using a simple finite difference of a slowly increasing analog value. For most sampling instants, the difference is zero. But when the digital value changes by 0.1, the finite difference changes by $0.1/\Delta t$, where Δt is the sampling interval. The output of the derivative calculation is a series of pulses whose magnitude is very large for short sampling intervals. The purpose of derivative mode smoothing is to address this behavior.

The analog-to-digital converters used in the analog input hardware of most digital systems provide a resolution of at least 1 part in 4000. But today, the trend is to replace this approach with transmitters that transmit the value in engineering units to a finite number of digits. Often the resolution is established based on what the process operators need to see. But when the input is the process variable for a controller, a greater resolution is normally required. The minimum resolution should be 1 part in 1000, and preferably higher if either the controller is to be tuned with a large gain or if a derivative is to be used in the controller.

To illustrate, consider a pressure transmitter with a measurement range of 0 to 10 psig. The resolution in the digital value is 0.1 psig or 1 part in 100 or 1% of span, which is adequate for the process operators. But consider the impact on the proportional mode. The controller gain of the pressure controller is 5%/%. Every time the pressure changes by 0.1 psig (or 1% of span), the controller output changes by 5%, which is a noticeable bump. Thinking about adding a derivative? With this resolution, forget it.

Linearity. With the introduction of electronic transmitters and the subsequent incorporation of microprocessors into the transmitters, the output of most transmitters varies linearly with the value of the measured variable.

The transmitter operating line (a plot of the transmitter output as a function of the process variable in engineering units) is usually a straight line. The departure from linearity rarely exceeds 1%, and from a process control point of view, it is not a source of problems. Consequently, the transmitter gain or sensitivity is as follows:

$$\text{Transmitter gain} = \frac{\text{Span of the output signal}}{\text{Engineering span for the measured variable}}$$

Although the output of the transmitter is linear, the output from the basic sensor within the measurement device is often nonlinear. Common examples include the following:

Orifice meter. Flow varies nonlinearly with pressure drop across the meter (applies to all head-type flow meters).

Thermocouple. Temperature varies nonlinearly with millivolts.

RTD. Temperature varies nonlinearly with resistance.

When a transmitter is installed, the linearization is normally provided within the transmitter. However, most control systems continue to provide the capability to perform the necessary linearizations.

Turndown Ratio. Although transmitters and controllers are very capable of performing the linearization, there are some consequences. An example is the limited turndown ratio for the orifice meter.

For a measurement device, the turndown ratio is defined as follows:

$$\text{Turndown ratio} = \frac{\text{Maximum value that can be accurately measured}}{\text{Minimum value that can be accurately measured}}$$

For a truly linear measurement device, the turndown ratio is usually 10:1 or better. However, for the orifice meter, the turndown ratio is around 4:1. The poor value for the turndown ratio is a consequence of the nonlinear relationship inherent within the meter.

For all head-type flow meters (including the orifice meter), the pressure drop across the meter varies with the square of the flow through the meter:

$$\frac{\Delta P}{\Delta P_{MAX}} = \left[\frac{F}{F_{MAX}} \right]^2$$

where

F = Volumetric flow through the meter

F_{MAX} = Volumetric flow when the pressure drop is ΔP_{MAX}

ΔP = Pressure drop across the meter

ΔP_{MAX} = Upper range value for the pressure transmitter (the relationship between F_{MAX} and ΔP_{MAX} is determined by the meter design)

Differential pressure transmitters commonly provide the option to perform the square root extraction.

Suppose the minimum measurable value for the ΔP is 0.1 ΔP_{MAX}. The corresponding flow is $0.1^{1/2} F_{MAX} = 0.316 F_{MAX}$ for a turndown ratio of about 3:1. What would be required of the ΔP measurement to attain a 10:1 turndown ratio on flow? For $F = 0.1 F_{MAX}$ (a 10:1 turndown ratio), then $\Delta P = 0.01 \Delta P_{MAX}$ (a 100:1 turndown ratio). But even if the transmitter is capable of this turndown ratio, any noise in the ΔP measurement is greatly amplified by the square root extraction required to convert to flow.

9.2. VERY SMALL PROCESS GAIN

When the controller gain K_C is expressed in %/% (as most controllers do), a measurement range is required for the process variable. In performing the loop calculations, the set point, the process variable, and the control error are expressed as a percent of span based on this measurement range.

The designers of control systems can select from the following three options regarding the measurement range for the loop PV:

1. When configuring the PID controller, a source is specified for the PV input. The measurement range of the source can be retrieved and used

as the measurement range of the PV. In most cases, the source is an input from the transmitter associated with the measurement device, so the measurement range of the PV for the loop is the same as the measurement range of the transmitter.

2. Specify the measurement range for the PV input as part of the configuration data for the PID controller. Most newer systems take this approach. When the source of the PV input is the measured variable from a transmitter, the measurement range of the PV does not have to be the same as the measurement range for the transmitter. However, in most cases, it will be.

3. Express the controller gain K_C in engineering units, such as %/°F for the temperature controller for the hot water process. Although this approach is definitely an option, designers choose this approach so infrequently that we shall say no more about it.

The transition from option 1 to option 2 is driven by measurement device technology. When the transmission medium is the 4–20 ma current loop, a measurement range is mandatory. But with digital communications, the measured value can be transmitted in engineering units. For example, a common approach with temperature measurements is to transmit the temperature in either °C or °F with a resolution of 0.1°. With this approach, no measurement range is required for transmitting the value. In fact, no measurement range is required until the value is used within the PID calculations.

Let us begin the discussion in this section by assuming that the measurement range for the PV within the PID controller is the same as the measurement range for the transmitter. Our first objective is to understand what problems can develop with this approach. We will then examine the options for addressing the consequences.

Effect of Wide Transmitter Span on the Process Gain. As the term is used by control engineers, the process gain K is the product of the valve gain K_V, the process unit gain K_E, and the transmitter gain K_T. A small value for any one of the gains would cause the process gain to be small (assuming neither of the other two gains is unusually large).

Of the three gains, the transmitter is the most frequent contributor of a small gain. A temperature transmitter with a measurement range of 80°C to 100°C has a gain of 5%/°C; a transmitter with a measurement range of 0°C to 100°C has a gain of 1.0%/°C.

Sometimes designers are just overly conservative (or just plain sloppy), but occasionally a wide span in the transmitter is required for reasons such as the following:

Startup. For normal operations, the 80°C to 100°C measurement range is appropriate. But on startup, the initial temperature is near ambient conditions, which requires the 0°C to 100°C measurement range.

Batch process. For product A, the 80 °C to 100 °C measurement range is perfect. But for product B, a measurement range of 0 °C to 20 °C is required. To cover all batches, the 0 °C to 100 °C measurement range is configured.

Extruder Example. One application where spans are routinely oversized is extruder temperatures. Multiple temperature controllers are required, six or so along the barrel and one or two more for the die. For a polypropylene extruder, the temperatures are above 500 °F when the extruder is in operation. A measurement range of 500 °F to 600 °F would suffice for normal control of extruder temperatures.

However, a temperature measurement range of 0 °F to 600 °F is normally provided. The reason is startup. Depending on the nature of the extrusion process, startup can be extremely critical. If the screw is started prematurely, a plug of solid material can develop in the die, which leads to extreme pressures within the extruder and a major malfunction referred to as "blowing the head off."

To prevent this, a predefined sequence is followed to heat the extruder before starting the screw. For each zone, the temperature is raised to a specified value and held at this value for a specified time. This is repeated until operating temperatures are attained. This necessitates the 0 °F to 600 °F measurement range.

Effect of a Small Process Gain on the Controller Gain. As discussed, the loop performance depends on the loop gain, which is the product of the process gain K and the controller gain K_C. A wide transmitter span has no effect on the appropriate value for the loop gain. Suppose a controller gain of 5%/% is required for a temperature measurement range of 500 °F to 600 °F (transmitter gain is 1%/°F). If the measurement range for the transmitter is 0 °F to 600 °F (transmitter gain is 0.167%/°F), a controller gain of 30%/% is required. At least theoretically, it is possible to compensate for a small process gain by tuning the controller with a large controller gain.

When conventional controllers were used for extruder temperatures, the adjustment range for the controller gain was typically 1.0%/% to 100%/% (for normal service, the adjust range was typically 0.1%/% to 10%/%). A few controllers provided a "gain times 10" switch so that the controller could be used for applications such as extruder temperature control as well as for normal applications.

In digital systems, the adjustment range for the controller gain is typically 0.01%/% to 100%/%. They are easily capable of the large gains required in applications such as extruders, but unfortunately, there are some side effects of a large controller gain (such as 50%/%). We usually can do this in temperature loops. Rarely can this be done in other loops. And sometimes we encounter problems even in temperature loops.

Noise in the Measured Variable. When the controller gain is large, any change in the measured variable is amplified and then sent to the final control element.

Suppose the controller gain is 50%/% and there is a measurement noise of 0.1% of the measured variable span. The measurement noise is amplified by a factor of 50 and appears in the controller output. That is, the signal to the final control element contains noise of 5.0% of the output span.

Large values for the controller gain can be used only when there is essentially no measurement noise. Temperature measurements are usually the only candidates. For the extruder, the thermocouple is inserted into a hole drilled into the barrel of the extruder. The thermocouple is sensing the temperature of a massive block of metal, so the measurement noise should be essentially zero. If noise is present, its origin is electrical, not process.

Vacuum Control Problem. Figure 9.2 illustrates a process vessel that is operated under high vacuum. A material is charged with the vessel open to the atmosphere, and then the vacuum is established. The vacuum is provided by a vacuum pump, and it is controlled by adjusting the recycle stream around the vacuum pump.

Once the vacuum has been established, a measurement range of 0 to 50 mm Hg absolute would be satisfactory. However, what about establishing vacuum? Suppose the material charged to the vessel contains a small amount of solvent that vaporizes under vacuum. To prevent foaming, the vacuum must be established in a gradual manner. To do so requires a pressure measurement range of 0 to 800 mm Hg absolute.

This process requires two pressure transmitters, one for control (0 to 50 mm Hg) and one for startup (0 to 800 mm Hg). If one attempts to use the

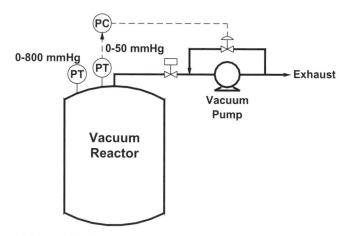

Figure 9.2. Pressure measurements for a vacuum process.

wide span for control, a large controller gain would be required, and would excessively amplify the noise in the pressure measurement.

Narrowing the Span. The accuracy of conventional analog transmitters was often expressed as a percent of span. When this is the case, changing the transmitter's measurement range from 0 °F to 600 °F to 500 °F to 600 °F significantly improves the accuracy of the measured value. But with microprocessor technology, the accuracy is often expressed as a percent of the upper range value. When this is the case, changing the measurement range from 0 °F to 600 °F to 500 °F to 600 °F has no effect on the accuracy of the measured value.

When the controller configuration parameters provide for specifying a separate span for the PV, the following possibility originates:

Measurement range for input from measurement device: 0 °F to 600 °F

Measurement range for PV for controller: 500 °F to 600 °F

If a controller gain of 30%/% is required for a measurement range of 0 °F to 600 °F, then a controller gain of 5%/% would be required for a measurement range of 500 °F to 600 °F. On the surface, this seems to be beneficial, since the controller gain is now within the range generally considered to be normal. However, this is more appearance than reality.

Suppose the noise on the measured value from the transmitter is ±0.1 °C. For a measurement range of 0 °F to 600 °F, the noise is 0.0167% of span. Multiplying the noise on the PV by the controller gain of 30%/% gives a noise on the controller output of 0.5% of span.

When the measurement range of the PV for the controller is 500 °F to 600 °F, will the noise on the measured value still be 0.0167% of span? Unfortunately, the answer is no. The noise is ±0.1 °C regardless of the measurement range. Therefore, for a measurement range of 500 °F to 600 °F, the noise is ±0.1 °C or 0.1% of span. Multiplying the noise on the PV by the controller gain of 5%/% gives a noise on the controller output of 0.5% of span. This is the same as before!

The level of noise on a signal is determined at the source, that is, within the measurement device itself. Instead of resorting to gimmicks within the controller, the problem must be addressed at its source.

9.3. TEMPERATURE MEASUREMENTS

Temperatures are key variables in most process facilities. Most consist of a temperature-sensitive probe inserted into a thermowell. The thermowell contributes to both the dynamic and the steady-state issues pertaining to the measurement.

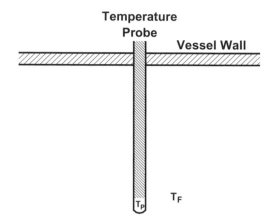

Figure 9.3. Bare probe for temperature measurement.

Bare Probe. To keep the configuration as simple as possible, let us start with a bare temperature probe, that is, a probe that is not inserted in a thermowell, as illustrated in Figure 9.3. In practice, such installations are limited to measuring the temperature of noncorrosive gasses at low pressures, such as air within ducts. The probe is inserted through the walls of a vessel or pipe into a fluid. Let the temperature of the probe be T_P; let the fluid temperature be T_F.

The supplier of the temperature probe will provide statements of how close the measured value is to the probe temperature T_P. The accuracy of the measurement is one of the factors taken into consideration in choosing among various temperature measurement technologies (thermocouples, RTDs, filled bulbs, etc.). We shall eliminate these issues by assuming that the temperature measurement of T_P is perfect.

What we really want to know is the fluid temperature T_F. Therefore, let us explore the factors that would lead to the probe temperature T_P being different from the fluid temperature T_F.

Steady-State Error. One potential source of steady-state error is heat conduction through the stem of the probe. In some applications, the probe is inserted through the walls of a vessel that is also outfitted with a jacket for heating or cooling the contents of the vessel.

If the vessel is being cooled, the wall temperature is less than the fluid temperature, and heat flows from the fluid to the probe and then to the wall of the vessel. If the vessel is being heated, the heat flow is in the opposite direction. But anytime there is heat flow, there is a temperature difference. We can state that the probe temperature will be between the fluid temperature and the wall temperature. But in general, to determine how close the probe temperature is to the fluid temperature requires that we somehow obtain another measurement of the fluid temperature that is not influenced by these errors. Conceptually, this is easy, but in practice, it proves to be far more difficult under process operating conditions.

Another situation that leads to static errors is where heat is being transferred by radiation. This is a potential source of significant errors in temperature measurements in combustion processes.

Suppose the static error is constant. For example, suppose the probe temperature is always 2 °C below the fluid temperature. What is the impact? First, static errors do not impair the ability of a regulatory control system to maintain constant conditions within the process. The problem is that these conditions are different from expected, which affects process performance.

Suppose we are measuring the temperature of a chemical reactor. If the probe reads 2 °C low, then the reactor temperature is 2 °C higher than we think. For most reactions, such a difference will impact the properties of the product. The quality control laboratory would determine that one or more product quality measures are not what is anticipated. Those knowledgeable in the product chemistry would then suggest that the reactor be run at a lower temperature. Once the target for the reaction temperature is reduced by 2 °C, the product properties would be as expected.

In applications where the target for the measured variable can be adjusted based on other information, we will compensate for static errors. But otherwise, static errors affect process performance.

Accuracy and Repeatability. Let us begin by explaining these two terms:

Accuracy: Degree of conformity of the measured value with a standard or accepted value. Perhaps you would prefer to know the degree of conformity of the measured value with the true value of the measured variable, but in practice, how can we determine its true value?

Repeatability: If the measured values are exactly the same on two different occasions, the difference in the process conditions on these two occasions reflects the nonrepeatability of the measurement device.

The scatter patterns for the targets in Figure 9.4 illustrate the concepts of accuracy and repeatability. Good accuracy requires good repeatability, but an inaccurate measurement can be very repeatable in that the result is consistently in error and in the same manner.

Since regulatory control is focused on maintaining constant conditions within the process, its performance depends on the repeatability of the measurement device, not on its accuracy. Endeavors such as process optimization, constraint control, and the like depend on the accuracy of the measurement device.

Dynamic Error. Suppose there is no steady-state error. Then suppose the fluid temperature increases abruptly by 5 °C. How does the probe temperature

Accuracy	Good	Poor	Poor
Repeatability	Good	Good	Poor

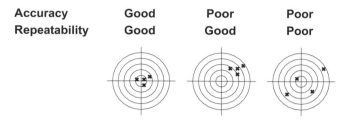

Figure 9.4. Accuracy and repeatability.

respond? The probe temperature will not increase as rapidly as the fluid temperature. How rapidly the probe responds depends on two factors:

Heat transfer coefficient. The greater the heat transfer coefficient, the greater the heat transfer for a given temperature difference. This improves the response of the probe, so it is recommended that the probe be installed at a location where the fluid flows rapidly across the probe, which increases the heat transfer coefficient.

Mass of the probe. The greater the mass of the probe, the more heat that must be transferred to increase the temperature of the probe by 1 °C. As this makes the probe respond more slowly, it is recommended that the mass of the probe be as small as possible.

For simple configurations such as in Figure 9.3, an expression can be developed for the time constant of the measurement device. An energy balance around the probe involves the following two terms:

$$\text{Heat transfer rate:} \quad U\,A\,(T_F - T_P)$$

$$\text{Accumulation of energy within the probe:} \quad M\,c_P\,\frac{dT_P}{dt}$$

where

A = Area for heat transfer
c_P = Probe heat capacity
M = Mass of probe
U = Heat transfer coefficient between probe and surrounding fluid

Equating these terms and rearranging gives the following differential equation:

$$M\,c_P\,\frac{dT_P}{dt} = U\,A\,(T_F - T_P)$$

$$\frac{Mc_P}{UA}\frac{dT_P}{dt} + T_P = T_F$$

This is the equation for a first-order lag, the time constant of the probe being the coefficient $M\, c_P/U\, A$. For fast response, we need a short time constant, which means a high heat transfer coefficient U and a small probe mass M.

Dynamic errors in the measurement device impair the ability of the control system to maintain constant conditions within the process. This suggests that the measurement device should respond as rapidly as possible.

In practice, the response required from the measurement device depends on the dynamics of the process. The usual rule of thumb is that the time constant of the measurement device should not exceed one tenth the time constant of the process. That is, the measurement device should respond at least ten times faster than the process. In practice, applying this rule of thumb is difficult because the time constant of most measurement devices under process operating conditions is not known.

Use of Thermowells. Most temperature probes are inserted into a thermowell that provides the following functions:

1. Protects the temperature probe from the process fluids.
2. Permits access to the temperature probe while the process is in operation.

Since air is not a good conductor of heat, the temperature probe is usually spring-loaded so as to force the probe against the tip of the thermowell. The alternative is to fill the thermowell with some type of heat conducting fluid.

Thermowells can be manufactured from a variety of materials and come in a variety of types, including standard class for normal services, thin-walled for low pressure gas service, heavy duty for high-pressure service, and glass-coated for corrosive service.

From a control standpoint, the thermowell is a necessary evil—inserting additional mass between the variable of interest and the probe temperature leads to a slower response. Using a standard class well adds substantial mass, especially when the probe is a thermocouple or RTD. Consequently, the response speed is determined primarily by the thermowell. Some corrosive services require a glass-coated well; the poor heat transfer of the glass slows the response. High-pressure applications require a heavy-duty well; the additional mass slows the response.

Suppose we replace the filled bulb with a thermocouple or RTD. The time constant of the bare probe will be significantly less. The response speed is still greater when inserted into a thermowell; however, the percentage improvement will be much less. Especially for heavy-duty and glass-coated wells, the dynamic characteristics are determined primarily by the thermowell.

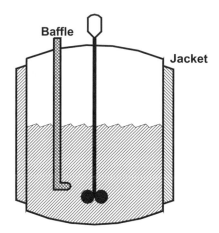

Figure 9.5. Glass-lined reactor.

Glass-Lined Reactors. Some reaction media are highly corrosive. For such services, glass-lined reactors such as illustrated in Figure 9.5 must often be used. All wetted surfaces are coated with glass, including the vessel itself, the agitator, and the baffles that are inserted to improve mixing.

How do you measure the temperature within such a reactor? Inserting a probe for temperature, measurement is not an easy endeavor. The more one studies this situation, the more the baffle begins to look like a thermowell. In some early installations, a filled bulb was inserted into the baffle, which was then filled with heat conducting liquid. Today, the reactor manufacturers incorporate a metallic plug (titanium, tantalum, or the like) at an appropriate location on the baffle, and embed either a thermocouple or an RTD within the plug. This is certainly an improvement; however, one is still basically reading the temperature of the baffle. The mass of the baffle coupled with relatively poor heat transfer for glass makes for a slow responding system. Following some event that is known to affect the temperature, it can be minutes before any change is observed in the measured temperature.

There is no question that inserting the temperature probe into the baffle will result in a slowly responding measurement device. The question is how slow. The best way to quantify the answer to this question would be to determine the value for the time constant of the measurement device.

How do we do this? With the vessel filled with reacting materials under reacting conditions, we have no way to determine the value for the time constant of the measurement device. With the vessel filled with water, there are some options. But usually the characteristics of the reacting medium are different from water. The heat transfer coefficient will be different, and anything that affects the heat transfer coefficient will affect the time constant of the measurement device.

In such applications, the time constant of the measurement device will surely impair the performance of the control system. But with no way to determine a value for the time constant of the measurement device, we have no way to quantify its effect on control system performance.

9.4. FILTERING AND SMOOTHING

Excessive use of filtering is a common practice in the process industries. Whenever an experienced control person is asked to investigate a control problem, one of the first steps is to ascertain where filtering is being applied. The next step is often to remove it. Given that processes are slow, a filter with a time constant of perhaps five or ten seconds is unlikely to be a problem. But when the filter time constant is a minute or more, it is time to ask questions. Often someone is trying to hide something that needs to be corrected.

Experienced control people like to tell "war stories" regarding filters or, actually, the misapplication of filters. And there seems to be no end of these, with new ones appearing regularly, which includes such things as the output of one filter being the input to another filter, filters on the output of a controller, etc. No wonder the first step is to ascertain where filters are being applied, and the second step is to turn them off.

Components of a Signal. Every input to a control system is a signal consisting of four components:

1. Controllable disturbances.
2. Uncontrollable disturbances.
3. Measurement noise.
4. Stray electrical pickup.

Any signal can be represented in terms of sinusoidal components. The bandwidth of the controller is the highest frequency of these sinusoidal components to which the controller can effectively respond. Disturbances whose frequencies are lower than this are controllable; disturbances whose frequencies are higher than this are uncontrollable.

The measured variable input may also contain measurement noise. This is especially prevalent in some measurements, such as weight gauges. However, it can also occur in flow, level, and pressure. When a thermowell is present, the temperature measurement should be free of noise, but when the measurement device is a pyrometer, noise is definitely a possibility.

Every analog signal contains stray electrical pickup, primarily at either 50 or 60 Hz. For current loop inputs, the analog input processing hardware in digital systems is specifically designed to remove this component. For digital inputs from smart transmitters, this must be dealt with in the transmitter itself.

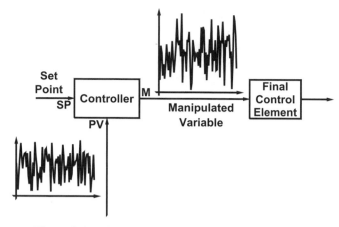

Figure 9.6. Effect of noise on the measured variable.

Effect of Measurement Noise. When noise is present in the measured variable, the controller translates this noise to the controller output, as illustrated in Figure 9.6. When the controller gain is 1%/%, any noise in the measured variable is simply transferred to the output of the controller. Controller gains greater than 1%/% amplify the noise, which must usually be avoided when significant noise is present. The major consequence of noise in the controller output is its impact on the final control element. Most are mechanical in nature, so noise leads to more rapid wear and increased maintenance.

Noise on the input to an integrator has little effect on the output of the integrator, which basically averages the noise. Consequently, the reset mode is not significantly impacted by noise in the measured variable. In a very noisy environment, the reset mode becomes the primary mode of control. However, this results in a loop that responds very slowly.

Filter. In control applications, we would like to remove all components completely except those to which the controller can effectively respond. This leaves the controllable disturbances, which are the lowest frequency components of the signal. This function is provided by a "filter" that is inserted, either in hardware or in software, between the measurement device and the process variable input to the controller. The location of the filter is illustrated in the block diagram in Figure 9.7.

As applied in process control, the filter is specifically a "low-pass" filter. Such filters attenuate the high-frequency components of a signal but not the low-frequency components. An ideal low-pass filter removes completely all components above a specified frequency (the cutoff frequency), and passes unaltered all components below this frequency. Such filters are unrealizable. Practical low-pass filters attenuate the components above the cutoff frequency, the attenuation factor increasing with frequency. Components at frequencies

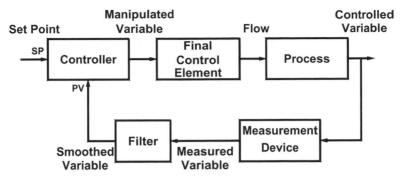

Figure 9.7. Block diagram of a control loop with a filter.

well below the cutoff frequency are passed basically unaltered, but some attenuation is applied to those components whose frequency is just below the cutoff frequency.

Exponential Smoothing. The most common digital filter, often referred to as the exponential filter, is the discrete counterpart to the RC-network analog filter that is described by the differential equation for a first-order lag:

$$\tau_F \frac{dY}{dt} + Y = X$$

where

X = Input to the filter
Y = Output of the filter (the smoothed value)
τ_F = Filter time constant

To obtain a digital filter, this differential equation is converted to a difference equation using backward differences:

$$\tau_F \frac{Y_i - Y_{i-1}}{\Delta t} + Y_i = X_i$$

where

X_i = Filter input at sampling instant i
Y_i = Smoothed value at sampling instant i
Δt = Time interval on which the filter calculations are executed (the filter calculations are sometimes executed on a shorter time interval than the time interval on which the control calculations are executed)

Finally, we solve the resulting equation for Y_i, the current value of the smoothed value:

$$Y_i = \frac{\tau_F}{\tau_F + \Delta t}\, Y_{i-1} + \frac{\Delta t}{\tau_F + \Delta t}\, X_i$$

The current filter output Y_i is computed from the previous filter output Y_{i-1} and the current value of the input X_i to the filter. Note that the two coefficients sum to unity. This permits the filter equation to be expressed as follows:

$$Y_i = k\, Y_{i-1} + (1-k)\, X_i$$

where $k = \tau_F/(\tau_F + \Delta t)$ = filter coefficient or smoothing coefficient.

The value of the filter coefficient is between 0.0 (no smoothing) and 1.0 (heavy smoothing). The filter coefficient k depends on the filter time constant τ_F and the time interval Δt on which the filter calculations are executed.

Because of the simplicity of the equation for the exponential filter, the smoothing provided by most digital control systems is the exponential smoothing. The smoothed value is retrieved (this is Y_{i-1}), the current value of the input is retrieved (this is X_i), the filter equation is executed, and the result is deposited in the location for the smoothed value.

Moving Average. To process people with no background in controls, the exponential filter is a bit mysterious. Arithmetic averages they understand. The following equation can be used for smoothing:

$$Y_i = \frac{1}{N} \sum_{j=0}^{N-1} X_{i-j}$$

This filter is the arithmetic moving average filter, which is sometimes referred to as a moving average filter.

The term "moving" reflects the nature of the calculations. The usual arithmetic average of N values would yield a new smoothed value after each N input values. That is, if the input is updated on a one-second interval and the average is of 10 values, a new smoothed value would be computed every 10 seconds. The "moving" average works a bit differently. On each execution interval, the arithmetic average of the previous N values of the input is computed. This requires a storage array for the previous N values of the input (actually, only N–1 values must be stored). On each execution, the calculations are as follows:

1. A new value of the input is retrieved.
2. The arithmetic average is computed from this value and the values from the storage array.

3. The oldest value in the storage array is overwritten by the new value of the input.

Basically, the moving average filter averages the values of the input over the previous time interval N Δt. The effective lag of the moving average filter is approximately ½ N Δt. Provided τ_F = ½ N Δt, the performance of the moving average filter is similar to, but not identical to, that of the exponential filter.

As compared with the exponential filter, the moving average filter is more demanding in both its computational burden and its storage requirements. In the early versions of digital systems, both had to be taken seriously. But with today's microprocessors, neither is an issue. Nevertheless, most smoothing is accomplished with the exponential filter.

Response of the Exponential Filter to a Step Change. Let there be a step change of 5 units in the measured variable at time = 2.0 minutes. When noise with a standard deviation of 1 unit is superimposed on the measured variable, the responses in Figure 9.8 present the smoothed variable for three values of the filter time constant. Increasing the filter time constant reduces the noise in the smoothed variable. This is achieved by slowing the response to all changes in the measured variable, including the controllable disturbances. This adversely affects the performance of the control loop.

The equations for the exponential filter are the same as the equations for a time constant; both are first-order lags. In a noise-free environment, the response to a step change would be an exponential lag. In this regard, adding a filter to a loop has about the same effect as inserting a temperature probe into a thermowell. Both insert an additional time constant into the loop, which always has an adverse effect on the performance of the loop. Of course, the adverse effect increases as the time constant increases.

Value for Filter Time Constant. In the process industries, the usual approach to determining the degree of smoothing is as follows:

> If the process variable is "bouncing around" too much, then increase the smoothing.

In effect, the degree of smoothing is determined from observations of the variance in the smoothed variable. This approach often leads to excessive smoothing and to degraded loop performance.

Use as little smoothing as possible, preferably none. But if filtering is applied, always do so in such a manner that a quantitative measure is available for the degree of smoothing. In practice, there are three possibilities for smoothing a measured variable:

Level. About 75% of level measurements use either pressure or differential pressure transmitters. But for demanding applications, technologies such as noncontact radar and guided wave radar provide superior performance.

Flow. Especially in chemicals and pharmaceuticals, the orifice meter is being displaced by the coriolis meter, the magnetic flowmeter, the vortex-shedding meter, and the ultrasonic meter.

Density. The coriolis meter senses both flow and density.

CHAPTER 10

Final Control Elements

The objective of this chapter is to examine the control issues pertaining to the various types of final control elements installed in the process industries:

Positioning valves. These continue to dominate, so they receive the most attention.

ON/OFF valves. These are simpler, less expensive, and more reliable than positioning valves.

Variable speed drives. When coupled to centrifugal pumps, this approach offers some advantages over valves.

10.1. VALVES AND FLOW SYSTEMS

A flow system embodies all equipment required to transport a fluid from a source to a destination. All flow systems require piping, fittings, and valves (hand valves, control valves, and/or block valves). Additional components may include pumps, compressors, exchangers, filters, and so on. Figure 10.1 illustrates a flow system.

Equations relating pressure drop and flow are available for piping, fittings, exchangers, and so on. For most components of the flow system, the pressure drop across the component varies with the square of the flow. In flow systems such as the one in Figure 10.1, the flow depends on the characteristics of all elements of the flow system, not just the characteristics of the control valve.

For control purposes, we would prefer that the flow vary linearly with the control valve position. We can purchase a valve with linear characteristics. However, installing such a valve in a flow system such as in Figure 10.1 does not result in the flow varying linearly with valve position. The nonlinear contributions of the other components distort the relationship.

Valves. A valve is a device that starts, stops, or regulates the flow of a fluid by adjusting the position of a movable part. Valves may be applied for either

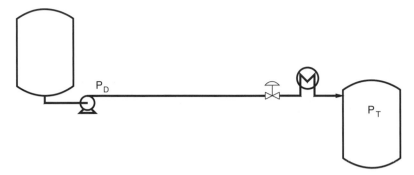

Figure 10.1. A flow system.

ON–OFF service (often called block valves) or for regulatory service (often called control valves). Our interest is in control valves.

Based on the nature of the moving part, valves are broadly classified into two categories:

Rising stem (sliding stem) valves. Examples include globe, angle, and diaphragm valves.

Rotary stem valves. Examples include ball and butterfly valves.

The stem is the movable part that affects the flow through the valve. The term "valve position" refers to the position of the stem. The valve position M is usually expressed in percent, but expressing M as a fraction is more convenient in some equations.

The choice regarding type of valve is based on factors such as cost, pressure recovery, minimum pressure within the valve (flashing?), tendency to leak when closed, and so on. This list rarely contains any item that directly pertains to control.

Actuators. A control valve requires an actuator that is capable of positioning the movable part to any value between the two extremes of fully open and fully closed. Positioning actuators are generally classified as to the source of power—pneumatic, electric, and occasionally hydraulic.

The pneumatic diaphragm actuator is usually preferred, primarily because of its characteristics on loss of power (that is, loss of supply air). Air pressure on one side of the diaphragm provides a force that opposes a spring. As air is supplied, the spring is either compressed or extended. On loss of either control air or supply air, the spring returns to its normal length. The behavior is as follows:

Air-to-open configuration. Supplying air to the diaphragm causes the valve to open; on loss of air, the spring drives the valve to the closed position.

Air-to-close configuration. Supplying air to the diaphragm causes the valve to close; on loss of air, the spring drives the valve to the open position.

Replicating this characteristic in electric actuators increases complexity and cost. Motor-driven valves can be configured to fail open or fail closed on loss of the control signal. But for loss of power, motor-driven valves fail in last position.

Other advantages are often cited for the pneumatic actuator, including reliability, low cost, and so on. However, it is their failure characteristics that really make them the actuator of choice.

Inherent Valve Characteristics. The inherent valve characteristics relate the flow through the valve to the valve position, but under a very important stipulation, the pressure drop across the valve is constant. That is, the pressure drop across the valve is not affected by valve position. For most control valves installed in process plants, this is not the case. However, the intent of the inherent valve characteristics is to specify how the valve itself behaves. This behavior has to be combined with the behavior of the other elements of the flow system to obtain the installed valve characteristics.

Normally the relationship for the inherent valve characteristics is expressed as follows:

$$\frac{F}{F_{MAX}} = f(M)$$

where

F $\quad$ = Flow through the valve at valve position M

F_{MAX} = Flow through the valve when fully open (M = 1)

M $\quad$ = Valve position as a fraction between 0 and 1

The value of F_{MAX} is determined by the size of the valve and the pressure drop across the valve.

The inherent valve characteristics depend only on the valve. Valve manufacturers supply this relationship either in graphical or tabular form.

Installed Valve Characteristics. In flow systems such as in Figure 10.1, the pressure drop across the valve is not constant. At low flows, the pressure upstream of the valve is essentially the pump discharge pressure P_D; the pressure downstream of the valve is essentially the pressure P_T in the destination tank. But as the flow increases, the pressure upstream of the valve drops because of the friction losses in the piping between the pump and the valve. In addition, the pressure downstream of the valve increases because the pressure drop across the exchanger increases with flow.

Since the pressure drop across the valve is not constant, the inherent valve characteristics do not describe the relationship between the flow through the valve and the valve position. All other elements of the flow system contribute to this relationship. The resulting flow versus valve position relationship is normally referred to as the installed valve characteristics. Since these characteristics depend on all of the components in the flow system, we cannot expect the valve manufacturer to supply this information. But knowing what components constitute the flow system, the installed valve characteristics can be computed from the characteristics of these components and the inherent valve characteristics.

10.2. VALVE SIZING

This is not a discourse on valve sizing, so we will not be examining the details. Our interest is the influence of valve sizing (or errors in valve sizing, or more specifically, oversized valves) on control loop performance.

The equations, correction factors, and other issues pertaining to valve sizing are delineated in a standard (1). Most valve manufacturers provide computer programs that perform the sizing calculations, as well as calculate related factors such as valve noise.

The sizing programs are very good at performing the calculations. But to size a valve, the following information is required:

1. Maximum flow through the valve.
2. Expected pressure drop across the valve.
3. Properties of the fluid flowing through the valve.

Give the program conservative values for these parameters, and guess what, the valve is oversized! The results of the valve-sizing calculations are no better than the data supplied to the program (the proverbial "garbage in, garbage out" problem).

Liquids. For liquid service (incompressible fluid), the flow through the valve is related to the pressure drop by the following equation:

$$Q = C_V \left[\frac{\Delta P_V}{G} \right]^{1/2}$$

Solving this equation for the valve flow coefficient C_V gives the usual equation for sizing valves:

$$C_V = \frac{Q}{(\Delta P_V / G)^{1/2}}$$

where

    C_V = Valve flow coefficient
    G   = Specific gravity of the fluid
    Q   = Volumetric flow through valve
    ΔP_V = Pressure drop across the valve

The flow coefficient can be thought of as the flow through the valve with a unit pressure drop across the valve. For liquids, the customary units are as follows:

	Flow	Pressure Drop
English units	gallons/min	psi
Metric units	liters/min	bar

To size a valve, we must have two quantities:

Flow through the valve. In most cases, the flow through the valve can be obtained directly from the process flowsheet. However, in some cases, a larger flow may be required under some other situation (startup, emergency response, or other). We need to know the maximum flow that the valve will be expected to deliver under any circumstances.

Pressure drop across the valve. There are various guidelines for the appropriate pressure drop ΔP_V across the control valve. The more common are as follows:

- 10 psi for globe valves; 5 psi for rotary valves.
- 20% to 50% of the pressure drop caused by friction in the remainder of the flow system.
- 5% of total pressure.

In recent years, the emphasis on energy conservation has tended to lower the pressure drops across control valves. Inputting energy to a pump only to dissipate it at the control valve is not very productive. However, a lower control valve pressure drop leads to a larger valve (higher capital costs) and to reduced maximum flow through the fully open valve (decreased rangeability).

It has previously been observed that oversized valves are common. Factors such as the following contribute to this:

- Some oversizing is incorporated into the valve-sizing procedure. A valve will be selected with a C_V larger than the value calculated from the sizing equation. Starting with an inflated value for the maximum flow compounds the excess.
- In the flow system in Figure 10.1, the valve pressure drop is one factor that determines the pump size. But conversely, the excess capacity due to

selecting the next larger pump size further increases the pressure drop across the control valve.

The sizing equation determines the value of the valve C_V required to obtain a specified flow Q for a specified pressure drop ΔP_V across the valve. This relationship can be applied at any valve opening, not just at the fully open position. The inherent characteristics supplied by the valve manufacturer describe how the flow varies with valve position M by providing either graphical or tabular data on the C_V at various values of M.

In the material that follows, we shall use the following notation:

C_V = Flow coefficient for the valve at the current valve position M

$C_{V,100}$ = Flow coefficient for the fully open valve. Note that $0 \le C_V \le C_{V,100}$.

$C_{V,SIZE}$ = Flow coefficient from the sizing calculations. Normally $C_{V,100} > C_{V,SIZE}$.

Limitations of the Sizing Equation for Liquids. The equation for sizing control valves for liquid service is used almost universally. The sizing equation presented previously is for turbulent flow. Laminar flow is handled by applying correction factors to the sizing equation.

The sizing equation also assumes liquid throughout the valve. The valve behaves like an orifice. Flow velocities increase within the valve, which in turn leads to lower pressures within the valve. Should the pressure fall below the vapor pressure of the liquid, flashing will occur within the valve. Flashing damages the wetted parts of the valve and must be avoided.

Standards also apply to the piping configuration for the test equipment used to determine the C_V for a valve. Correction factors are also required when the field piping differs from the test configuration.

Gases and Vapors. Gases and vapors are compressible fluids, which complicates the sizing equations and leads to different preferences. A very common vapor is steam, so special design equations are provided for steam. The computer program provided by a valve manufacturer embodies that manufacturer's preferred equations for sizing valves for gas and vapor services. We shall not present any of the equations.

For gases and vapors, the maximum possible velocity within the valve is the velocity of sound at the temperature and pressure within the valve. This sonic velocity limitation is the basis for choked flow for gases and vapors.

Installed Valve Characteristics. How the control valve affects loop performance is determined by how the flow varies with valve position for the valve as installed. These are generally referred to as the installed valve characteristics; however, every component of the flow system has an influence. The greater the pressure drop across a component, the greater its influence on the installed valve characteristics.

In any flow system, there are two components of the total flow system pressure drop:

Static pressure head. This is the difference between the static pressure at the destination and the static pressure at the source, including any gravity heads resulting from different elevations.

Friction head. This head is due to the resistance to fluid flow. This head increases approximately with the square of the fluid flow rate.

The pressure drop in the valve is entirely friction head. Let us define $\%\Delta P_V$ as the pressure drop across the control valve as a percentage of the total friction head for the flow system. Generally $\%\Delta P_V$ varies with the fluid flow rate, so the value of interest for $\%\Delta P_V$ is its value at the design flow.

If $\%\Delta P_V = 100$, then

- The installed characteristics are the same as the inherent valve characteristics.
- The pressure drop across the valve is constant (assuming the flow system pressure drop is constant).

If $\%\Delta P_V < 100$, then

- The installed characteristics will differ from the inherent valve characteristics.
- The pressure drop across the valve will vary with valve position.

The installed valve characteristics depend on the following factors:

1. Inherent valve characteristics.
2. $\%\Delta P_V$ at the design flow.
3. $C_{V,100}$.
4. The pressure drop versus flow relationships for the other components of the flow system.

Of the factors in the above list that affect the installed valve characteristics, $\%\Delta P_V$ and $C_{V,100}$ pertain to valve sizing. Increasing the degree to which a valve is oversized has two consequences that impair control performance:

1. The gain or sensitivity at the design flow increases. This amplifies the effect of valve positioning errors on the flow.
2. You may have heard an operator say "once this valve is more than 50% open, it has no effect on the process." Oversized valves cause this type of behavior.

10.3. INHERENT VALVE CHARACTERISTICS

The inherent valve characteristics describe how the flow through the valve varies with valve position. This relationship is supplied by the manufacturer of the valve as either a graph or a set of tabulated values. The data can be presented in various forms:

1. C_V as a function of valve position M.
2. $C_V/C_{V,100}$ or %C_V as a function of valve position M.
3. F/F_{MAX} as a function of valve position M.

The inherent valve characteristics are always presented with the stipulation that the pressure drop across the valve is constant, that is, does not vary with valve position. When ΔP_V is constant, the relationships depend only on the valve.

Valve manufacturers obtain the inherent characteristics for a valve by conducting tests using a valve test stand. A constant pressure drop is imposed across the valve, and the flow through the valve is measured for various values of the valve position. A graph of the results would be the inherent characteristics for the valve.

For a flow system, the pressure drop available for fluid flow is the total pressure drop across the flow system less any static pressure head and gravity head. A good measure of the degree to which the valve dominates is the ratio of the pressure drop across the valve to the pressure drop available for fluid flow, or equivalently, the percent of the pressure drop available for fluid flow that is dissipated at the valve. Let us designate this percentage as %ΔP_V.

Categories of Inherent Valve Characteristics. With regard to how the flow changes with valve position, valves are divided into the following categories:

1. Linear.
2. Increasing sensitivity.
3. Decreasing sensitivity.

Figure 10.2 illustrates the variation of flow with valve position for each type of valve.

Control valves are of either the linear or the increasing sensitivity type. Decreasing sensitivity valves (also called quick opening valves) are applied where the flow must increase rapidly once the valve begins to open. This type of valve is commonly used for ON–OFF or block valves. As decreasing sensitivity valves are rarely used as control valves, this type of characteristic will receive no further attention.

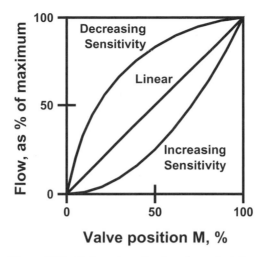

Figure 10.2. Categories of valve characteristics.

Linear Characteristics. If the inherent valve characteristics are linear, the flow increases linearly with valve position. This is represented by the following relationship:

$$\frac{C_V}{C_{V,100}} = \frac{F}{F_{MAX}} = f(M) = M$$

where M is the valve position as a fraction between 0 and 1.

We have previously stressed the problems that originate from nonlinear components of a control loop. One would naturally assume that linear valves would be desirable. However, very few are installed. Most components of a flow system are nonlinear, with the pressure drop varying with approximately the square of the flow. Coupling these nonlinear elements with a linear valve will result in a nonlinear relationship. Most control valves are of the increasing sensitivity type. In a sense, the nonlinearities within the valve at least partially compensate for the nonlinearities in the other components of the flow system.

Equal-Percentage Characteristics. This valve is described by the following statement:

A change in valve position gives a proportional percentage change in flow.

Mathematically, a small change in valve position is dM. A percentage (actually a fractional) change in flow is dF/F. The differential equation for the equal-percentage valve is as follows:

$$\frac{dF}{F} = k \, dM$$

where k is a proportionality constant, with a typical value being 4. This differential equation can be integrated to give the theoretical equation for an equal-percentage valve:

Differential equation: $\dfrac{dF}{F} = d(\ln F) = k\,dM$

Integrated equation: $\ln F = kM + C$, where C is the constant of integration

Boundary condition: At $M = 1$, $F = F_{MAX}$

Value for constant: $\ln F_{MAX} = k + C$ or $C = F_{MAX} - k$

Theoretical equation: $\ln F - \ln F_{MAX} = k(M - 1) = -k(1 - M)$

$$\frac{F}{F_{MAX}} = e^{-k(1-M)}$$

But there is a problem with the theoretical equation. As the graph of the theoretical equal-percentage equation in Figure 10.3 illustrates, the flow at $M = 0$ (valve closed) is not zero:

$$\frac{F}{F_{MAX}} = e^{-k} > 0$$

For $k = 4$, the flow at $M = 0$ is about 2% of the maximum flow. Therefore, a small modification to the theoretical characteristics is necessary to obtain the practical equal-percentage characteristics presented in Figure 10.3. The inherent valve characteristics must be provided either as a table of values or as a graph.

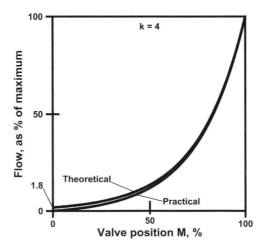

Figure 10.3. Equal-percentage inherent valve characteristics.

TABLE 10.1. Effect of k on Equal-Percentage Valve
Characteristics

M, %	%C_V, k = 3	%C_V, k = 4	%C_V, k = 5
0	0	0	0
10	2	1	0.4
20	4	2	1
30	8	4	2
40	12	7	4
50	18	12	8
60	26	19	13
70	38	29	22
80	53	44	36
90	73	66	60
100	100	100	100

Two equal-percentage valves do not necessarily have the same inherent valve characteristics. Table 10.1 presents the values of %C_V for k = 3, k = 4, and k = 5. To define precisely the characteristics of an equal-percentage valve requires a table or graph for the inherent valve characteristics.

Construction. The type of valve characteristics (linear, equal-percentage, or other) is determined by the shape of the wetted parts (trims) within the valve. For globe valves and angle valves, the valve characteristics are determined by the shape of the plug. For ball valves, the valve characteristics are determined by the shape of the port through the ball.

The valve manufacturers intentionally shape one or more of the internal parts so that the desired valve characteristics are realized. For most valves, the manufacturers can supply trims that provide linear characteristics, equal-percentage characteristics, quick opening, and others.

When ordering a valve, the desired characteristics must be specified, or in some cases, the characteristics of the valve must be considered in selecting which valve to order. For most valves, it is possible to change the characteristics by changing the wetted parts. However, it is better to get it right the first time.

Butterfly Valves. In line sizes of 6″ or more, the economy of the butterfly valve is very appealing. As for all valves, the valve characteristics are determined by the wetted parts. With globe valves and ball valves, the internal parts can be shaped; however, the shapes of the internal parts of the butterfly valve are fixed.

Figure 10.4 presents the inherent characteristics for a butterfly valve manufactured by Masoneilan (2). Except when nearly fully open, the valve exhibits increasing sensitivity characteristics. However, its characteristics depart considerably from those of an equal-percentage valve.

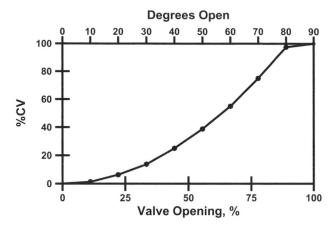

Figure 10.4. Inherent characteristics of a butterfly valve.

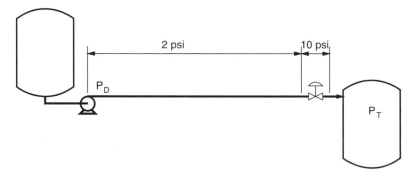

Figure 10.5. A flow system dominated by a valve.

10.4. FLOW SYSTEM DOMINATED BY CONTROL VALVE

In the flow system in Figure 10.5, material with a specific gravity G of 1.0 is being pumped from one vessel to another. At the design flow of 60 GPM, the pressure drops are as follows:

1. Piping section from pump to control valve: 2 psi.
2. Across the control valve: 10 psi.

The total pressure drop available for fluid flow ΔP_T is 12 psi. From this pressure drop, we are taking 83% across the valve ($\%\Delta P_V = 83\%$ at design conditions).

To keep this example as simple as possible, we shall assume that the pump discharge pressure is constant. In practice, the analysis should include the pump curves that relate the pump discharge pressure to flow.

Maximum Flow. Let us apply the orifice equation to the piping section:

$$Q = C_o(\Delta P/G)^{1/2}$$

The orifice coefficient C_o for the piping section is computed from the flow (60 GPM) and the pressure drop (2 psi):

$$C_o = \frac{Q}{(\Delta P/G)^{1/2}} = \frac{60}{(2/1)^{1/2}} = 42.4 \, \text{GPM}/\text{psi}^{1/2}$$

The maximum flow is when there is no pressure drop across the valve (a valve with an infinite C_V). The pressure drop across the piping section is the total pressure drop ΔP_T of 12 psi for the flow system. The maximum flow Q_{MAX} is computed as follows:

$$Q_{MAX} = C_o(\Delta P_T/G)^{1/2} = 42.4 \, (12/1)^{1/2} = 147 \, \text{GPM}$$

As the control valve is oversized more and more, the maximum flow will approach this value.

Valve Sizing. A flow of 60 GPM and a pressure drop of 10 psi across the control valve give the following value for $C_{V,SIZE}$:

$$C_{V,SIZE} = \frac{Q}{(\Delta P_V/G)^{1/2}} = \frac{60}{(10/1)^{1/2}} = 19.0 \, \text{GPM}/\text{psi}^{1/2}$$

If the value of $C_{V,100}$ is 19.0 GPM/psi$^{1/2}$, then when the control valve is fully open, the flow would be 60 GPM. Obviously the value of $C_{V,100}$ must be greater than 19.0 GPM/psi$^{1/2}$; the question is how much greater. Increasing the value of $C_{V,100}$ means that we are oversizing the valve. We need to examine the following consequences of increasing $C_{V,100}$:

- A larger flow can be attained when the valve is fully open. However, the flow can never exceed the value of Q_{MAX} (147 GPM).
- A flow of 60 GPM will be attained at a smaller value of the valve opening.

Let us examine these two questions, first for a linear valve and then for an equal-percentage valve.

Linear Characteristics. For a linear valve, Figure 10.6 presents the installed valve characteristics for four cases:

$C_{V,100}$	Q for M = 100%	M for Q = 60 GPM	Comments
$C_{V,SIZE}$	60 GPM	100%	Exhibits only a small departure from linearity.
2 $C_{V,SIZE}$	98 GPM	50%	Exhibits a larger but still small departure from linearity.
5 $C_{V,SIZE}$	134 GPM	20%	Exhibits a significant departure from linearity.
10 $C_{V,SIZE}$	143 GPM	10%	Relationship is very nonlinear. If the valve is more than 50% open, the sensitivity is nearly zero. The operator would observe that "once this valve is more than 50% open, it has no effect on the process."

As the control valve provides most of the pressure drop, the installed characteristics are essentially linear for reasonable values of $C_{V,100}$. But if the valve is oversized by more than a factor of 2, the installed valve characteristics are nonlinear. There is a lesson here: If you install a linear valve, make sure it is properly sized.

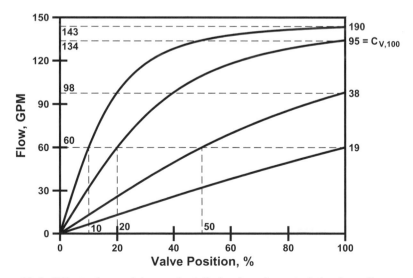

Figure 10.6. Effect of oversizing on installed valve characteristics for a linear valve; flow system dominated by the valve.

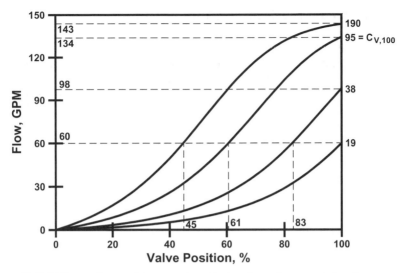

Figure 10.7. Effect of oversizing on installed valve characteristics for an equal-percentage valve; flow system dominated by the valve.

Equal-Percentage Characteristics.

For an equal-percentage valve, Figure 10.7 presents the installed valve characteristics for four cases:

$C_{V,100}$	Q for M = 100%	M for Q = 60 GPM	Comments
$C_{V,SIZE}$	60 GPM	100%	Exhibits the nonlinear behavior typical of equal-percentage characteristics.
2 $C_{V,SIZE}$	98 GPM	83%	Exhibits the nonlinear behavior typical of equal-percentage characteristics.
5 $C_{V,SIZE}$	134 GPM	61%	Decreasing sensitivity behavior appears at high valve openings.
10 $C_{V,SIZE}$	143 GPM	45%	Decreasing sensitivity behavior is more pronounced at large valve openings. An equal-percentage valve that operates at less than 50% open is grossly oversized.

When the control valve provides most of the pressure drop, the installed characteristics for a properly sized equal-percentage valve exhibit the nonlinearities characteristic of the equal-percentage relationship. But when the valve is significantly oversized, the installed characteristics of the equal-percentage valve are preferable to those of the linear valve. In a sense, equal-percentage valves are more forgiving with respect to oversizing.

10.5. FLOW SYSTEM DOMINATED BY PROCESS

In the previous examples of flow systems, the valve has taken most of the pressure drop ($\%\Delta P_V > 50\%$). Let us add another source of pressure drop to the flow system, namely a heat exchanger between the control valve and the destination tank. At the design flow of 60 GPM, let the pressure drops be as illustrated in Figure 10.8:

1. Piping section from pump to control valve: 2 psi.
2. Across the control valve: 10 psi.
3. Across the exchanger: 28 psi.

The total pressure drop available for fluid flow is 40 psi. From this pressure drop, we are now taking 25% across the valve ($\%\Delta P_V = 25\%$ at design conditions).

As in previous examples, we shall assume that the pump discharge pressure is constant.

Maximum Flow. The orifice coefficient C_o for the combined piping and exchanger is computed from the flow (60 GPM) and the combined pressure drop (30 psi):

$$C_o = \frac{Q}{(\Delta P/G)^{1/2}} = \frac{60}{(30/1)^{1/2}} = 11.0\,\text{GPM}/\text{psi}^{1/2}$$

The maximum flow is when there is no pressure drop across the valve (a valve with an infinite C_V). The pressure drop across the piping and the exchanger is the total pressure drop ΔP_T of 40 psi for the flow system. The maximum flow Q_{MAX} is computed as follows:

$$Q_{MAX} = C_o(\Delta P_T/G)^{1/2} = 42.4\,(40/1)^{1/2} = 69.6\,\text{GPM}$$

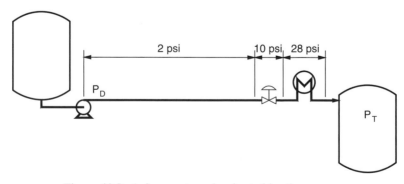

Figure 10.8. A flow system dominated by the process.

As the control valve is oversized more and more, the maximum flow will approach this value.

Valve Sizing. A flow of 60 GPM and a pressure drop of 10 psi across the control valve give the following value for $C_{V,SIZE}$:

$$C_{V,SIZE} = \frac{Q}{(\Delta P_V / G)^{1/2}} = \frac{60}{(10/1)^{1/2}} = 19.0 \, \text{GPM}/\text{psi}^{1/2}$$

If the value of $C_{V,100}$ is 19.0 GPM/psi$^{1/2}$, then when the control valve is fully open, the flow would be 60 GPM. The consequences of increasing $C_{V,100}$ include the following:

- A larger flow can be attained when the valve is fully open. However, the flow can never exceed 69.6 GPM.
- A flow of 60 GPM will be attained at a smaller value of the valve position (smaller valve opening).

Let us examine these two questions, first for a linear valve and then for an equal-percentage valve.

Linear Characteristics. For a linear valve, Figure 10.9 presents the installed valve characteristics for four cases:

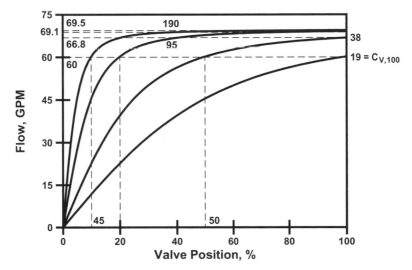

Figure 10.9. Effect of oversizing on installed valve characteristics for a linear valve; flow system dominated by the process.

$C_{V,100}$	Q for M = 100%	M for Q = 60 GPM	Comments
$C_{V,SIZE}$	60 GPM	100%	Exhibits decreasing sensitivity characteristics.
2 $C_{V,SIZE}$	66.8 GPM	50%	Exhibits even stronger decreasing sensitivity characteristics.
5 $C_{V,SIZE}$	69.1 GPM	20%	Above an opening of 50%, the valve has almost no effect on flow.
10 $C_{V,SIZE}$	69.5 GPM	10%	Above an opening of 25%, the valve has almost no effect on flow.

The source of the nonlinearities in the installed valve characteristics is that the process pressure drop varies with the square of the flow. Even for reasonably sized valves, these nonlinearities dominate the installed characteristics. The more the valve is oversized, the more pronounced the nonlinearities.

Equal-Percentage Characteristics. For an equal-percentage valve, Figure 10.10 presents the installed valve characteristics for four cases:

$C_{V,100}$	Q for M = 100%	M for Q = 60 GPM	Comments
$C_{V,SIZE}$	60 GPM	100%	Reasonably linear from a valve opening of 50% up.
2 $C_{V,SIZE}$	66.8 GPM	83%	Quite linear between M = 40% and M = 80%.
5 $C_{V,SIZE}$	69.1 GPM	61%	Reasonably linear below 61%, but above 61%, the effect on flow decreases rapidly.
10 $C_{V,SIZE}$	69.5 GPM	45%	The operators observe that "once the valve is more than 50% open, it has no effect on the process."

Each of the curves in Figure 10.10 has a reasonably wide linear region. However, normal procxess operations are at the upper end of the linear region. Because the $\%\Delta P_V$ is only 25%, the flow cannot be increased very much above its design value. This leads to the flat portion on the installed characteristics, especially when the valve is grossly oversized.

Selecting Valve Characteristics. Selecting the appropriate inherent valve characteristics is often summarized as follows:

Linear valve. Choose if at design conditions the valve is taking more than half the pressure drop available for fluid flow.
Equal-percentage valve. Choose if at design conditions the valve is taking less than half the pressure drop available for fluid flow.

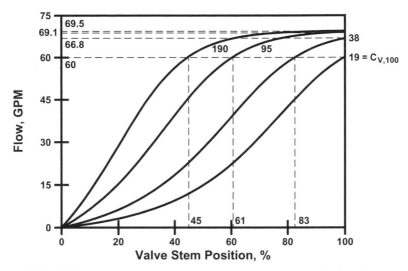

Figure 10.10. Effect of oversizing on installed valve characteristics for an equal-percentage valve; flow system dominated by the process.

This is a broad generalization, and there are certainly exceptions.

The only way to be sure how the valve will behave is to determine its installed characteristics. Unfortunately, this is not routinely done. Computer programs to perform the calculations are readily available. However, these programs require detailed specifications on all components of the flow system. The above simple guidelines usually get it right, which makes it difficult to justify the expenditure of effort to calculate the installed characteristics of every control valve.

In most organizations, there is a predisposition to use only equal-percentage valves. If you propose to install a linear valve, you will have to make a very good case for it (and you still might not win!).

Effect of %ΔP$_V$. In flow systems such as the one in Figures 10.5 or 10.8, the energy dissipated across the control valve originates with the pump. Especially for applications with large flows, energy considerations suggest that the pressure drop across the valve should be as small as possible; that is, %ΔP$_V$ should be reduced as much as possible. But the downsides to reducing %ΔP$_V$ must also be recognized:

- The control valve size and cost increase. Reducing %ΔP$_V$ reduces the pressure drop across the valve. To maintain the same flow at a lower pressure drop requires a larger valve.
- The rangeability is reduced. At a %ΔP$_V$ of 75%, the maximum flow (control valve fully open) is approximately twice the design flow. At a

%ΔP_V of 25%, the maximum flow is approximately 15% more than the design flow.

- The design flow will be near the upper end of the linear range of the valve. The controller can significantly reduce the flow, but only a small increase is possible. The flat region of the installed characteristics commences just above the design flow.

10.6. VALVE NONIDEALITIES

Ideally, the valve stem position should always agree with the control signal to the valve. But being mechanical devices, valves can exhibit a variety of nonidealities that generally increase with age and wear on the mechanical parts. Of these, we shall only examine two:

- Hysteresis.
- Stiction.

The presence of these and other nonidealities is normally suspected from the performance of the control loop. But to quantify the nature of the nonideality, one needs a measurement of the actual stem position. Smart valves are usually equipped with such a measurement that can be transmitted to the control system. Older valves are not so equipped, so one must be acquired to conduct the test. The test usually consists of stroking the valve from fully closed to fully open, and then vice versa. The measured valve for the stem position is then plotted versus the control signal to reveal the nature and severity of the nonideality.

Hysteresis. Hysteresis is normally associated with mechanical fittings that are not tight. Rotary valves are especially prone to this problem. The actuator is connected to the valve stem by a coupling that fits over the valve stem much like a wrench on the head of a bolt. When new, the coupling fits tightly to the valve stem. But with age and wear, the fit loosens, resulting in hysteresis.

The effect of hysteresis is felt primarily when the valve reverses direction. Suppose the control valve has a hysteresis of 2% centered about the position signal to the valve. The following statements apply:

- When the valve is opening, the actual valve position is 1% less than the position signal to the valve.
- When the valve is closing, the actual valve position is 1% greater than the position signal to the valve.
- When the valve reverses direction (opening to closing or closing to opening), the position signal to the valve must change by 2% before there is any change in the actual valve position.

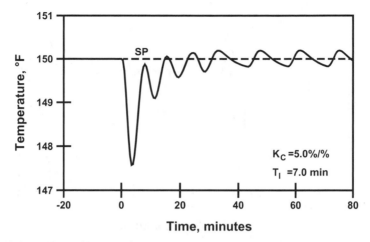

Figure 10.11. Effect of hysteresis on the temperature loop for the hot water process.

Figure 10.11 illustrates the effect of a 1% hysteresis in the steam control valve for the hot water process. The response is to an increase in hot water demand from 500 lb/min to 600 lb/min.

A modest hysteresis has little effect on performance parameters such as the decay ratio, maximum departure from the set point, and so on. Its main effect is that the loop does not line out at the set point. Instead, the hot water temperature continues to cycle about the set point. Such cycles are called limit cycles, and they are the result of some nonideal component within the control loop. In this case, the source is the hysteresis in the control valve. Limit cycles are frequently irregular; the limit cycle in Figure 10.11 increases more rapidly than it decreases.

The amplitude and period of the limit cycle are affected by both the process characteristics and the controller tuning. Dead time in the process increases both the amplitude and the period of the cycle. Although tuning affects the cycle, limit cycles cannot be eliminated by tuning. Besides, the controller should be tuned to the process, not to the valve nonidealities.

Stiction. Where one mechanical part must slide against another, a force is required to overcome friction. All valves require packing around the stem to reduce leaks. In a sliding stem valve, the stem rise is opposed by friction due to the packing. In a rotary stem valve, turning the stem is opposed by friction due to the packing. In both valves, the impact is primarily on small movements in stem position. To cause the stem to move requires a sufficient force to overcome the friction, which in turn requires some difference between the control signal and the current stem position (that is, some positioning error). The effective result is that there is some minimum stem position movement that the valve is capable of making. The term normally applied to this movement is "stiction."

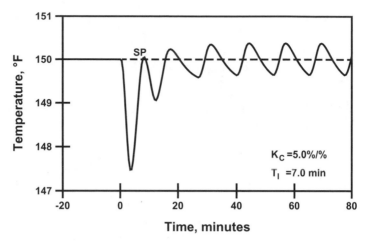

Figure 10.12. Effect of stiction on the temperature loop for the hot water process.

A simplistic view of stiction is that a certain positioning error is required to overcome the friction, but once the stem begins to move, it will move to the position specified by the position signal to the control valve. Suppose the position signal is slowly increasing to a valve with a stiction of 1%. The valve stem does not move until the actual valve position is 1% less than the position signal to the valve. But when the position error attains 1%, the valve quickly moves to the position specified by the input signal. Instead of changing smoothly as called for by the position signal to the valve, the actual valve position changes in a sequence of small step changes of 1%. Similar behavior is observed when the control signal is decreased.

Figure 10.12 illustrates the effect of stiction in the steam control valve for the hot water process. The response is to an increase in hot water demand from 500 lb/min to 600 lb/min. The control valve has a stiction of 1%. As for hysteresis, stiction also has little effect on performance parameters such as the decay ratio, maximum departure from the set point, and so on. However, stiction also leads to a limit cycle in the hot water temperature.

The existence of a limit cycle in the response suggests a nonideality somewhere in the loop. Its amplitude and period will depend on the nature of the nonideality. But from the limit cycle in the response, it is not possible to suggest the nature of the nonideality or the component from which it originates. The first step should be to put the controller on manual and to fix its output. If the limit cycle continues, it originates within the process or possibly the measurement device. If the limit cycle disappears, the control valve is very likely the source. Additional tests are then required to determine the nature of the nonideality.

Use of a Flow Controller. Figure 1.10 presented a temperature-to-flow cascade configuration as an alternative to the simple feedback configuration for controlling the hot water temperature for the hot water process.

Figure 10.13 illustrates the performance of the temperature loop in the temperature-to-flow cascade. The response is to an increase in hot water demand from 500 lb/min to 600 lb/min. Even though the control valve has a

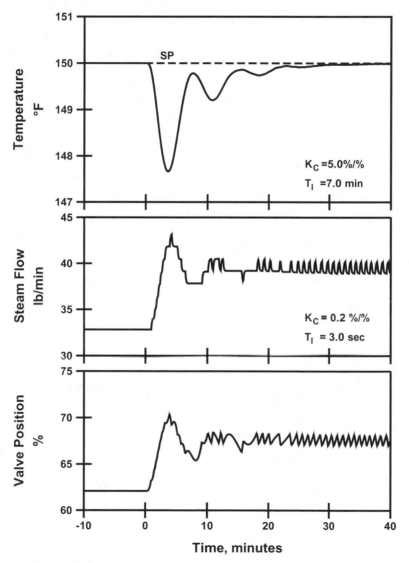

Figure 10.13. Performance of temperature-to-flow cascade with 1% stiction in the control valve.

stiction of 1%, there is no noticeable effect on the temperature control performance. Specifically, there is no limit cycle in the hot water temperature. The flow controller effectively isolates the temperature loop from the nonidealities in the control valve.

Figure 10.13 also illustrates the steam flow and steam valve position. The limit cycle now appears in the steam flow. Since the flow controller is fast, the period of the limit cycle is much shorter, in fact, so short that it has no effect on the hot water temperature. A similar cycle is also present in the output signal to the control valve. The increased wear on the valve has maintenance implications.

If you are only concerned about control of the primary process variable (in this case, the hot water temperature), installing a flow controller is a viable alternative to correcting the problems in the control valve. But unfortunately, this proves to be a short-term fix. At some point in the future, the consequences of the increased wear on the valve will have to be addressed.

10.7. VALVE POSITIONER

The equilibrium position of the diaphragm actuator is determined by a force balance on the stem. Figure 10.14 illustrates the three forces:

1. Force exerted by the air against the diaphragm.
2. Force exerted by the spring.
3. Force exerted on the plug by the fluid flowing through the valve.

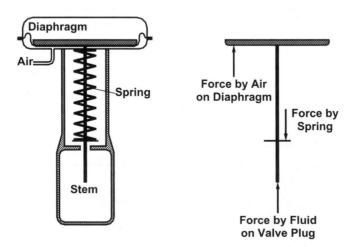

Figure 10.14. Force balance on the diaphragm actuator.

As the pressure of the air increases, the force exerted against the diaphragm increases, which in turn extends the spring. If the force exerted on the plug is constant, the relationship between actuator position and air pressure (the control signal) is reproducible. However, the force exerted on the plug depends on line pressure, fluid characteristics, and other factors that are not necessarily constant. By affecting the equilibrium position of the actuator, these factors become disturbances to the control loop.

Valve Positioner. The valve positioner senses the stem position (tradition-ally by a mechanical linkage to the valve stem), compares the stem position with the control signal, and adjusts the air to the diaphragm so as to make the stem position agree with the control signal. Basically, the valve positioner is a proportional-only controller whose configuration is as follows:

- Measured variable: the valve stem position
- Set point: the position signal from the controller
- Manipulated variable: air to diaphragm

Some valve positioners are entirely pneumatic, and they require a pneumatic control signal. Others are electro-pneumatic, with the control signal being a 4-20 ma current loop. Finally, the smart valves also incorporate a valve posi-tioner. Most accept either a 4-20 ma current loop or a digitally transmitted value.

Use of Positioners. Positioners are optional on diaphragm actuators. The main argument against positioners is purchase cost, but they are also one more item to maintain.

From a control standpoint, all valves should be equipped with a valve posi-tioner. They are most needed in slow loops such as temperature; they provide the least benefit in fast loops such as flow.

Getting valve positioners installed on control valves usually precipitates a confrontation with the project manager. As soon as the project manager runs short of money (as they always do), the proposal to eliminate all valve posi-tioners is floated. Control engineers have to fight this proposal, but they do not always win. Some plants have few, if any, valves with positioners. Some plants have positioners on all but the flow loops (usually the result of a com-promise with the project manager). Very few plants have positioners on all valves.

Smart valves inherently include a valve positioner (although it can be dis-abled if desired). As these valves gain universal acceptance, hopefully the fights with the project manager will become a thing of the past.

Valve Positioners in Flow Loops. Although rarely the case, a valve posi-tioner can potentially degrade the performance of a flow loop.

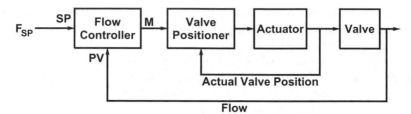

Figure 10.15. Flow loop with valve positioner.

The block diagram in Figure 10.15 illustrates a flow loop with a valve that is equipped with a valve positioner. The valve positioner is basically a proportional-only feedback loop within the valve. The resulting configuration is a cascade configuration, with the flow loop being the outer loop and the valve position loop being the inner loop.

For a cascade configuration to work, the inner loop must be faster than the outer loop, preferably by a factor of five. The major dynamics are between the air to the diaphragm (the output of the valve positioner) and the valve stem position, that is, within the actuator itself. A change in valve stem position almost instantly results in a change in flow (negligible dynamics in the valve). If the flow controller were aggressively tuned, the valve positioner loop would be the slow loop and would degrade the performance of the flow loop. However, very few flow loops are aggressively tuned. Typical tuning is a gain of 0.2%/% and a reset time of 3 sec. With this very conservative tuning in the flow controller, the valve positioner will not cause problems.

Cams in the Valve Positioner. Suppose you have installed an equal-percentage valve, but conclude that a linear valve is more appropriate. You could replace the existing trims (wetted parts) with linear trims. In addition to the cost of the new trims, maintenance will have to dismantle and reassemble the valve with the new parts.

Another alternative is to "fake out" the valve positioner. The stem position sensed via the linkage to the valve stem is the stem position of the equal-percentage valve. Suppose we translate this to an equivalent stem position for a linear valve. Many valve positioners provide a mechanism for doing this. The linkage to the valve stem rotates a cam within the positioner. The position input to the valve positioner logic is taken from the cam. For a 1-to-1 cam, the position input to the valve positioner logic is the stem position sensed through the linkage. But with a cam that translates equal-percentage stem positions to equivalent linear stem positions, the valve seems to behave like a linear valve.

Consider preparing a cam that makes the valve positioner believe that an equal-percentage valve is a linear valve. The cam must do the following:

1. The input is the current stem position of the equal-percentage valve.
2. The %C_V for the equal-percentage valve at this stem position can be determined from the inherent characteristics of the equal-percentage valve.
3. What position of a linear valve would give the same %C_V? For a linear valve, this is simply %$C_V/100$. For other types of valve characteristics, this would have to be determined from the inherent characteristics for that valve.

Suppose the actual valve is a butterfly valve that is 70% open. If its inherent characteristics are as presented in Figure 10.4, the %CV at 70% open is 61%. Suppose the control valve is to seem to have equal-percentage characteristics with a k of 4. From Table 10.1, an equal-percentage valve that is approximately 87% open would have a %CV of 61%. Therefore, the input to the valve positioner should be 87% instead of 70%.

Smart Valves. A smart valve is the result of incorporating a microprocessor into the valve positioner. This permits several functions to be added. As compared with conventional valves, the advantages of smart valves include:

1. Configuring, calibrating, and troubleshooting are far easier.
2. Valve positioner function is provided.
3. Interface options are as follows:
    a. Current loop.
    b. Digital.
4. With a digital interface, the actual valve position can be transmitted to the control system.
5. Multiple characterization functions are provided (basically a digital implementation of the cam).

If desired, the valve positioner function can be disabled. However, there are no cost savings to attract the attention of project managers.

Smart valves are being purchased for most new control valves. However, the transition from "dumb" valves to smart valves is lagging behind the transition from dumb transmitters to smart transmitters. The reason is simple—control valves are costly items, at least as compared with most measurement devices. Replacing installed "dumb" valves with smart valves is difficult to justify.

10.8. ON–OFF CONTROL

Figure 10.16 illustrates on/off control applied to the hot water process. An on/off temperature controller generates the output to a two-position valve on

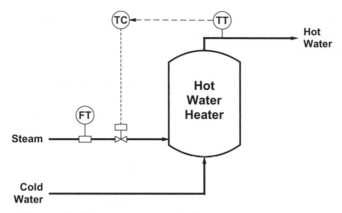

Figure 10.16. ON–OFF control of hot water temperature.

the steam. The simplest ON/OFF control logic generates the discrete signal to the final control element as follows:

Direct Acting	Reverse Acting
if (PV > SP)	if (PV < SP)
M = TRUE	M = TRUE
else	else
M = FALSE	M = FALSE

The thermostat in most residential heating/cooling systems includes ON/OFF control logic (the thermostat is a temperature sensor and controller combined into one unit). In the summer or "Cool" mode, the controller is direct acting, and it activates the cooling unit if the temperature is above the target. In the winter or "Heat" mode, the controller is reverse acting, and it activates the heating unit if the temperature is below the target.

Limit Cycle. ON/OFF controllers do not "line out." Their "steady state" is a limit cycle that persists indefinitely. Figure 10.17 illustrates this for ON–OFF control of hot water temperature.

The limit cycles from ON/OFF control are not necessarily centered about the set point. The limit cycle in Figure 10.17 varies from approximately 4 °F below the set point to approximately 5.2 °F above the set point. This is also reflected in the percent of the time that the steam valve is open. For the limit cycle in Figure 10.17, the period is just over four minutes. The steam valve is open for approximately 44% of the cycle.

The amplitude of the cycle depends on the process dynamics. The higher the process gain is, the larger is the amplitude of the limit cycle. Both the amplitude and the period of the limit cycle increase with the process dead time.

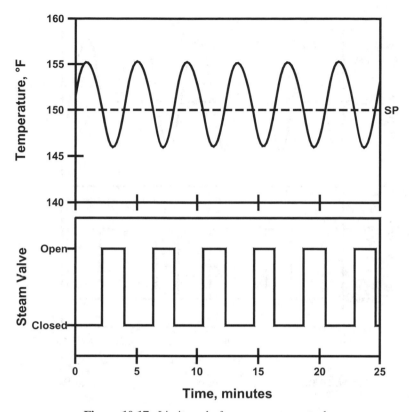

Figure 10.17. Limit cycle from ON–OFF control.

Deadband. Many ON/OFF controllers include an adjustment generally referred to as the deadband (DB). Incorporating the deadband gives the following switching logic:

Direct Acting	Reverse Acting
if (PV > SP + DB)	if (PV < SP – DB)
M = TRUE	M = TRUE
else if (PV < SP – DB)	else if (PV > SP + DB)
M = FALSE	M = FALSE
else	else
M is unchanged	M is unchanged

Increasing the deadband lengthens the time between switches (less wear on the final control element), but it increases the amplitude. The limit cycle in Figure 10.18 is for the hot water temperature controller with a deadband of $1.0\,°F$. The limit cycle now varies from $5.8\,°F$ below the set point to $7.5\,°F$ above the set point with a period of approximately five minutes.

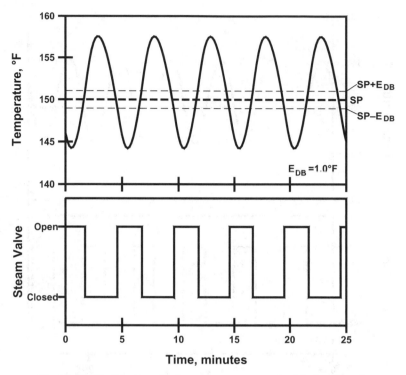

Figure 10.18. Limit cycle from ON–OFF control with deadband.

Derivative or Pre-act. Derivative or pre-act has the opposite effect of dead-band, specifically; it reduces the amplitude of the limit cycle but at the expense of more frequent switching. With derivative or pre-act, the switching logic is based on

$$PV + T_D \frac{dPV}{dt}$$

Basically, a projected value for the PV is obtained by assuming that the current rate of change will persist for one derivative time.

The response in Figure 10.19 illustrates the effect of the derivative in the hot water temperature controller. With a derivative time of 0.2 min, the limit cycle varies from 1.3 °F below the set point to 1.6 °F above the set point with a period of approximately 2.5 min.

Applications. Since limit cycles are inevitable, ON/OFF controllers can only be used in applications where some variance is acceptable in the controlled variable. This normally eliminates their consideration for the important loops in

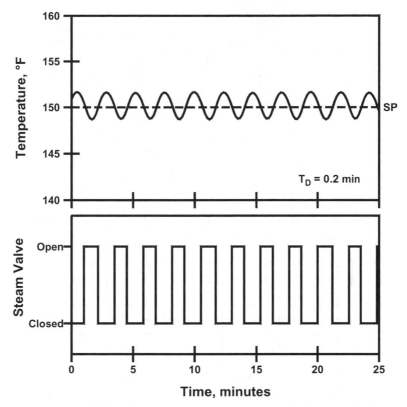

Figure 10.19. Limit cycle from ON–OFF control with pre-act.

most processes. However, their use in the auxiliary loops (such as hot water temperature) is appealing.

Their main advantages are their simplicity and economy. The controller itself is simple, and the adjustment of the deadband and/or derivative is relatively easy. ON/OFF final control elements are invariably simpler and less expensive than modulating final control elements.

10.9. TIME PROPORTIONING CONTROL

Processes are low-pass filters. That is, they do not respond to high-frequency signals or to any short-duration disturbance. In such situations, the process basically responds to the long-term average of that signal.

The schematic in Figure 10.20 illustrates the use of time proportioning control for the hot water process. The temperature controller is a standard PID controller. The output M from the PID controller is the input to the time proportioning logic that provides modulating control using an ON–OFF

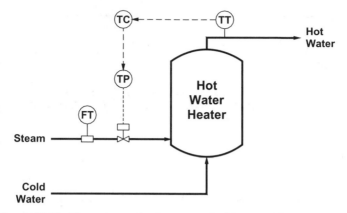

Figure 10.20. Time proportioning control of hot water temperature.

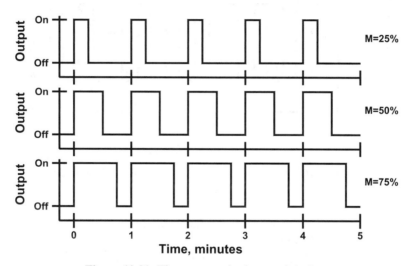

Figure 10.21. Time proportioning modulation.

actuator. The input to the time proportioning logic is an analog signal; the output is a discrete signal.

Time Proportioning Modulation. Modulation is achieved by varying the percent of the cycle time that the output is energized. Let M be the input to the time proportioning logic that specifies the percent of the cycle time that the output is to be energized. Figure 10.21 illustrates the modulation for $M = 25\%$, $M = 50\%$, and $M = 75\%$.

Time proportioning provides linear behavior regardless of the characteristics of the valve. When the valve is closed, the flow is zero. When the valve is fully open, let us designate the flow through the valve as F_{MAX}. The process responds to the average flow F_{AVG} over the cycle time. The value of F_{AVG} varies linearly with M:

M	F_{AVG}
0%	0
25%	0.25 F_{MAX}
50%	0.5 F_{MAX}
75%	0.75 F_{MAX}
100%	F_{MAX}

These values are exact only if the actuator changes states instantly. Control valves move at a finite rate called the slewing rate. Changing states requires a finite time called the travel time or transition time. Consequently, M = 50% will not exactly give a flow of $0.5 F_{MAX}$.

Effect of Cycle Time. With time proportioning modulation, the output cycles from OFF to ON and back to OFF on each time proportioning cycle. When the actuator is a mechanical device (such as a valve), decreasing the cycle time increases the wear on the actuator. Decreasing the cycle time from 1 minute to 30 seconds doubles the cycles per unit time made by the actuator, which means more wear and maintenance.

The controlled variable will have a cycle whose period is the same as the cycle time. Its amplitude depends on the process time constants (dead time has no effect on the amplitude of this cycle). Where mechanical actuators are used, the cycle time is usually increased until the amplitude of the cycle is the maximum that can be tolerated.

Temperature Control. Using time proportioning control has no appreciable effect on the temperature controller. The controller is tuned in the usual way. For the hot water process, we previously determined that a controller gain of 5.0%/% and a reset time of 7.5 min give approximately a quarter decay ratio. This same tuning can be used in the time proportioning configuration. Figure 10.22 presents the response to a set-point change from 150 °F to 160 °F using a time proportioning cycle time of 1 minute.

Extruder Temperature Control. Extruders require multiple temperature controllers, six or so along the barrel and one or two more for the die. The final control element for each temperature loop is an electrical heater band. Time proportioning logic is used to switch the heater ON and OFF, with the temperature controller specifying the percent of time that the heater is to be on.

When mechanical switches were used to switch the heaters ON and OFF, the cycle time for the time proportioning logic was typically 30 seconds. If the cycle time were shortened to 15 seconds, the number of switches per unit time would

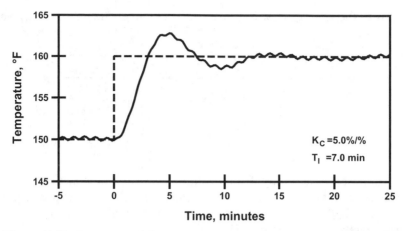

Figure 10.22. Response of time proportioning control to a set-point change.

double, and the lifetime of the switches would be cut in half. But with the advent of solid state switching circuits, the lifetime of the switches is largely independent of the number of switches. This has permitted the cycle time to be reduced to 10 seconds (or even less if one so desires).

Positive Displacement Pumps. Reactors often require the addition of very small amounts of materials such as additives, catalyst, defoaming agents, or other. When this has to be done continuously, a common approach is to use a small positive displacement pump. The issue is how to vary the feed rate for these positive displacement pumps. Some can be equipped with variable stroke lengths. A variable speed drive is another alternative.

Yet another alternative is to use time proportioning logic. Usually there is no flow measurement. Instead, the feed rate (volume per stroke) of the positive displacement pump is known. To feed at half this rate, the pump is switched ON for half the time proportioning cycle. When time proportioning is used, the pump motor is started once each time proportioning cycle. The concern is overheating the motor by starting too frequently. Fortunately, most such applications can tolerate a relatively long cycle, such as 2 or even 5 minutes.

10.10. VARIABLE SPEED PUMPING

When a centrifugal pump is providing the driving force for fluid flow, two alternatives are available for regulating flow (3):

1. Constant speed drive and control valve, usually on the pump discharge.
2. Variable speed drive.

The former provides smooth flow even at very small valve openings, although in some cases, low flow protection consisting of recirculation back to the source vessel or pump suction is required. But with variable speed pumping, smooth flow is not always provided at reduced speeds. There are two possible culprits:

1. The drive.
2. The centrifugal pump.

To obtain variable speed capabilities, early installations used DC drives. However, today most are variable-frequency AC drives or VFDs. VFDs are designed for applications that require precise speed control, and relative to these, variable speed pumping is not very demanding.

When problems develop with variable speed pumping, the centrifugal pump needs to be analyzed in light of the pump performance curves. Certain aspects of the pump curves can definitely lead to variations in the flow at reduced speeds.

An Application. Low flow protection for a centrifugal pump with a constant speed drive normally entails recirculation back to either an upstream vessel or the suction of the pump. The configuration in Figure 10.23 is the surest way to achieve low flow protection. The flow through the pump is measured, with the output being the measured variable for a flow controller that positions a control valve in the recirculation line.

The set point for the flow controller is the minimum required pump flow. Under normal process operating conditions, the pump flow should exceed the minimum required flow, so the flow controller will close the valve in the recirculation line. But should the pump flow drop below the minimum, the flow controller will quickly open the control valve in the recirculation line to provide the required flow.

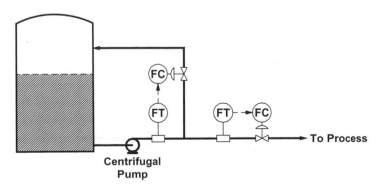

Figure 10.23. Low flow protection for centrifugal pump.

For a constant speed drive, a control valve is also required in the line to the process in order to attain the flow required for the process. Figure 10.23 illustrates a flow controller for this purpose. In some configurations, this flow loop is the inner loop for a level-to-flow cascade, a temperature-to-flow cascade, or other cascade arrangement. But in some cases, the flow is not measured, and the output of the level, temperature, or other controller is connected directly to the control valve. The analysis is basically the same for all configurations, but in what follows, a flow controller will be assumed.

Often some modifications are made to reduce costs:

1. As pressure measurements are less expensive than flow measurements, either a discharge pressure measurement or a pump differential pressure measurement is substituted for the pump flow measurement. The success of this approach depends on the nature of the pump performance curves; it only works if the pressures change significantly with pump flow.
2. One control valve can be eliminated by inserting a fixed orifice into the recirculation piping. This provides some recirculation flow at all times, even when the process flow exceeds the minimum flow required for low flow protection. Extra energy is required to provide the flow, and it is possible that a larger pump will be required as well.

Figure 10.24 presents the configuration for meeting these requirements by using a VFD. The "SC" element is a speed controller that is included with the drive electronics, but as will be discussed subsequently, it could be a torque controller instead.

The drive manufacturers tend to stress energy savings as the justification for VFDs. But especially in industries such as specialty chemicals, there are other possibilities. Using a VFD potentially makes the following equipment unnecessary:

1. Measurement for the pump flow.
2. Recirculation piping.

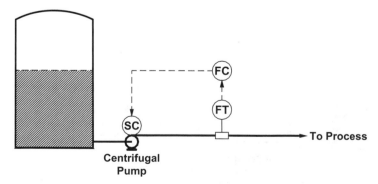

Figure 10.24. Centrifugal pump with VFD.

3. Control valve in the recirculation piping.
4. Control valve in the line to the process.

The controller for pump flow is also eliminated, but this component is likely to be in the software of the digital controls and therefore "free."

"Potentially eliminated" must also be stressed. The above only considers the normal process operations. The minimum flow for the VFD will be less than the minimum flow for the constant speed drive, but it will not be zero. There are always startup issues to be addressed, and usually other considerations. For example, fluid backflows through a centrifugal pump, so if positive shutoff is required, a block valve is required in the line to the process.

Although energy savings in industries like specialty chemicals are likely to be small, other factors can be very significant:

Equipment costs. When the items of equipment must be manufactured from special materials, elimination of only one item of equipment can easily offset the additional cost of the VFD. And these savings come upfront, which appeals to project managers.

Maintenance costs. Control valves are high-maintenance items. Even replacing a control valve by a block valve reduces maintenance costs. When the fluid is corrosive or toxic, maintenance costs escalate, and raise issues pertaining to the exposure of maintenance personnel to potentially harmful materials. Life-cycle costs need to be taken seriously.

From a control perspective, VFDs have another distinct advantage. Specifically, they do not exhibit stiction, hysteresis, and other aberrations resulting from the mechanical nature of control valves. Today's drives incorporate either speed control or torque control, thus providing the counterpart to the positioner advocated by control engineers for control valves. But with a control valve, the flow regulation capability usually meets the requirements of the process. Justifying VFDs on improved flow regulation continues to be elusive.

Flow Regulation. The flow through a pump is determined by the following curves:

Pump performance curve (or just "pump curve"). This curve describes the performance of the pump, relating pump flow, pump head, and pump speed. This curve is provided by the pump manufacturer.

System performance curve (or just "system curve"). This curve relates the head and flow for all components in the fluid flow path (the flow system), excluding the pump. There are two components that determine this curve:

Static head. This is the difference in the head across the flow system when the flow is zero. This difference in head includes the pressure head and the gravity head. The pump must work against this head even at zero flow.

Friction head. As the flow increases, pressure drop due to the fluid's resistance to flow increases by approximately the square of the flow. All components in the flow system contribute to the friction head.

Suppose the friction head is known for a specific flow (this data point should be available from the design calculations). Assuming the friction head varies with the square of the flow, the system curve can be approximated as follows:

$$H_P = H_S + H_D(Q_P/Q_D)^2$$

where

Q_P = Process flow, GPM
H_P = Head at process flow Q_P, ft
H_S = Static head, ft
Q_D = Design flow, GPM
H_D = Friction head at flow Q_D, ft

The operating point is always where the pump curve intersects the system curve.

Control valves regulate the flow by varying the friction head. That is, they work entirely through the system curve. This is illustrated by the example in Figure 10.25. System curves are presented for a fully open control valve and for a control valve that is 75% open. For the fully open control valve, the flow is 177 GPM. As the control valve closes, the resistance to fluid flow increases. This increases the friction head component of the system curve, which makes the system curve intersect the pump curve at a lower flow. When the control valve is 75% open, the flow is 127 GPM.

With variable speed pumping, the system curve is fixed, but the pump curve shifts with pump speed. Some pump suppliers provide pump curves for certain selected speeds; others provide a pump curve only for the rated speed of a constant speed drive. In either case, the affinity laws can be applied to obtain pump curves at other speeds:

$$Q = Q_C(N/N_C)$$
$$H = H_C(N/N_C)^2$$

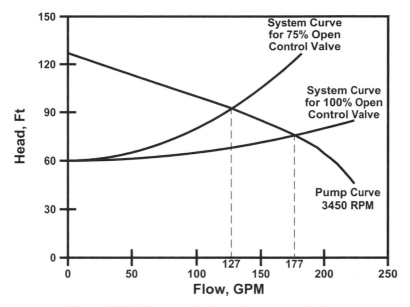

Figure 10.25. Flow regulation using a control valve.

where

N = Speed, RPM

N_C = Speed for the pump curve supplied by the manufacturer, RPM

Q = Volumetric flow at speed N, GPM

Q_C = Volumetric flow at speed N_C (from the pump curve), GPM

H = Head at speed N, ft

H_C = Head at speed N_C (from the pump curve), ft

Flow decreases in proportion to the pump speed; head decreases in proportion to the square of the pump speed. The affinity laws also state that power decreases with the cube of the pump speed, which strengthens the arguments by the VFD manufacturers for energy savings.

Figure 10.26 illustrates regulating the flow by varying the pump speed. The system curve is fixed. Pump curves are illustrated for 3450 RPM, 2850 RPM, and 2470 RPM. For a pump speed of 3450 RPM, the flow through the pump will be 177 GPM. For a pump speed of 2850 RPM, the flow through the pump will be 98 GPM. For a pump speed of 2370 RPM, there will be no flow through the pump. In fact, 2370 RPM is the minimum pump speed for flow. That is, there will be flow through the pump only for pump speeds in excess of 2370 RPM.

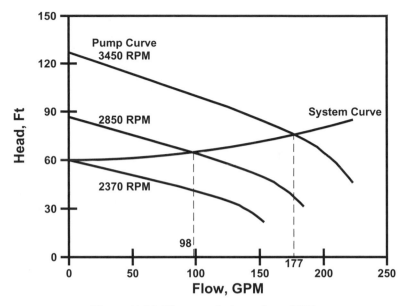

Figure 10.26. Flow regulation using a VFD.

The minimum pump speed to obtain flow is easily computed. When there is no flow, the system head is the static head. From the pump curve, the head delivered by a pump with no flow and running at speed N_C can be determined. The minimum pump speed to obtain flow is computed using the affinity laws as follows:

$$N_{MIN} = N_C(H_S/H_{C0})^{1/2}$$

where

N_{MIN} = Minimum pump speed for flow, RPM
N_C = Speed for the pump curve supplied by the manufacturer, RPM
H_S = Static head (for the system curve), ft
H_{C0} = Head at speed N_C and zero flow (from the pump curve), ft

For the above example, $N_C = 3450\,$RPM, $H_S = 60\,$ft, and $H_{C0} = 127\,$ft. The minimum pump speed is 2370 RPM. The VFD must operate over the range of 2370 RPM to 3450 RPM. The minimum speed is 69% of the maximum speed; the turndown ratio is 1.46:1. VFDs are easily capable of meeting these requirements.

Propagation of Variance. Small variations in both the static head and the friction head occur in any pumping installation. These variations in head lead

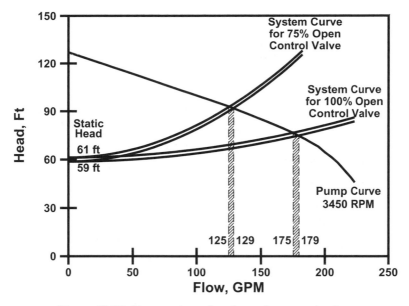

Figure 10.27. Propagation of variance for control valve.

to variations in flow; that is, variations in head propagate to variations in flow. The magnitude of this propagation of variance can also be determined from the performance curves.

Figure 10.27 illustrates the propagation of variance from head to flow. Suppose the static head is 60 ± 1 ft. In Figure 10.27, system curves are drawn for a static head of 59 ft and for a static head of 61 ft. The intersection of each with the pump curve is determined. For the fully open control valve, the flow is 177 ± 2 GPM; for the 75% open control valve, the flow is 127 ± 2 GPM. In each case, a variance of ± 1 ft in head is propagated to a variance of ± 2 GPM. Closing the control valve does not significantly affect the propagation of variance from head to flow.

Figure 10.28 illustrates the propagation of variance from static head to flow for pump speeds of 3450 RPM and 2450 RPM. For a pump speed of 3450, a static head of 60 ± 1 ft gives a flow of 127 ± 2 GPM (same as obtained previously for the fully open control valve). But at 2450 RPM, a static head of 60 ± 1 ft gives a flow of 19 ± 6 GPM. As the pump speed is reduced, the variance in flow increases by a factor of 3.

Even with this increase in variance, the pump would probably still perform satisfactorily. At low flows, the pump curve in Figure 10.28 exhibits a significant sensitivity of head to flow. This is not the case for all pumps. For some, the pump curve at low flows is essentially flat (slope is zero). That is, at low flows, the pump head is independent of flow. For such pump curves, the propagation of variance from static head to pump flow would be much larger.

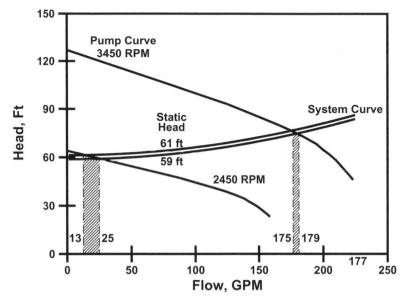

Figure 10.28. Propagation of variance for a VFD.

If the pump curve is flat at low flows, one can expect to experience cycling between no flow and some positive flow. If one expects a pump with a VFD to perform smoothly at low flows, pumps with flat pump curves should be avoided. Pumps with pump curves such as in Figure 10.26 would not be expected to exhibit the cycling.

Flow Regulation with a VFD. The amount of energy transferred from the motor to the pump is determined by two variables:

Speed. Traditionally, a tachometer or encoder is required to measure the speed. But with the latest VFD technology, speed is often calculated from other motor variables.

Torque. Torque can be calculated from other motor variables.

VFDs can now be configured to control either of these variables. If speed is controlled, the input signal to the drive electronics is the desired speed. When torque is controlled, the input signal to the drive electronics is the desired torque.

For pumps, the variable of interest is the flow through the pump. This variable can be influenced by either varying the speed of the drive or varying the torque delivered by the drive. The pump curves are traditionally drawn with pump speed as a parameter. This might imply that the flow through the pump is best controlled by varying the pump speed, but this is not necessarily the case.

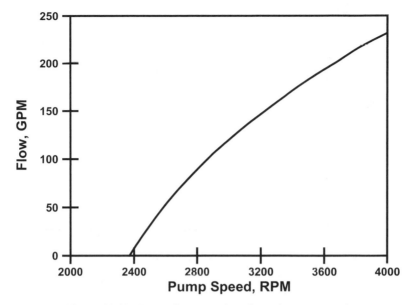

Figure 10.29. Pump flow as a function of pump speed.

When the input signal to the drive electronics is the target for the pump speed, the relationship between flow through the pump and the pump speed is important. This can be computed from the pump curve and system curve, with the result as shown in Figure 10.29. The departure from linearity is noticeable, which means that the sensitivity of flow to pump speed is not constant.

When the input signal to the drive electronics is the target for the torque, the relationship between flow through the pump and the torque is important. This can also be computed from the pump curve and system curve, with the result as shown in Figure 10.30. The graph exhibits only a slight departure from linearity. As the PID control equation is a linear equation, a linear relationship between the pump flow and the signal to the drive electronics is preferred.

The graphs in Figures 10.29 and 10.30 were computed from the pump curve in Figure 10.26. This pump curve exhibits significant sensitivity of head to flow at low pump flows. For some pumps, the pump curve is flat at low flows; that is, the head is almost constant at low pump flows. For such pumps, the departure from linearity is much greater, with the graph of pump flow versus pump speed being almost vertical at low flows. But even for a pump with a flat pump curve, the relationship between pump flow and torque is nearly linear.

Based on the nature of the relationship of pump flow to torque, the control signal to the drive electronics should preferably be the target for the torque. This also seems consistent with the future directions for VFD technology.

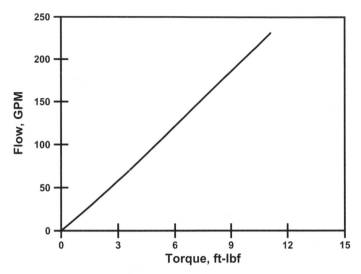

Figure 10.30. Pump flow as a function of torque.

10.11. SUMMARY

Perhaps a third of all process control problems originate in the control valve. The common culprits include

Oversized valve. This gives a high sensitivity at the normal operating conditions, but the sensitivity approaches zero at large valve openings.

Valve nonidealities. Valves are mechanical devices that wear with age. The resulting behavior from the valve leads to a limit cycle in the control loop. "Fix the valve" is always the best approach, but in difficult services (such as a caustic slurry), one's expectations have to be realistic.

Installed valve characteristics. Except for butterfly valves, most valves are purchased with equal-percentage characteristics. This is not always appropriate.

The advantages of the "smart" valve are generally recognized, although some object to the additional cost in small valves. There are two distinct advantages:

1. The smart valve provides a superior response to changes in the signal to the valve.
2. The additional information available from the valve also permits problems to be identified and resolved more expeditiously.

Applications of variable speed pumping continue to increase. Today's drive technology is clearly up to the task, and it now provides options such as torque control as well as speed control. If problems originate when a constant speed drive is replaced with a variable speed, the natural tendency is to blame the drive. However, changing the pump speed causes the operating point to shift to a different location on the pump performance curves. The nonlinear nature of these curves can definitely cause problems to develop.

LITERATURE CITED

1. "Flow Equations for Sizing Control Valves", ANSI/ISA 75.01.01 (IEC 60534-2-1 Mod)-2007.
2. Masoneilan 39003 High Performance Butterfly Valves (HPBV), Specification Data, CM9003, April, 2000, p. 10.
3. Smith, C. L., "Watch Out with Variable Speed Pumping", *Chemical Processing*, Vol. 71, No. 5 (May 2008), p. 33.

CHAPTER 11

Process and Instrumentation Diagrams

The purpose of the process and instrumentation (P&I) diagram is to specify the control strategy for a process. Although controller tuning gets far more attention, developing the P&I diagram is really the essence of process control.

Tuning is a procedure for getting the characteristics of a controller "in tune" with the characteristics of the process. In a similar sense, the P&I must also be "in tune" with the process. Where it is not, the affected controllers will be untunable and the controllers will remain on manual.

This chapter uses examples to illustrate three considerations pertaining to P&I diagrams:

1. Certain aspects of process behavior are subtle and if not understood can lead to loops that cannot fulfill their purpose. A chlorine vaporizer is used to illustrate.

2. Because of their simplicity, simple feedback configurations are the logical starting point. However, functions such as cascade can significantly reduce the variance in the process variable. The hot water process is used to illustrate.

3. In some cases, there are two or more possible locations for the control valve, which has consequences on aspects such as the limits imposed on process operations. A steam-heated exchanger is used to illustrate.

11.1. DEVELOPING P&I DIAGRAMS

If you examine several P&I diagrams, you will get the impression that the guiding principle is summarized by the following statement:

Control each variable with the nearest valve that has a significant influence on that variable.

Practical Process Control: Tuning and Troubleshooting, by Cecil L. Smith
Copyright © 2009 John Wiley & Sons, Inc.

Those that develop P&I diagrams certainly do not follow such a simplistic approach; however, it turns out that way most of the time.

A more probable explanation is our tradition of focusing on dynamic issues. Dynamically, a near valve usually influences a variable faster than a distant valve. Dynamic considerations suggest we should control each variable with the nearest valve that has a significant influence on that variable.

When a P&I diagram is deficient, the problem usually originates in the steady-state relationships. A valve might seem to have a significant influence on a controlled variable when in fact it does not. Even though most modern designs are based on models, the developers of the P&I diagram base their decisions largely on their understanding of the process. There are two pitfalls in this:

1. As process designs become more complex, the cause-and-effect relationships become more difficult to sort out.
2. The emphasis on shortening the design and construction cycle reduces the time one has to reflect on how a process really works.

This section will emphasize that a control problem just might be the result of a deficiency in the P&I diagram. Errors in the P&I diagram almost always result in one or more untunable controllers. Developing the P&I diagram is really the most crucial step to achieving a control system that delivers the performance required by the process.

Process Design Procedure. For a process plant, the design effort yields two "deliverables":

Process flowsheet. This is produced by the process design team to meet the stated objectives for the plant.

P&I diagram. This is produced by the controls and instrumentation group based on the process flowsheet from the design team.

Who developed the P&I diagram for a specific plant? We are referring to the intellectual content. The name of the engineering company that provided the detailed design services usually appears on the drawing. They provided the drafting services and certainly filled in some of the details. But traditionally, the in-house controls group at central engineering made (or at least blessed) all major decisions pertaining to a P&I diagram. But right-sizing and outsourcing has produced a smaller central engineering with younger people. Even so, some companies expect the remaining people to do the work, whereas others look outside to technology suppliers, engineering companies, consultants, and so on.

The point is as follows: Do not assume that the control configuration on the P&I diagram is appropriate to the process. A smaller controls group with less experienced people is being given less time to develop the P&I diagram for processes of ever increasing complexity.

When digital controls began to compete with conventional controls, one of the arguments for installing digital systems was the flexibility to implement changes to the control configuration. But for whatever reason, this has not transferred to implementing modifications to the P&I diagram. If the plant cannot produce a salable product, changes to the P&I diagram will be made. But if the plant can "live with" the consequences of the deficiency, correcting a P&I diagram once a plant is operating is usually a frustrating experience. The usual attitude is to continue with the devil that we know instead of risking the unknown. The lesson here: Get it right the first time!

Keep it Simple. Most loops on a P&I diagram will be simple feedback control loops consisting of three components:

- A measurement device.
- A PID controller (although the derivative mode will not be used in most).
- A final control element, typically a control valve.

The appealing attributes of this approach include

1. Simple to design. It is based largely on a qualitative understanding of process behavior.
2. Widely known (by both control engineers and instrument technicians).
3. Requires only one process measurement.
4. Can be implemented using a variety of commercial control products, including single-loop controllers, distributed control systems (DCS), programmable logic controllers (PLC), and so on.

But in process applications, simple PID control has its limitations, with the most frequently encountered limitations as follows:

1. The PID controller is a linear controller, but most processes are non-linear. Consequences occur when the process must be operated over a wide range.
2. The response of the PID controller is limited by the process dynamics. Especially when significant dead time is present, the PID controller has to be tuned so that it responds slowly.
3. The simple PID controller works best when only one of the manipulated variables has a significant influence on the controlled variable. This is not always the case.

Digital technology offers many capabilities that can be advantageously used to address these and other problems. But there is a downside, which is probably best referred to as "creeping elegance." At the design stage, the tendency is to ask "what if" questions regarding the controls, and then to answer each of these questions by incorporating something into the controls to address the issue.

Often the argument is made that we will get the plant running with the simplest configuration required just to operate the plant. Then we will ascertain what control configuration enhancements will improve the performance of the process. Actually this is a sound approach 3/4 solve the real problems and not the hypothetical ones. But will we have the discipline to actually do it?

Engineers versus Instrument Technicians. In most plants, the job description for instrument technicians includes the responsibility for tuning the controllers. However, developing and modifying the P&I diagram is not in their job description. When deficiencies in the P&I diagram are causing tuning problems, the instrument technicians will not be able to tune the loop satisfactorily, but they do not have the authority (and probably not the expertise) to identify the cause of the problem and to correct it.

To identify and correct problems with the P&I diagram, the engineers must be involved. In most cases, the engineers will also attempt to tune the loop. However, their objective is to understand the nature of the problem being experienced, and to make sure that their understanding of the process behavior is consistent with observed behavior. Often the problems with the P&I diagram occur because our understanding of process behavior is inadequate.

11.2. P&I DIAGRAM FOR A CHLORINE VAPORIZER

To illustrate the issues in developing a P&I diagram, let us consider the chlorine vaporizer illustrated in Figure 11.1. The chlorine gas from the vaporizer is a feed to the main process. The chlorine is delivered in cylinders that are equipped with a steam coil. The chlorine vapor line is connected to the process and steam is supplied to the coil.

Two variables are measured and controlled, as follows:

Chlorine flow. Control of this flow is critical for the process to operate properly.

Vaporizer pressure. The cylinder is outfitted with a relief device that we do not wish to actuate. But as long as this does not occur, the vaporizer pressure is acceptable.

Both variables are measured. There are two control valves, one in the chlorine vapor line to the process and one in the steam supply to the vaporizer.

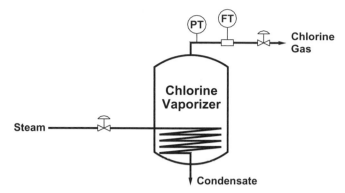

Figure 11.1. Chlorine vaporizer with measurements and control valves.

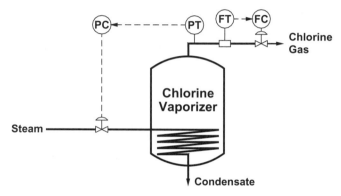

Figure 11.2. Control chlorine flow by manipulating the control valve in the chlorine vapor line; control vaporizer pressure by manipulating the control valve on the steam supply.

Proposed P&I Diagram. Even those who have spent years developing P&I diagrams have difficulties explaining exactly how they go about it. Certainly a significant component is their past experiences, both with what worked and what did not work. To some extent, this is "pattern recognition." Unfortunately, what initially seems to be the same or similar will occasionally turn out to be very different.

For most flow control configurations, the flow measurement senses the flow through the control valve that is manipulated to control the flow. In Figure 11.1, the measurement for chlorine flow is immediately upstream of the control valve for the chlorine vapor feed to the process. This suggests the flow controller for the chlorine flow as illustrated in Figure 11.2.

We are now left with only one option for controlling the vaporizer pressure, namely, by manipulating the control valve on the steam supply. The steam flow certainly affects the vaporizer pressure. Adding the pressure controller completes the P&I diagram in Figure 11.2 for the chlorine vaporizer.

One observation: If your approach is to "control each variable with the nearest valve," the result is the P&I diagram in Figure 11.2.

Unfortunately, there is a serious problem with the configuration in Figure 11.2. The chlorine vapor valve has a short-term effect on the chlorine flow, but its long-term effect is very small. To understand how the process really works, we need to return to the configuration in Figure 11.1, that is, with the measurements and control valves but no controls.

With no controls (as in Figure 11.1), consider the response of both chlorine flow and vaporizer pressure to a change in the chlorine vapor valve position. On opening the chlorine vapor valve, there will be a significant increase in the chlorine flow. Unfortunately, this increase is temporary. Removing more chlorine vapor causes the pressure within the vaporizer to decrease. Does this cause more chlorine to be vaporized?

For the chlorine vaporizer, the chlorine flow is determined by the heat transferred from the steam:

$$F = Q/\lambda$$

where

   F = Chlorine flow

   Q = Heat transfer rate within the vaporizer

   λ = Latent heat of vaporization of the chlorine

Opening the vapor valve decreases the pressure (and temperature) within the vaporizer, which increases the heat transfer rate Q in the above equation. But on the other hand, the value of the latent heat of vaporization increases as the pressure decreases. Both of these are essentially secondary effects—the chlorine vaporization rate is determined by the steam flow, which depends primarily on the position of the control valve in the steam supply. Consequently, the steady-state sensitivity of chlorine flow to the chlorine vapor valve position is very small. This causes difficulties for the chlorine flow controller as configured in Figure 11.2.

When the chlorine vaporizer is at steady state, the following statements must be true:

1. The vaporization rate of chlorine within the vaporizer must be the same as the chlorine flow through the control valve in the chlorine vapor line.
2. The condensation rate of steam must be the same as the steam flow through the control valve on the steam supply.

3. The condensation rate of steam determines the rate of heat transfer from steam to chlorine, which in turn determines the vaporization rate of the chlorine.

Configurations in Which One Loop Depends on Another. The process relationships suggest that the control valve in the chlorine vapor line affects the vaporizer pressure, but not the chlorine flow. The process relationships assume that no control actions are being taken, so in effect the vaporizer pressure controller is on manual.

But normally we would control the vaporizer pressure. Consider using only the vaporizer pressure controller (the flow controller in Figure 11.2 is on manual). Unlike the chlorine flow controller, the vaporizer pressure controller will function if used alone. So we could tune the vaporizer pressure controller first and then tune the chlorine flow controller.

With the pressure controller in automatic, consider opening the control valve in the chlorine vapor line. The response is as follows:

1. Opening the control valve in the chlorine vapor line increases the chlorine flow.
2. The increased chlorine flow decreases the vaporizer pressure.
3. The pressure controller responds by opening the control valve on the steam supply.
4. Opening the control valve on the steam supply increases the steam flow and the heat transfer rate.
5. The increased heat transfer rate increases the vaporization rate of the chlorine.
6. The increased vaporization rate of the chlorine increases the vaporizer pressure (the pressure controller adjusts the position of the control valve on the steam supply so as to maintain the vaporizer pressure at its target).
7. With the same pressure in the vaporizer and a larger opening of the control valve in the chlorine vapor line, the increase in the chlorine flow will be sustained.

This response suggests that the chlorine flow can be controlled by manipulating the control valve in the chlorine vapor line provided the vaporizer pressure controller is on automatic. Should the vaporizer pressure controller be placed in manual, the chlorine flow loop will not perform properly and must be placed on manual. With digital controls, logic can easily be added such that the chlorine flow controller is permitted to be in automatic only if the vaporizer pressure controller is in automatic.

However, another requirement also must be met. When one loop is dependent on another loop, this must be a slow loop that is dependent on a fast loop. Just keep the fast loop in automatic, and the configuration will function.

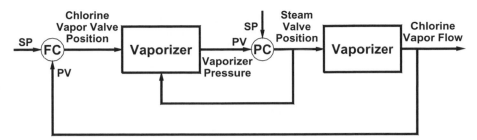

Figure 11.3. Relationship between flow controller and pressure controller.

However, when a fast loop is dependent on a slow loop, this configuration will not function satisfactorily. Unfortunately, this is the case for the P&I diagram in Figure 11.2.

The relationship between the vaporizer pressure controller and the chlorine flow controller is illustrated by the block diagram in Figure 11.3. The basis for each component of the block diagram is as follows:

1. The chlorine flow controller manipulates the control valve in the chlorine vapor line.
2. The long-term effect of the control valve in the chlorine vapor line is only on the vaporizer pressure; its long-term effect on chlorine flow is insignificant.
3. The vaporizer pressure controller responds to changes in the vaporizer pressure by manipulating the control valve on the steam supply.
4. The control valve on the steam supply affects both the chlorine flow and the vaporizer pressure.

The block diagram in Figure 11.3 contains two loops, one for the chlorine flow and one for the vaporizer pressure. Note that the vaporizer pressure loop is totally contained within the chlorine flow loop. Such loops are said to be "nested," with the outer loop (the chlorine flow loop) being dependent on the inner loop (the vaporizer pressure loop).

A similar but not identical loop structure applies to cascade control configurations. The structure is similar in that there are two "nested" loops. The structure is different in the arrangement for the controllers. In the block diagram in Figure 11.3, external set points are available for both loops (that is, the chlorine flow set point and the vaporizer pressure set point can be individually specified). In the cascade configuration, an external set point is only available for the outer loop (the set point for the inner loop is determined by the outer loop).

For cascade configurations to function properly, the inner loop must be faster than the outer loop, preferably by a factor of five. This requirement applies to nested loops in general, so it also applies to the block diagram in

Figure 11.3. Is the vaporizer pressure loop (the inner loop) faster than the chlorine flow loop (the outer loop)? Definitely not. Flow loops are generally the fastest loops in a process, and certainly the chlorine flow loop will be faster than the vaporizer pressure loop. As a result, the control configuration in Figure 11.2 will not function properly.

Tuning the Loops. Prior to startup, a decision must be made as to the order in which the loops are to be tuned. The issues include the following:

1. Can any loops be tuned before putting chemicals into the process? We often test the equipment by pumping water, blowing air, or similar endeavors. As part of this effort, is there an opportunity tune any of the loops? Sometimes we can, but there are no such opportunities for the chlorine vaporizer.
2. Tune the fast loops first. For the chlorine vaporizer, this suggests to tune the flow loop and then to tune the pressure loop.
3. Tune the most important loops first. For the chlorine vaporizer, maintaining the proper chlorine flow to the process is crucial, so this also suggests to tune the flow loop and then to tune the pressure loop.

Tune Flow Then Pressure. Let us assume that the flow loop in Figure 11.2 is to be tuned first. The customary tuning coefficients for a flow loop, namely, a gain of about 0.2%/% and a reset time of 3 sec or so, should be satisfactory for this loop. We manually adjust the steam valve to first heat the chlorine to its boiling point. We then open the chlorine valve to admit chlorine to the process, switch the flow controller to automatic, and specify the desired set point for the chlorine flow. And now we encounter a problem. Depending on where we position the control valve on the steam, one of the following occurs:

Steam valve not sufficiently open (or flow controller set point too high). The chlorine flow controller drives the vapor valve fully open, but the chlorine flow remains below its set point. The heat being supplied by the steam is insufficient to vaporize the amount of chlorine required for the current flow controller set point.

Steam valve open too much (or flow controller set point too low). The chlorine flow controller is able to maintain the desired chlorine flow, but only by gradually closing the chlorine vapor valve. This causes the vaporizer pressure to gradually increase. If we allow this to continue, the relief device will eventually react. We are vaporizing more chlorine than required for the current flow controller set point.

By continuing to adjust the opening of the control valve on the steam, we can get the flow loop to line out. However, it is very sensitive to the opening of this control valve, and any upset to the process or change in the chlorine flow set point initiates one of the above two reactions from the flow controller.

For a flow controller, whenever the customary tuning coefficients (0.2%/% for the controller gain and 3 sec for the reset time) do not provide at least some measure of control, one should examine that flow loop very carefully. Hopefully this will at least lead to the conclusion that the chlorine flow controller in Figure 11.2 can only function if the vaporizer pressure controller is in automatic.

Tune Pressure Then Flow. The vaporizer pressure controller will function properly when used alone. That is, we are able to tune the vaporizer pressure controller in Figure 11.2. With the vaporizer pressure controller in automatic, we can change the opening of the chlorine vapor valve, and the pressure controller will respond so as to maintain the vaporizer pressure at its set point.

Once we have tuned the vaporizer pressure controller, we can attempt to tune the chlorine flow controller. But again the usual flow controller tuning coefficients do not provide acceptable performance. With these tuning coefficients, the chlorine flow controller responds too rapidly, leading to oscillations in both chlorine flow and vaporizer pressure. To eliminate the oscillations, the chlorine flow controller must be tuned to respond very slowly. A controller gain of 0.2%/% is very low, so further decreasing the controller gain will not be effective. Instead, the reset time must be increased from 3 sec to several minutes. Eventually this will eliminate the oscillations, but the chlorine flow controller responds so slowly that its performance is very poor.

As noted in the block diagram in Figure 11.3, the control configuration consists of two nested loops, the inner loop being the vaporizer pressure loop and the outer loop being the chlorine flow loop. The outer loop must be slower than the inner loop. We have achieved this by tuning the chlorine flow loop to respond very slowly. Although we can eventually eliminate the oscillations in this manner, rarely is the resulting performance acceptable.

The only hope for success is first to tune the vaporizer pressure controller to respond as rapidly as the process dynamics permit. Then slow the response of the chlorine flow controller until the oscillations are eliminated. Unfortunately, this is contrary to the relative importance of the two loops. The most important controlled variable is the chlorine flow to the process. Provided the relief device is not actuated (and chlorine is vented), the performance of the vaporizer pressure loop is acceptable.

Alternate P&I Diagram. In multivariable terminology, the chlorine vaporizer in Figure 11.1 is referred to as a 2×2 process—there are two controlled variables (chlorine flow and vaporizer pressure) and two manipulated variables (chlorine vapor valve position and steam valve position). The configuration in Figure 11.2 proposes the following configuration:

1. Control the chlorine flow by manipulating the chlorine vapor valve position.
2. Control the vaporizer pressure by manipulating the steam valve position.

The term "pairing" is often used to designate the selection of which manipulated variable will be used to control each controlled variable.

In a 2×2 process, there are only two possible pairings. In addition to the one for the configuration in Figure 11.2, the following pairing is possible:

1. Control the chlorine flow by manipulating the steam valve position.
2. Control the vaporizer pressure by manipulating the chlorine vapor valve position.

The P&I diagram for this configuration is illustrated n Figure 11.4.

The two loops in Figure 11.4 will perform individually:

Chlorine flow controller on auto; vaporizer pressure controller on manual (fixed chlorine vapor valve position). The output of the chlorine flow controller is the steam valve position. This determines the steam flow, which in turn determines the chlorine vaporization rate. At steady state, the chlorine flow must equal the chlorine vaporization rate. The vaporizer pressure increases or decreases until the flow through the chlorine vapor valve equals the chlorine vaporization rate.

Vaporizer pressure controller on auto; chlorine flow controller on manual (fixed steam valve position). The chlorine vaporization rate and the chlorine flow are also fixed (at steady state, these must be equal). To

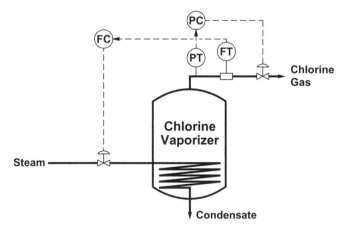

Figure 11.4. Control chlorine flow by manipulating the control valve on the steam supply; control vaporizer pressure by manipulating the control valve in the chlorine vapor line.

achieve the desired vaporizer pressure requires a certain pressure drop across the control valve. As the flow through the valve is fixed, the valve position can be adjusted until this pressure drop is achieved.

Consequently, it would be possible to tune each of these loops individually, with the other loop on manual. But as will be explained in the next chapter on interaction, this does not assure that the two loops will function properly when both are in automatic.

11.3. SIMPLE PID CONTROL CONFIGURATION

The simple feedback control configuration for temperature control of the hot water process in Figure 1.9 consists of a PID controller, a measurement device for the hot water temperature, and a control valve on the steam supply. If the performance of the process is acceptable, this simple configuration is perfectly adequate. But should the performance prove unsatisfactory, the deficiencies need to be identified and control configurations proposed to address these deficiencies.

Deficiencies in process performance are usually caused by variance in one or more key process variables. This variance may in turn be caused by variance in other process variables. The hot water temperature is potentially one of these other process variables. The variance in hot water temperature results from some combination of the following:

- Disturbances, specifically, changes in hot water demand and changes in steam supply pressure.
- Steam valve sizing and related issues.
- Controller tuning.

Changes in Hot Water Demand. As for most utility processes, the hot water process experiences frequent changes in hot water demand. There are times of high hot water demand and times of low hot water demand. The transition from high demand to low demand (and vice versa) is not in one large change in demand, but instead is the cumulative effect of many small changes in hot water demand. Large changes in hot water demand occur very infrequently and are always associated with other major disruptions to process operations.

Let us assume that this is a major source of the variance in hot water temperature. The ability of a feedback controller to respond to these is limited by the dynamics of the hot water process. This limits the potential of activities such as tuning to address this source of variance.

Changes in Steam Supply Pressure. Significant changes in the steam supply pressure occur occasionally. This change in steam supply pressure affects the steam flow, which in turn affects the hot water temperature. Changes in steam supply pressure occur far less frequently than changes in hot water demand, but the changes in steam supply pressure tend to have a larger effect on the hot water temperature.

Let us assume that this is a contributor to the variance, but not the major source. Reducing this source would be beneficial, but only this would not be sufficient.

Valve Issues. Control valve sizing is always a potential issue. Check the steam valve opening for the higher hot water demands. If the valve is an equal-percentage valve and is more than 50% open, sizing is probably satisfactory.

Another possibility is nonidealities in the control valve. When the process is operating at a relatively steady hot water demand, is there a cycle in the hot water temperature? If so, does it stop when the controller is placed in manual?

Let us assume the sizing is satisfactory and that there is only a small cycle in the hot water temperature that is not currently contributing significantly to the variance in hot water temperature.

Controller Tuning. In any utility process, a valid issue is the controller performance at different throughputs. Let us assume the controller has previously been tuned under the following conditions:

High hot water demand. The performance is good when the hot water demand is high, but the loop cycles when the hot water demand is low.

Low hot water demand. The tuning is more conservative, so the controller responds more slowly.

As cycling is unacceptable, the tuning for the low hot water demand is in use. But when tuned this way, the response to the frequent changes in hot water demand is inadequate. The desire is for a control configuration that provides good control at both low and high hot water demands.

Solution. Scheduled tuning is one possibility to achieve more consistent performance. However, this approach relies entirely on simple feedback control to respond to the changes in hot water demand. Let us consider two cascade configurations that use additional measurements to provide enhanced response to changes in hot water demand and steam supply pressure.

The process operating lines will be used in analyzing these configurations. For the simple PID configuration in Figure 1.9, the controlled variable is the hot water temperature T_w. The manipulated variable is the controller output M to the steam control valve. The operating line for the process is a plot of T_w versus M. Figure 11.5 presents the operating lines for three different hot water demands. The sensitivity or gain of the process is as follows:

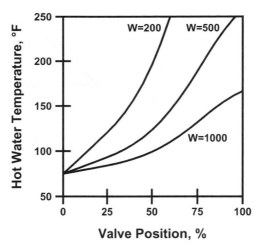

Figure 11.5. Operating lines for simple feedback control.

$$\frac{\partial T_w}{\partial M}$$

This is the slope of the operating lines in Figure 11.5, which is strongly affected by the hot water demand.

11.4. TEMPERATURE-TO-FLOW CASCADE

One potential enhancement to the control configuration for the hot water process is to add a steam flow controller. The result is the temperature-to-flow cascade configuration illustrated in Figure 1.10. As compared with the simple PID control, this configuration requires an additional measurement for the steam flow.

The temperature-to-flow cascade configuration requires two PID controllers:

Steam flow controller. The process variable is the steam flow F; the controller output is the steam valve position M. The process gain or sensitivity is

$$\frac{\partial F}{\partial M}$$

This process sensitivity is determined by the characteristics of the steam valve and flow system for the steam supply.

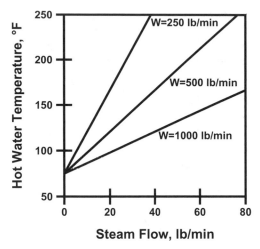

Figure 11.6. Operating lines for temperature-to-flow cascade.

Hot water temperature controller. The process variable is the hot water temperature T_w; the controller output is the set point for the steam flow controller. But at steady state, the steam flow and the steam flow set point are equal, so the manipulated variable is the steam flow F. The operating lines are plots of T_w versus F, and they are presented in Figure 11.6 for three different hot water demands. The process sensitivity or gain is

$$\frac{\partial T_w}{\partial F}$$

Since the sensitivity is with respect to steam flow instead of steam valve position, the steam valve and flow system characteristics have no effect on this sensitivity. Each operating line is essentially linear; however, the slope of the operating line clearly depends on the hot water demand.

Changes in Hot Water Demand. Like the simple PID configuration, the temperature-to-flow cascade configuration relies on feedback control to respond to load changes. The hot water temperature controller provides the primary response to changes in hot water demand. This temperature controller will still have to be tuned for a low hot water demand, the result being an unacceptably slow response when the hot water demand is high.

Changes in Steam Supply Pressure. With the temperature-to-flow cascade configuration, the flow controller provides the primary response to changes in steam supply pressure. The steam flow controller responds as soon as the steam supply pressure affects the steam flow. Since the change in flow occurs much sooner than the change in hot water temperature, the temperature-to-flow cascade responds far faster to changes in steam supply pressure than the simple PID configuration. This will essentially eliminate this source of variance in the hot water temperature.

Valve Issues. The steam flow controller isolates the hot water temperature controller from all issues regarding the control valve. The valve nonidealities now cause the steam flow to cycle about the set point provided by the hot water temperature controller. But since the flow loop is fast, the period of the cycles will be very short. Being much slower, the hot water process basically averages the cycles in steam flow, and no cycles will be observed in the hot water temperature. However, the faster response of the flow controller increases the wear on the control valve, which has long-term implications for valve maintenance.

Controller Tuning. Tuning the hot water temperature controller in the temperature-to-flow cascade will be very similar to tuning the simple PID controller. Problems will still be experienced with performance at different hot water demands. The temperature controller will have to be tuned at a low hot water demand. At a high hot water demand, the controller will respond too slowly. With the temperature-to-flow cascade, the variance in hot water temperature due to changes in hot water demand will still be a problem.

11.5. TEMPERATURE-TO-FLOW-RATIO CASCADE

Another potential enhancement to the control configuration is to control the ratio of steam flow to cold water flow. The result is the temperature-to-flow-ratio cascade configuration illustrated in Figure 11.7. As compared with simple PID control, this configuration requires two additional measurements:

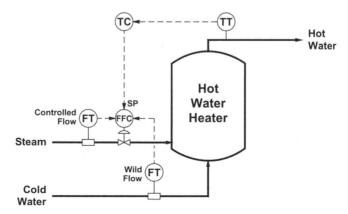

Figure 11.7. Temperature-to-flow-ratio cascade control of hot water process.

- Steam flow.
- Cold water flow.

This configuration also requires two PID controllers:

Steam-flow-to-cold-water-flow ratio. Sometimes called a flow-to-flow controller (FFC); this controller adjusts the signal to the control valve so as to maintain two flows in a specified ratio. The flow through the valve is called the controlled flow; the other flow is called the wild flow. The set point to the flow-to-flow controller is the desired ratio of controlled flow to wild flow. For the hot water process, the controlled flow is the steam flow F; the wild flow is the cold water flow Q. The set point is the desired value of the ratio F/Q of steam flow to cold water flow. The process sensitivity is determined by the characteristics of the valve and flow system, and it is expressed as follows:

$$\frac{\partial (F/Q)}{\partial M} = \frac{1}{Q}\frac{\partial F}{\partial M}$$

Hot water temperature controller. The process variable is the hot water temperature T_w; the controller output is the set point for the flow-to-flow controller. But at steady state, the flow ratio and the flow ratio set point are equal, so the manipulated variable is F/Q. The process sensitivity or gain is

$$\frac{\partial T_w}{\partial (F/Q)}$$

The process operating line is a plot of the hot water temperature T_w versus the steam-to-cold water ratio F/Q. Figure 11.8 presents the process operating line for the temperature loop. The process operating line exhibits only a modest departure from linearity, and the operating line is independent of the hot water demand.

Changes in Hot Water Demand. Any change in the hot water demand is immediately detected in the cold water flow measurement, which causes the flow-to-flow controller to change the steam flow. Flow-to-flow controllers respond very rapidly (basically the same as flow controllers), so there will be an almost immediate response to any change in hot water demand. For the simple PID control configuration (Figure 1.9) and for the temperature-to-flow cascade configuration (Figure 1.10), the hot water temperature controller provides the primary response to changes in hot water demand. In the temperature-to-flow-ratio cascade in Figure 11.7, the primary response is provided by the flow-to-flow controller, which is far faster than the temperature control-

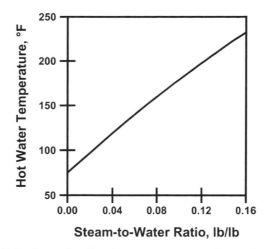

Figure 11.8. Operating line for temperature-to-flow-ratio cascade.

ler. For changes in hot water demand, the performance of the temperature-to-flow-ratio cascade is far superior.

Changes in Steam Supply Pressure. In the temperature-to-flow-ratio cascade, the flow-to-flow controller provides the primary response to changes in steam supply pressure. The change in steam supply pressure affects the steam flow (the controlled flow), which affects the ratio. The flow-to-flow controller responds to such changes very quickly. For steam supply pressure changes, the performance of the temperature-to-flow-ratio cascade is about the same as the performance of the temperature-to-flow cascade. Both are far superior to the performance of simple PID control.

Valve Issues. The flow-to-flow controller also isolates the hot water temperature controller from any nonidealities in the control valve. The benefits are essentially the same as for the temperature-to-flow cascade.

Controller Tuning. With both the simple PID control and the temperature-to-flow cascade, the temperature controller had to be tuned at a low hot water demand. This is not the case for the temperature-to-flow-ratio cascade. The operating line in Figure 11.8 is a plot of hot water temperature as a function of the steam-to-water ratio. The graph exhibits essentially a constant slope that does not depend on the hot water demand. Therefore, we can tune the hot water temperature controller at any hot water demand, and it will provide essentially the same performance at other hot water demands.

Benefits. The temperature-to-flow-ratio cascade in Figure 11.7 addresses the major deficiencies experienced with the simple PID loop but at a price—two

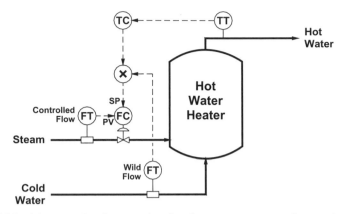

Figure 11.9. Alternate implementation for the temperature-to-flow-ratio cascade.

additional flow measurements are required. Rarely are they are already installed; usually their purchase, installation, and maintenance costs have to be justified.

Figure 11.7 is the flow-to-flow controller implementation of the temperature-to-flow-ratio cascade. An alternative is the configuration in Figure 11.9, which implements the temperature-to-flow-ratio cascade using a steam flow controller whose set point is computed from the cold water flow measurement and the target for the steam-to-flow ratio.

11.6. STEAM HEATER WITH CONTROL VALVE ON STEAM

The exchanger illustrated in Figure 11.10 heats the liquid flowing through the tubes by condensing steam on the shell. The liquid is a hydrocarbon fluid that enters at 150 °F. The typical flow rate is 1000 lb/min. The steam supply pressure is 75 psig (saturation temperature is slightly less than 320 °F). The control objective is to maintain the liquid outlet temperature at target.

No control valves are shown in Figure 11.10. To control the liquid outlet temperature, a final control element must be installed that influences the heat transfer rate in the exchanger. We shall examine three possibilities:

- Control valve on steam supply (1).
- Control valve on condensate.
- Liquid bypass with one or two control valves.

In the configuration illustrated in Figure 11.11, the final control element is a control valve on the steam supply. The condensate is discharged through a trap into the condensate return system. We shall examine the other possibilities shortly.

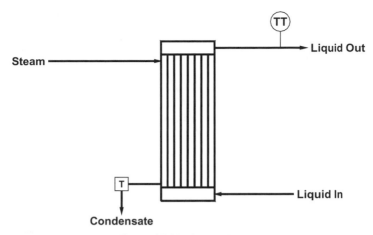

Figure 11.10. Steam heater.

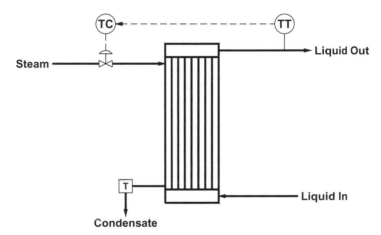

Figure 11.11. Control valve on steam supply.

Shell Pressure. What problems can originate when the control valve is installed on the steam supply to the steam heater? The answer to this question can be understood from a graph of the shell steam pressure as a function of the steam valve position. Figure 11.12 provides graphs for both a linear valve and an equal-percentage valve. Two features are of interest:

1. The pressure of the condensing steam in the shell of the exchanger is less than atmospheric pressure when
   - The linear control valve is less than 6% open.
   - The equal-percentage control valve is less than 29% open.

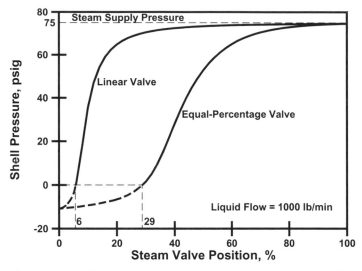

Figure 11.12. Shell pressure as a function of steam valve position.

When the pressure is below atmospheric, the condensate does not flow out of the shell of the exchanger. Ultimately, this imposes a minimum on the heat transfer rate that can be sustained by the exchanger.

2. Before the control valve is fully open, the pressure of the condensing steam approaches the steam supply pressure of 75 psig. Ultimately, this imposes the maximum on the heat transfer rate than can be sustained by the exchanger.

The tendency is to assume that the minimum and maximum heat transfer rates are imposed by the control valve, specifically, valve fully closed and valve fully open. For the steam heater, this is not the case.

Minimum Heat Transfer Rate. Let us assume that the condensate from the trap flows into a drain. When the steam valve position is less than 29% (equal-percentage) or 6% (linear) open, the shell pressure is below atmospheric. Under these conditions, the condensate will not flow out of the shell.

The minimum allowable shell pressure imposes a minimum heat transfer rate on the exchanger. Atmospheric pressure in the shell corresponds to a shell temperature of 212 °F. Since the liquid enters at 150 °F, heat will be transferred to the liquid. This heat transfer rate is the minimum that can be continuously sustained by the exchanger.

If the required heat transfer rate drops below the minimum that can be sustained, a cycling condition originates. For an equal-percentage valve, the behavior is as follows:

1. The controller positions the valve to less than 29% open.
2. The shell begins to fill with condensate, which reduces the effective heat transfer area and the heat transfer rate.
3. The liquid outlet temperature drops below target, causing the controller to increase its output to the control valve.
4. When the shell pressure exceeds atmospheric pressure, the condensate is forced from the shell, exposing the entire heat transfer area.
5. The heat transfer rate increases, which drives the liquid outlet temperature above its target.
6. The controller decreases its output to the control valve, eventually giving a valve position less than 29% open. This causes the cycle to repeat.

Maximum Heat Transfer Rate. The maximum possible steam pressure in the shell is the steam supply pressure, which is 75 psig for this exchanger. Theoretically, increasing the steam valve position always increases the steam pressure in the shell. But under some situations, the increase in shell steam pressure will be very small. As already noted on several occasions, control valves are commonly oversized in the process industries.

The graph of shell pressure as a function of steam valve position in Figure 11.12 illustrates the consequences. Long before the control valve is fully open, the steam pressure in the shell is essentially equal to the steam supply pressure. This is the case for valve openings in excess of 60% (equal-percentage) or 15% (linear). Beyond these valve openings, the effect of the control valve on the heat transfer is so small that the liquid outlet temperature cannot be effectively controlled using the control valve on the steam supply.

The net result is that the control valve is not usable over its full range of 0% to 100%. Instead, its usable range is from about 30% to about 60% (equal-percentage) or 6% to 15% (linear).

Upper Limit on Heat Transfer Rate. For exchangers and other heat transfer processes, the upper limit for the heat transfer rate can be imposed by two factors:

Media limited. The process is capable of condensing more steam than can flow through the control valve. In this case, the control valve would be effective up to 100%.

Heat transfer limited. The maximum steam flow is imposed by heat transfer, that is, by U A ΔT.

When the control valve is oversized (as it commonly is), the maximum will be imposed by the heat transfer limit. Control is generally effective for heat transfer rates up to about 90% of the heat transfer limit. Above that value, changes in the control valve opening have little effect on the heat transfer rate.

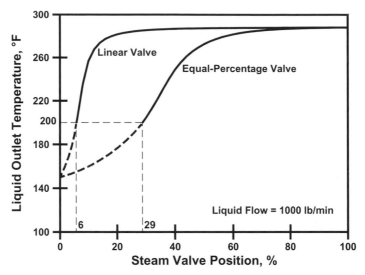

Figure 11.13. Operating lines for steam heater.

Operating Lines for Steam Heater. With the control valve on the steam supply as in Figure 11.11, the operating line is a plot of the liquid outlet temperature as a function of the steam valve position. Figure 11.13 presents the process operating lines for a linear valve and for an equal-percentage valve.

Let us begin by examining the limits on process operations. The minimum and maximum on the steam flows translate into corresponding limits on the liquid outlet temperature:

Minimum. The minimum controllable value for the liquid outlet temperature is about 200 °F.

Maximum. The absolute maximum on the liquid outlet temperature (corresponding to a shell pressure equal to the steam supply pressure) is just under 290 °F. However, the practical controllable limit is less. About 90% of the heat transfer capability is attained when the liquid outlet temperature is about 280 °F.

The valve position at which these limits are imposed differs for the two types of valves. However, the limits in terms of the liquid outlet temperature (and the steam flow through the valve) are exactly the same.

Equal-Percentage Valve. For the equal-percentage valve, the minimum and maximum limits correspond to valve positions of about 30% and 60% respectively. But for valve openings above 50%, the effect of the steam valve on the liquid outlet temperature decreases rapidly. The slope of the process operating line is the process gain or process sensitivity. A decrease in the process sensitivity has the same effect on loop performance as a decrease in the controller gain; specifically, the loop responds more slowly. Between valve positions of 30% and 50%, the controller can be effectively tuned to give consistent performance. But for valve positions between 50% and 60% (liquid outlet temperatures between 270 °F and 280 °F), the loop will respond more slowly. If the controller gain is increased to compensate for the reduced process gain, the higher gain will give problems at lower liquid outlet temperatures.

This is one case where some will propose characterization functions, scheduled tuning, or other approaches commonly referred to as adaptive control (but in reality, these are nonlinear controls). For example, the controller can be tuned to conditions experienced when the valve position is between 30% and 50%. But when the valve position is between 50% and 60%, a higher controller gain can be used. At a constant throughput, this would likely prove successful. Shortly we will consider variations in throughput, which greatly complicate this logic.

Linear Valve. For the linear valve, the operating line in Figure 11.13 suggests that this is not a good choice for the valve characteristics. The minimum corresponds to a valve position of about 6%; the maximum corresponds to a valve position of about 14%. This range is simply too narrow for the control to be effective. One contributing factor is that the control valve is oversized by approximately a factor of four.

Effect of Throughput. Some processes operate at essentially constant throughput; others do not. Refining processes are probably the best examples of processes with relatively small changes in throughput. Power generation processes are examples of processes that are designed with the expectation of significant changes in throughput. Most utility processes are designed to provide a turndown ratio of 4:1 or more.

With the control valve on the steam supply, Figure 11.14 presents the process operating lines (equal-percentage valve only) for liquid flows of 1000 lb/min, 2000 lb/min, and 4000 lb/min. The heat transfer generally increases as the flow increases, but not in a linear fashion. The general shape of the operating lines is basically the same for all.

Tuning Issues. For operating lines with curvatures such as the ones for the steam heater, the temperature controller is generally tuned to give acceptable performance in the region where the slope of the operating line is the steepest. This is where the process will have its highest sensitivity, which in turn requires the smallest value for the controller gain. The steepest slope decreases in magnitude as the throughput increases. Consequently, the higher the throughput, the lower the process sensitivity, and the higher the controller gain required to achieve consistent loop performance.

This is another potential application for scheduled tuning. A measurement is required for the liquid flow through the exchanger. For low liquid flows, the

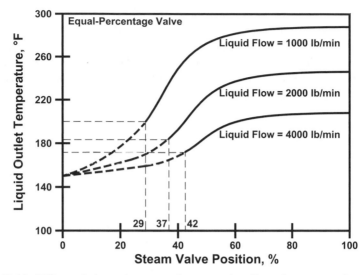

Figure 11.14. Effect of throughput on the operating lines for an equal-percentage valve.

process gain is high, so a low controller gain is appropriate. As the liquid flow increases, the controller gain should be increased (approximately proportional to the increase in liquid flow). These applications of scheduled tuning are usually successful. But attempting to also compensate for the curvature in the operating line greatly increases the complexity of the logic.

Batch Processes. Batch processes can also exhibit extreme variations in throughput. Consider a batch reactor with a jacket for removing heat. The dynamics of a production-scale reactor are determined by the contents of the reactor, that is, the reacting media. As compared with the reacting media, the dynamics associated with the jacket are far shorter, which means that the jacket is essentially at an equilibrium state that reflects the conditions within the reactor. As the conditions within the reactor change, the jacket basically tracks conditions within the reactor.

For batch reactors, the turndown ratio pertains to the heat transfer rate between the reactor and the jacket. For many batch applications, this heat transfer rate varies substantially during the batch. Typically the heat transfer rate is highest in the early stages of the batch. During the later stages, the heat transfer rate usually drops off considerably. Turndown ratios of 50:1 are experienced in practice, which has major repercussions on all aspects of the process (jacket design, valve sizing, cooling/heating media flow measurement, controller tuning, etc.).

Temperature-to-Flow Cascade. The control configuration illustrated in Figure 11.15 is the temperature-to-flow cascade configuration for the exchanger. This configuration consists of two control loops:

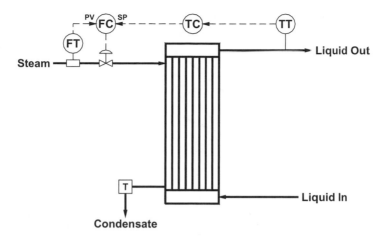

Figure 11.15. Temperature-to-flow cascade.

Steam flow controller. The controlled variable is the steam flow; the manipulated variable is the position of the steam control valve. This loop is the inner loop of the cascade.

Liquid outlet temperature controller. The controlled variable is the liquid outlet temperature; the manipulated variable is the set point of the steam flow controller. This loop is the outer loop of the cascade.

Cascade configurations are usually justified on the improved dynamic response to disturbances to the inner loop. Suppose there are significant variations in the steam supply pressure. For the simple feedback configuration in Figure 11.11, the temperature controller responds only after the change in steam supply pressure affects the liquid outlet temperature. But for the cascade configuration in Figure 11.15, the flow controller responds when the change in steam supply pressure affects the steam flow. The cascade configuration responds to this disturbance far sooner than the simple feedback configuration.

Operating Lines for Temperature Loop. Although cascade is commonly justified based on dynamic considerations, the steady-state characteristics sometimes (but not always) offer additional advantages for the cascade configuration. For three liquid flow rates, Figure 11.16 presents the operating lines for the temperature loop in the cascade control configuration. These operating lines are very different from the operating lines for simple feedback control in Figure 11.13.

The process operating lines for the liquid outlet temperature loop in the cascade configuration are essentially linear. Furthermore, they are not affected by the valve characteristics; the flow controller has to contend with these (and other issues associated with the valve).

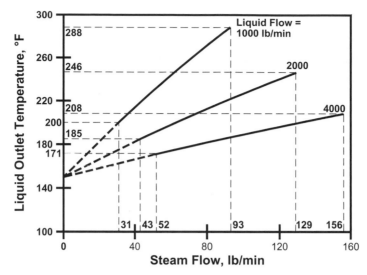

Figure 11.16. Operating lines for temperature loop of temperature-to-flow cascade.

Limits on the Heat Transfer Rate. The operating lines presented in Figure 11.16 also reflect the limits on process operations. Consider the operating line for a liquid flow of 1000 lb/min:

Minimum. Below a steam flow of 31 lb/min, the operating line is indicated by a dashed line. The steam flow of 31 lb/min corresponds to a valve position of 29% and to a liquid outlet temperature of 200 °F. Should a set point less than 31 lb/min be specified, condensate will build up within the shell of the exchanger, leading to the cycling described earlier. This cycle will be most evident in the steam flow, and it will have a higher frequency than the cycle that occurs for simple feedback control.

Maximum. The operating line terminates at a steam flow of 93 lb/min. The steam flow of 93 lb/min corresponds to a fully open steam valve; higher steam flows may occur temporarily, but they will not be sustained. This creates the potential for windup to commence when the temperature controller specifies a flow set point in excess of this value. Digital control systems provide mechanisms (external reset, integral tracking, and/or inhibit increase/decrease) to prevent this windup, and it is essential that they be configured properly.

11.7. STEAM HEATER WITH CONTROL VALVE ON CONDENSATE

In the configuration illustrated in Figure 11.17, the control valve is in the condensate return line instead of in the steam supply line (2). This completely changes the mechanism for varying the heat transfer rate:

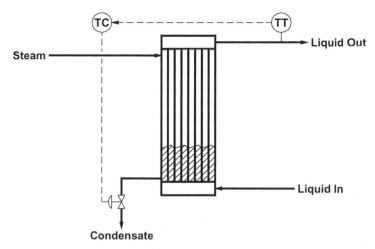

Figure 11.17. Control valve on condensate.

Valve on steam supply. Changing the steam valve position affects the shell pressure, which affects the shell temperature and the ΔT for heat transfer.

Valve on condensate. Changing the steam valve position affects the level of condensate within the exchanger, which affects the effective heat transfer area and the A term for heat transfer. The effective area for heat transfer is the heat transfer surface exposed to condensing vapors. The submerged heat transfer area mainly cools the condensate with little contribution to the total heat transfer rate.

With regard to condensate return, the pressure within the shell of the exchanger is always the steam supply pressure.

Dynamics. The exchanger in Figure 11.17 responds more slowly to control actions than when the control valve is on the steam supply. Changes in the condensate valve position must first affect the level within the exchanger. This results in a slower response, but usually it is sufficiently rapid to provide acceptable performance.

One casualty of the slow flow dynamics is the temperature-to-flow cascade. With the flow measurement on the steam supply, the temperature-to-flow cascade configuration is unlikely to have the 5:1 dynamic separation required for the cascade configuration to perform properly.

Valve Characteristics. Although equal-percentage valves are generally installed in the control valve on the condensate, an argument can be made for linear valve characteristics. The driving force for fluid flow is provided by the steam supply pressure. The hydrostatic head of the condensate within the exchanger is negligible. The pressure drop in the condensate return system is also usually small. Consequently, the pressure drop across the control valve in the condensate line is essentially constant. Most guidelines suggest linear valve characteristics for installations where the pressure drop across the control valve is constant.

Heat Transfer Rates. Let us examine the minimum and maximum heat transfer rates:

Minimum. Theoretically, the minimum heat transfer rate is zero. With the condensate valve closed, the exchanger completely fills with condensate, resulting in no heat transfer.

Maximum. The maximum heat transfer occurs when the exchanger is completely drained of condensate. The value for the maximum steam flow with the control valve on the condensate is exactly the same as when the control valve is on the steam supply.

However, the consequences of attempts to exceed the maximum are very different. With the control valve on the condensate, the exchanger has the potential to "blow steam" into the condensate return system. As the condensate valve opens, the condensate level within the exchanger drops. If the condensate valve is opened too much, the level drops entirely out of the exchanger and steam flows into the condensate return system.

Steam Trap. There is a very simple approach that will prevent blowing steam into the condensate return. In the configuration in Figure 11.18, a steam trap is inserted into the condensate line upstream of the control valve. As long as

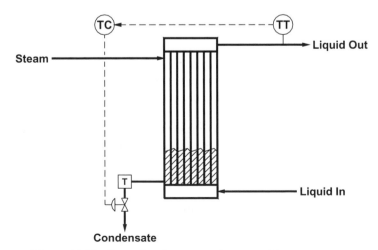

Figure 11.18. Control valve on condensate with a steam trap.

there is some condensate within the exchanger, the trap has no effect on the condensate flow. But should the condensate drain completely from the exchanger, the trap prevents steam from flowing into the condensate return system.

Unfortunately, the configuration in Figure 11.18 exposes the liquid outlet temperature controller to windup. Digital control systems provide a variety of windup prevention mechanisms. Activation of any of these windup protection mechanisms must be based on an event. In the default configuration, the event is the control valve being driven to a limit (as specified by the lower output limit or the upper output limit). But in the configuration in Figure 11.18, the event is the trap being activated to block steam from flowing into the condensate return. How does the control system know that this event has occurred? There is no indication to the control system, so the windup protection mechanisms cannot be activated at the appropriate times. One can try to specify a lower output limit that corresponds to the stream trap starting to block steam flow, but this is imperfect at best.

Temperature Control with Level Override. Another way to prevent the exchanger from "blowing steam" is to impose a minimum on the condensate level within the exchanger. The override configuration illustrated in Figure 11.19 incorporates this function into the simple feedback control configuration of Figure 11.17.

Since override configurations are beyond the scope of this book, only a general explanation will be provided. The key components are as follows:

- A measurement of the condensate level within the exchanger.
- A PID controller to control the level of the condensate within the exchanger. The set point for the condensate level corresponds to the

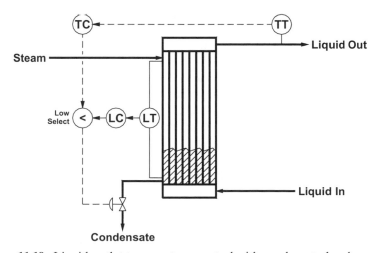

Figure 11.19. Liquid outlet temperature control with condensate level override.

physical elevation of the bottom of the exchanger (or, in practice, slightly above this elevation).
- The output of the condensate level controller and the output of the liquid outlet temperature controller are "auctioneered" by the low select, with the output being the position signal to the condensate valve.

The control logic switches between the following two:

- If the condensate level is above its set point, the output of the temperature controller determines the position of the control valve on the condensate. The liquid outlet temperature is maintained at or near its set point.
- When the condensate level drops to its set point, the output of the level controller determines the position of the control valve on the condensate. The liquid outlet temperature will be above its set point.

For the switch between these two to be smooth, certain windup issues must be properly addressed. Although beyond the scope of this book, this and other aspects of override controls must be understood before implementing an override control configuration.

There is another advantage of installing a level measurement on the condensate level within the exchanger. For most exchangers, the heat transfer equations impose the upper limit on the capacity. What percent of the available heat transfer capacity is being used? The answer is the exposed heat transfer area expressed as a percentage of the total heat transfer area.

This can be determined from the condensate level measurement. For vertical exchangers, the heat transfer area per unit of height is constant. If the exchanger is 25% filled with condensate, then 75% of the heat transfer capacity is used. The percent utilization of the heat transfer capacity is linearly related to the condensate pot level measurement. For horizontal exchangers, the relationship between exposed heat transfer area and condensate pot level is more complex, but a characterization function can be computed from the design parameters for the exchanger.

Shared Condensate Pot. The design for the condensate pot illustrated in Figure 11.20 permits more than one exchanger to discharge condensate into the same condensate pot. Dedicating a condensate pot to a single exchanger is generally not economical. However, we shall only illustrate a single exchanger, as the issues are the same for all exchangers that discharge to the condensate pot. There are other designs for condensate pot arrangements, but the control issues are basically the same.

This design in Figure 11.20 consists of two control loops:

- The liquid outlet temperature is controlled by manipulating a control valve on the condensate flowing from the exchanger to the condensate

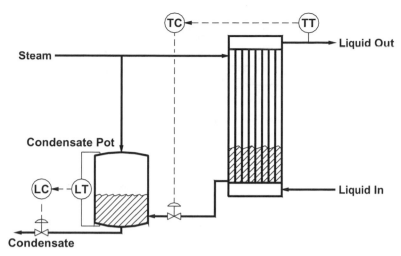

Figure 11.20. Shared condensate pot.

pot. A comparable control loop must be provided for each exchanger that discharges into the condensate pot.

• The condensate pot level is controlled by manipulating a control valve in the condensate return line. There is only one condensate pot level controller, regardless of the number of exchangers that discharge into the condensate pot.

The pressure of the condensing steam in the exchanger and the pressure within the condensate pot are both the steam supply pressure. The only driving force for condensate to drain from the exchanger to the condensate pot is gravity. The hydrostatic head for fluid flow is the difference between the level in the exchanger and the level in the condensate pot. Either the exchanger must be elevated or the condensate pot must be in a pit below grade level. The process designers must be very careful with regard to physical elevations.

With the condensate pot, it is not possible for the exchanger to "blow steam" into the condensate return system. Should the temperature controller attempt to exceed the maximum heat transfer rate, all of the condensate will drain from the exchanger and the maximum heat transfer rate is attained.

However, the liquid outlet temperature controller is susceptible to windup. Windup begins the instant all condensate is drained from the exchanger. Further opening the control valve between the exchanger and the condensate pot has no effect on the heat transfer rate, so the natural response of the temperature controller is to continue increasing the control valve opening. The normal windup protection mechanisms are not effective. Windup protection should be invoked the instant all condensate is drained from the exchanger, but the control configuration in Figure 11.20 provides no indication of such an event.

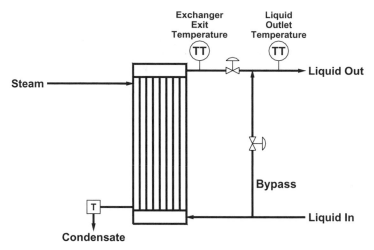

Figure 11.21. Liquid bypass.

If only the measurements indicated in Figure 11.20 are available, the capacity utilization for the exchanger cannot be determined. If this is required for an optimization strategy, a measurement of the condensate level in the exchanger must be added.

11.8. LIQUID BYPASS ARRANGEMENTS

Another approach to control the liquid outlet temperature is to bypass some of the liquid around the exchanger, as illustrated in Figure 11.21. One must distinguish the following two temperatures:

Liquid outlet temperature. This is the temperature sensed by the temperature transmitter located downstream of the point where the bypass flow and the flow through the exchanger are combined.

Exchanger exit temperature. This is the temperature of the liquid exiting the exchanger. The liquid flowing through the exchanger is heated to a higher temperature than the liquid outlet temperature. If the flow is sufficiently reduced, the temperature of the liquid leaving the exchanger approaches the temperature of the condensing steam within the exchanger.

In Figure 11.21, both temperature measurements are indicated, but usually only the liquid outlet temperature is measured.

Two normal control valves are indicated in Figure 11.21. However, there are two other options:

- A control valve in the bypass only. This is usually an economy measure. However, this configuration is unable to bypass all liquid around the exchanger, and consequently, it cannot reduce the heat transferred in the exchanger to zero.
- A three-way valve. As a three-way valve costs about the same as two normal control valves, they are not frequently installed. Furthermore, configurations such as split range (to be described shortly) cannot be implemented with a three-way valve.

There are no control valves on either the steam supply or the condensate. Only a trap is required in the condensate, so the full steam supply pressure is available for condensate return.

Bypass configurations respond very rapidly. Changes in the valve position quickly translate to changes in the liquid flows (bypass flow vs. flow through the exchanger), which in turn quickly affect the liquid outlet temperature.

The capacity utilization decreases with the bypass flow. That is, increasing the bypass flow decreases the outlet temperature and the heat transferred to the liquid. If the bypass flow and the flow through the exchanger were known, the percent utilization of the heat transfer capability could be calculated. However, these two flows are rarely measured.

Flow Simulation. To analyze the behavior of liquid bypass configurations thoroughly and accurately, a flow simulation is required. Performing a flow simulation is further complicated because any of the following are possible:

- The pressure drop across the exchanger-plus-bypass is constant.
- The total liquid flow (bypass flow plus flow through exchanger) is constant.
- The exchanger-plus-bypass is part of an even larger flow system. Ideally, the flow simulation should encompass the complete flow system. However, the results of the complete flow simulation will be somewhere between the extremes of the previous two cases.

The examples presented subsequently are for a constant pressure drop across the exchanger-plus-bypass.

Control Valve in Bypass Only. The configuration in Figure 11.22 provides only a control valve in the bypass. The performance of this arrangement depends on the sizing of the bypass valve, specifically, on what fraction of the total liquid flow bypasses the exchanger when the control valve is fully open. For the examples presented shortly, the bypass valve is sized such that 80% of the liquid bypasses the exchanger when the control valve is fully open.

The controller action for the bypass configuration in Figure 11.22 is opposite that of the previous configurations. If the liquid outlet temperature is

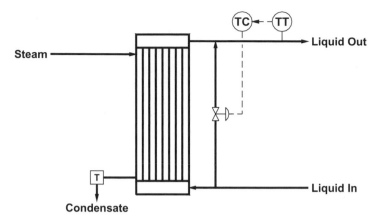

Figure 11.22. Control valve in bypass only.

increasing, the controller should increase its output to the bypass valve so as to bypass more liquid around the exchanger. This requires direct action in the liquid outlet temperature controller. In the previous configurations, the controller should be reverse acting. This change in directionality is also reflected in the slope of the process operating lines.

The maximum heat transfer rate occurs with the bypass valve completely closed. All liquid flows through the exchanger, which gives the maximum ΔT for heat transfer and consequently the maximum heat transfer rate. The maximum liquid outlet temperature is the same as for the previous configurations.

The operating lines in Figure 11.23 apply to the case where the pressure drop across the exchanger is constant. The observations are as follows:

Equal-percentage valve. The operating line exhibits only a modest departure from linearity, and certainly not enough to cause controller tuning difficulties.

Linear valve. The operating line exhibits significant nonlinearities of the decreasing sensitivity type. The sensitivity is highest at low valve openings, and it is lowest at large valve openings.

The equal-percentage valve is clearly preferred.

Two Control Valves. With two control valves as illustrated in Figure 11.21, the flow through the exchanger can be blocked so that the entire liquid stream bypasses the exchanger. This gives a heat transfer rate of zero (the minimum heat transfer rate). When the bypass valve is blocked, the entire liquid stream flows through the exchanger, which gives the maximum heat transfer rate.

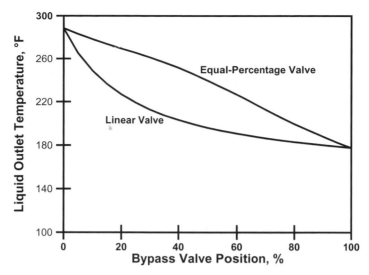

Figure 11.23. Operating lines for single valve in bypass (constant pressure drop from liquid in to liquid out).

For all examples presented herein, the two valves are sized as follows:

Valve in series with exchanger. With the bypass blocked and the valve in series with the exchanger fully open, the control valve is taking 20% of the pressure drop available for fluid flow.

Bypass valve. The valve is sized so that the flow is the same for the following situations:
- Bypass blocked and valve in series with exchanger fully open.
- Bypass fully open and the flow through the exchanger blocked.

Although there are two control valves, the only controller is the liquid outlet temperature controller. There is only one manipulated variable. In all previous examples, the number of manipulated variables was the same as the number of control valves. Configurations such as the following drive two (or possibly more) control valves using a single controller output:

- "See-saw" arrangements.
- Split range configurations.

We shall examine both.

"See-Saw" Arrangement. In the configuration presented in Figure 11.24, the output from the temperature controller positions both valves. One of the control valves must be "fail closed" and the other must be "fail open." Which is "fail-open" depends on the hazards analysis—for the configuration

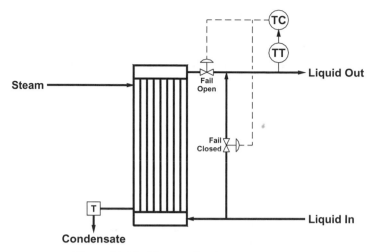

Figure 11.24. "See-saw" valve arrangement.

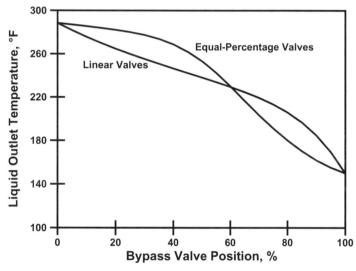

Figure 11.25. Operating lines for see-saw valve arrangement (constant pressure drop from liquid in to liquid out).

in Figure 11.24, the failure mode is for all liquid to flow through the exchanger. A controller output of 0% passes all flow through the exchanger. Reversing the "fail-open" and "fail-closed" would bypass all liquid on failure, with a controller output of 0% bypassing all of the liquid.

The process operating line for the temperature controller is a graph of the liquid outlet temperature as a function of the controller output. For a constant pressure drop across the exchanger-plus-bypass, Figure 11.25 presents the operating lines for linear valves and equal-percentage valves. Both exhibit significant changes in the slope and thus in the process gain.

Using equal-percentage valves in the see-saw arrangement presents considerable resistance to fluid flow. For an equal-percentage valve, the %CV at a valve opening of 50% is often around 20%. In the see-saw arrangement, if one valve is half open, the other is also half open. Under these conditions, these valves are providing considerable resistance to flow. The consequences are as follows:

- With a constant pressure drop across the exchanger-plus-bypass, the flow is greatly restricted when both valves are 50% open.
- To maintain a constant total liquid flow, a very large pressure drop is required when both valves are 50% open.

Split-Range Configuration. The configuration in Figure 11.26 is known as a split-range configuration. With conventional controls, split-range configurations were implemented through the zero and span adjustments at the control valves. But with digital systems, split-range configurations are normally implemented using characterization functions within the controls.

The split-range configuration in Figure 11.26 does not restrict the flow (bypass flow plus flow through exchanger), even if both valves are equal-percentage. The split-range logic is as follows:

Controller Output	Exchanger Valve Opening	Bypass Valve Opening
0%	100%	0%
50%	100%	100%
100%	0%	100%

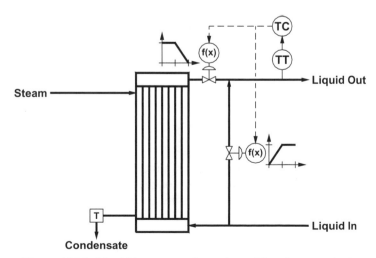

Figure 11.26. Liquid bypass configuration with split-range logic.

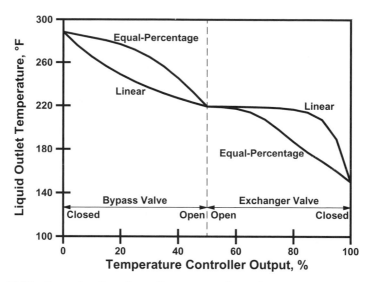

Figure 11.27. Operating lines for split-range configuration, constant pressure drop.

One of the valves is fully open at all times. When the controller output is at mid-range (50%), both valves are fully open. As the controller output increases above mid-range, the bypass valve remains fully open and the exchanger valve closes. As the controller output decreases below mid-range, the exchanger valve remains fully open and the bypass valve closes.

Figure 11.27 presents the operating lines for a constant pressure drop across the exchanger. The following three points on the process operating line do not depend on the valve characteristics:

- Temperature controller output of 0%. The bypass valve is fully closed; the exchanger valve is fully open. This gives the maximum liquid outlet temperature.
- Temperature controller output of 50%. Both valves are fully open.
- Temperature controller output of 100%. The bypass valve is fully open; the exchanger valve is fully closed. No liquid flows through the exchanger, so the liquid outlet temperature is the same as the liquid inlet temperature.

When the temperature controller output is less than 50%, the operating line depends only on the characteristics of the bypass valve (the exchanger valve is fully open). When the temperature controller output is greater than 50%, the operating line depends only on the characteristics of the exchanger valve (the bypass valve is fully open).

Both of the operating lines in Figure 11.27 exhibit a flat region just above mid-range. Flat regions on the process operating line are equivalent to a dead zone; changes in the controller output have little or no effect on the controlled variable (the liquid outlet temperature). The region is much wider for the linear valve than for the equal-percentage valve; this makes the linear valve characteristics undesirable in the exchanger valve.

For the equal-percentage valve, one solution might be to start closing the bypass valve at a controller output of 40% instead of at mid-range. When the split-range logic is implemented using a characterization function within the controller, such changes are very easy to implement.

11.9. SUMMARY

A major advantage of digital control technology is that changes are much easier. The control configuration from the P&I diagram is implemented via software ("softwiring" or its equivalent) as opposed to hardware.

But just try to get a plant to implement a change in the P&I diagram. In most cases, this advantage is not pursued. Making wiring changes, cutting holes in panels, and so on, made changes to the older hardware essentially impractical. Either it worked from the start, or it remained on manual for the life of the plant (or, at best, the operators learned to live with its deficiencies).

With digital systems, the result is too often the same. It does not have to be this way. But if you propose a change in the P&I diagram, be prepared for an endurance test. Reviews are appropriate, and they are mandatory where management of change procedures are followed. But the reviews often go on and on and on. Every conceivable objection is raised. In some cases, no decision is ever made. In others, a favorable decision is finally made, but then nothing is ever done. It remains on the "to be done" list until the sponsors give up. Some call this a "pocket veto."

LITERATURE CITED

1. Smith, C. L., "Achieve Effective Heat Exchanger Control", *Chemical Processing*, Vol. 71, No. 3 (March 2008), p. 33.
2. Smith, C. L., "Succeed with Condensate Control", *Chemical Processing*, Vol. 71, No. 7 (July 2008), p. 24.

CHAPTER 12

Loop Interaction

Most processes are multivariable in nature. The customary approach is to provide a single-loop PID controller for each variable to be controlled. Usually this works, but occasionally one or more controllers cannot be satisfactorily tuned. The objective of this chapter is to provide enough understanding of multivariable control that you can recognize when interaction between loops is causing tuning difficulties. The treatment herein of the solutions for interaction problems is rather superficial, as this is considered beyond the scope of this book.

12.1. MULTIVARIABLE PROCESSES

Figure 12.1 presents a generic representation of a multivariable process. The inputs are divided into two categories:

Manipulated variables: The values of these variables are at the discretion of the control system.

Disturbances: The values of these variables are determined by external factors.

The outputs are also divided into two categories:

Controlled variables: The values of these variables are to be maintained at or near their targets.

Dependent variables: The values of these variables are influenced by the manipulated variables and by the disturbance variables. They are not to be controlled to specified targets, but in some applications, constraints may apply to the dependent variables.

The dimensionality of a multivariable process is usually designated as n × m, where n is the number of manipulated variable inputs and m is the number of

---

Practical Process Control: Tuning and Troubleshooting, by Cecil L. Smith
Copyright © 2009 John Wiley & Sons, Inc.

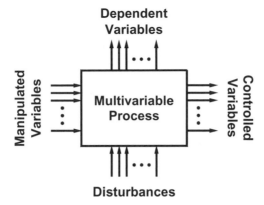

Figure 12.1. Multivariable process.

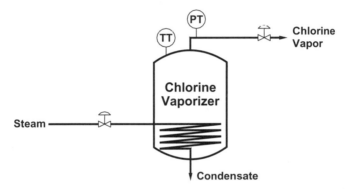

Figure 12.2. Controlling temperature and pressure for a chlorine vaporizer.

controlled variable outputs. For example, a 6×4 multivariable process has 6 manipulated variable inputs and 4 controlled variable outputs. A process with a single manipulated variable input and a single controlled variable output is a 1×1 process, and it is often referred to as a "single input, single output" process.

Controlled Variables. In a multivariable process, the controlled variables must meet one criterion: Each controlled variable must not be dependent on the other controlled variables.

Figure 12.2 illustrates a chlorine vaporizer with an internal steam coil. The chlorine leaves at the saturation temperature for the current pressure within the vaporizer (that is, the chlorine is not superheated). Measurements are provided for the temperature and pressure of the chlorine vapor. Valves are provided on the chlorine vapor stream and on the steam.

Both valves affect both chlorine temperature and pressure. However, it is not possible to specify targets independently for both chlorine temperature and chlorine pressure. These two variables are related by the vapor pressure relationships for chlorine. The temperature of the chlorine vapor is the boiling point of chlorine at the current pressure within the vaporizer. It is possible to control either chlorine temperature or chlorine pressure, but not both.

Square Configurations. A multivariable process is said to be square if the number of controlled variables equals the number of manipulated variables. Provided no constraints on the manipulated variables are encountered (such as a valve fully open or fully closed), it is possible for the control system to drive each controlled variable to its respective target.

Single-loop controllers can only be applied to square configurations. If there are four manipulated variables and four controlled variables, then four single-loop controllers are required. "Pairing" refers to how the controllers are arranged. A manipulated variable must be selected to control each controlled variable. The controlled variable is the measured variable for that controller; the controller output drives the selected manipulated variable. In some applications, the selection is obvious. In others, the choice is not clear at all. In yet others, subtle characteristics of the process make the "obvious" selection to be not as good as initially thought.

Skinny Configurations. A multivariable process is said to be "skinny" if the number of controlled variables exceeds the number of manipulated variables. In such configurations, it is not possible for the control system to drive all of the controlled variables to their targets. Skinny configurations are not commonly encountered in process applications.

It is only possible to control such a process in some "best" sense. This raises the question of what is meant by "best." A generic definition might be to control so as to minimize the sum of squares of the deviations from target. That is, "best" means

$$\min\left\{\sum E_k^2\right\}$$

where $E_k = SP - PV$ for controlled variable k.

This weighs all errors equally, which is probably not appropriate. A weighting factor for each error could be incorporated. Alternatively, an objective function specific to the application could be developed.

Fat Configurations. A multivariable process is said to be "fat" if the number of manipulated variables exceeds the number of controlled variables. Many combinations of the manipulated variables would drive all of the controlled variables to their targets. Process applications often prove to be fat.

An objective function is required to determine which of the many combinations of the manipulated variables is the "best" (minimum energy, maximum throughput, or the like). One approach to accomplish this is to "fix" (that is, set to a constant value) the necessary number of the manipulated variables to obtain a square configuration. The remaining manipulated variables are then adjusted so as to drive the selected controlled variables to their targets. Steady-state optimization routines then adjust the "fixed" manipulated variables so as to attain the "best" process performance. When approached in this manner, the regulatory control issues for the fat configuration are the same as for the square configuration.

Herein, we shall only address multivariable processes with a square configuration.

12.2. OFF-GAS SYSTEM

Figure 12.3 presents the schematic for the equipment to remove contaminants from a gas stream generated by the main process (1). This gas stream contains three materials that together present difficulties:

1. Small amount of particulate matter.
2. Sufficient water vapor that the dew point is above ambient temperature.
3. Sulfur dioxide (SO_2) or similar component that would cause any condensate to be acidic.

The gas stream is processed first through a bag house to remove the particulate matter. The gas is then sent to the scrubbers to remove the sulfur dioxide.

Any condensate would be acidic and, therefore, corrosive. This can be avoided if the gas stream remains above its dew point until arriving at the scrubbers. This will be achieved by adding air that has been heated by a furnace. To remain above the dew point through the bag house, some of the hot air is added upstream of the bag house. The remaining hot air is added

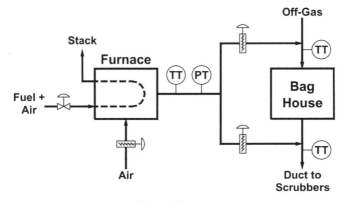

Figure 12.3. Off-gas process.

downstream of the bag house so that the gas stream remains above its dew point until it arrives at the scrubbers.

There are four variables to be controlled:

Bag house inlet temperature. This temperature must exceed the dew point of the off-gas stream plus the temperature drop due to heat losses in the bag house.

Scrubber duct inlet temperature. This temperature must exceed the dew point of the off-gas stream plus the temperature drop in the ducts to the scrubbers.

Hot gas pressure. The pressure ratings of the ducts must not be exceeded.

Hot gas temperature. If this temperature is too low, it will not be possible to maintain the required temperature of the off-gas stream entering the ducts to the scrubbers.

There are four final control elements:

1. Hot gas damper to bag house inlet.
2. Hot gas damper to bag house outlet (scrubber duct inlet).
3. Fresh air damper.
4. Furnace fuel control valve.

This is a 4×4 multivariable process.

Basis for Loop Pairing. Pairing refers to the selection of the final control element to be used as the manipulated variable for each controlled variable. The logic followed in selecting the pairing of manipulated variables with controlled variables is not easily expressed. The pairing has traditionally been developed by experienced control engineers, who undoubtedly draw on their past experiences as much as following any rigorous logical rules. But if one examines enough P&I diagrams, the common denominator can be summarized as follows:

Control each variable with the nearest final control element that significantly affects that variable.

From a dynamic perspective, this rule is probably appropriate. But as we shall see shortly, the steady-state effect of a final control element on the variable to be controlled can be much less that it appears.

For each controlled variable in the P&I diagram in Figure 12.4, the final control element is selected as follows:

Bag house inlet temperature. The bag house inlet temperature will be controlled using the hot gas damper to bag house inlet. Opening this

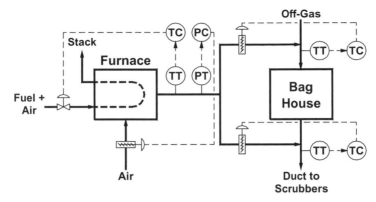

Figure 12.4. Proposed control configuration for off-gas process.

damper increases the flow of hot gas and, thus, increases the bag house inlet temperature.

Scrubber duct inlet temperature. Let us control the scrubber duct inlet temperature using the hot gas damper to bag house outlet. Opening this damper increases the flow of hot gas to the bag house outlet, which in turn increases the scrubber duct inlet temperature (as we shall see, this is not right).

Hot gas pressure at furnace exit. The remaining final control elements are the fresh air damper and the fuel control valve. The furnace has a heat exchanger so that the products of combustion do not flow into the hot gas stream (the water vapor resulting from combustion elevates the dew point). The fuel control valve has no effect on the hot gas pressure, so we must control the hot gas pressure using the fresh air damper.

Hot gas temperature at furnace exit. The only remaining controlled variable is the hot gas temperature; the only remaining final control element is the fuel control valve. The fuel flow definitely affects the hot gas temperature.

Unfortunately, there is a serious defect in the above logic, specifically, the cause-and-effect relationship between the scrubber duct inlet temperature and the hot gas damper to bag house outlet is far more complex that it initially appears. We will examine this in more detail shortly.

Sequence for Tuning the Controllers. To commission the control system, one must develop a strategy for tuning the loops one at a time. The following order is proposed:

1. Hot gas pressure controller. This controller can be tuned prior to lighting the furnace. As soon as the various blowers are started, the hot gas pressure controller can be tuned.

2. Hot gas temperature controller. Before tuning the temperature loops on the off-gas stream, hot gas must be available at the desired temperature.
3. Bag house inlet temperature controller. This temperature affects the bag house outlet temperature, which in turn affects the scrubber duct inlet temperature.
4. Scrubber duct inlet temperature controller.

This process is somewhat unusual in one regard. All four controllers can be tuned while blowing clean air through the process, that is, before the off-gas stream with the contaminants is being generated by the main process. For most processes, as many loops as possible are tuned before putting chemicals into the process.

This tuning sequence works very well up to the last step, namely, tuning the scrubber duct inlet temperature controller. Placing this controller on automatic induces upsets and cycling into the remaining loops.

Scrubber Duct Inlet Temperature Loop. To understand the problems with tuning the scrubber duct inlet temperature controller, the best approach is to remove all other controllers and examine how the process responds to control actions taken by this controller. Figure 12.5 illustrates the off-gas process with only the scrubber duct inlet temperature loop. In analyzing this loop, we want to understand what the process does, not what results when an effect is propagated from one loop to another.

Suppose the scrubber duct inlet temperature is too low. The scrubber duct inlet temperature controller should open the hot gas damper to bag house outlet. This increases the flow of hot gas to bag house outlet. However, where does the additional hot gas come from? It is some combination of the following:

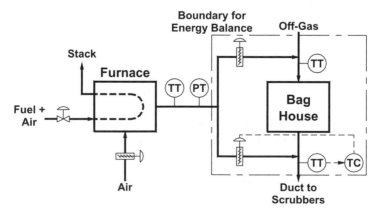

Figure 12.5. Off-gas process—scrubber duct inlet temperature loop only.

1. Hot gas is "stolen" from the bag house inlet. That is, opening the damper redirects some of the hot gas from the bag house inlet to the bag house outlet.
2. More fresh air enters the furnace.

Let us start by examining each individually.

Hot Gas Redirected from Bag House Inlet to Outlet. Let us assume that the fresh air flow is constant. If the scrubber duct inlet temperature is too low, the sequence of events is as follows:

1. The scrubber duct inlet temperature controller responds by opening the hot gas damper to bag house outlet.
2. This increases the hot gas flow to bag house outlet, but there is a corresponding decrease in the hot gas flow to bag house inlet.
3. Decreasing the hot gas flow to bag house inlet causes the bag house inlet temperature to drop, which causes the temperature of the gas leaving the bag house to also drop. Now it has to be heated more in order to attain the desired scrubber duct inlet temperature.
4. More gas is flowing through the hot gas damper to bag house outlet, but now a larger increase is required in the temperature of the bag house exit gas.

If the heat loss in the bag house is constant, redirecting hot gas has no effect at all. Consider an energy balance around the bag house. Let the boundary for the energy balance be the hot gas stream from the furnace as illustrated in Figure 12.5. The terms in the energy balance are as follows:

Energy in with the off gas: Redirecting air has no effect on this term.

Energy in with the hot gas from the furnace: We are currently assuming that redirecting air has no effect on the air flow through the fresh air damper. With this assumption, the flow and temperature of the hot gas stream are constant.

Heat loss from the bag house: For the moment, let us assume constant heat loss.

Energy out with the exit stream to the scrubber ducts: If the other three terms are constant, this term must be constant also. Therefore, redirecting the hot gas has no effect on the scrubber duct inlet temperature.

Of course, the heat loss from the bag house is not constant. Redirecting the hot gas affects the temperature in the bag house and, consequently, the heat loss in the bag house. Surely the bag house is insulated, which reduces the effect of the bag house temperature on heat loss. The net effect of redirecting hot gas on the scrubber duct inlet temperature is not zero, but it is too small for the scrubber duct inlet temperature controller to function properly.

Additional Fresh Air. If the scrubber duct inlet temperature is too low, the scrubber duct inlet temperature controller responds by opening the hot gas damper to bag house outlet. Let us assume that this has no effect on the flow through the hot gas damper to bag house inlet. With this assumption, opening the hot gas damper to bag house outlet has the following effect:

1. The flow through the hot gas damper to bag house outlet increases.
2. This increased gas flow is supplied entirely by additional flow through the fresh air damper.
3. Increasing the fresh air flow at constant fuel flow causes the hot gas temperature to decrease.
4. Decreasing the hot gas temperature decreases the bag house inlet temperature, which in turn decreases the temperature of the gas leaving the bag house.
5. More gas is flowing through the hot gas damper to bag house outlet, but it is cooler and has to cope with the lower temperature of the gas leaving the bag house.

The net consequence of this is to decrease the scrubber duct inlet temperature. This is opposite to the result expected from opening the hot gas damper to bag house outlet; that is, the controller is moving the damper in the wrong direction!

To understand this better, let us examine an energy balance encompassing the entire off-gas process. There are five terms in this energy balance:

Energy in with fresh air: Let us use the fresh air temperature as the reference temperature for the terms in the energy balance. By doing this, this term is zero regardless of the flow.

Energy in with the fuel: The fuel flow is constant, so this term in the energy balance is constant.

Energy in with the off gas: Increasing the fresh air flow has no effect on this term.

Heat loss from the bag house: The lower temperature in the bag house will reduce the heat loss slightly. This is a minor influence, so let us assume that the heat loss is constant.

Energy out with the exit stream to the scrubber ducts: If the other four terms are constant, this term must be constant also. But the flow is larger (the increase is the same as the increase in the fresh air flow). Higher flow and constant enthalpy means a lower enthalpy per unit mass, which means a lower scrubber duct inlet temperature.

The scrubber duct inlet temperature controller opens the hot gas damper to bag house outlet with the expectation of increasing the scrubber duct inlet temperature. But if all of the additional gas is supplied by fresh air, the result is the opposite.

Loop Interaction. For the scrubber duct inlet temperature controller, the manipulated variable (hot gas damper to bag house outlet) has little direct effect on its controlled variable (scrubber duct inlet temperature). But when all four loops are configured as in Figure 12.4, arguments are sometimes made that such configurations will work. Starting with a lower than desirable temperature at the inlet to the scrubber duct, the argument proceeds as follows:

1. To increase the scrubber duct inlet temperature, the scrubber duct inlet temperature controller opens the hot gas damper to bag house outlet.
2. To the extent that this affects the bag house inlet temperature, the bag house inlet temperature controller opens the hot gas damper to bag house inlet.
3. Both of these actions drop the hot gas pressure at furnace exit, so the hot gas pressure controller opens the fresh air damper.
4. Increasing the fresh air flow causes the hot gas temperature at furnace exit to drop, causing the hot gas temperature controller to open the fuel control valve to supply more heat.

The conclusion is that everything will be just fine. But starting with the scrubber duct inlet temperature controller, control actions are propagated from one loop to another. This works only when a slow loop propagates an effect to a fast loop. The desire is for the fast loop to be five times faster than the slow loop. None of the loops in Figure 12.4 meets this criterion.

Alternative Configuration. Figure 12.6 presents an alternative configuration that is based on the following logic:

Scrubber duct inlet temperature. If the bag house inlet temperature is at its target but the scrubber duct inlet temperature is below its target, then additional energy must be input at the furnace; that is, we must open the fuel control valve. Based on this reasoning, the fuel control valve is the logical choice for the manipulated variable for controlling the scrubber duct inlet temperature.

Bag house inlet temperature. The bag house inlet temperature will be controlled using the hot gas damper to bag house inlet.

Hot gas temperature at furnace exit. With the scrubber duct inlet temperature controller manipulating the fuel control valve, the only choice for controlling the hot gas temperature is to manipulate the fresh air damper.

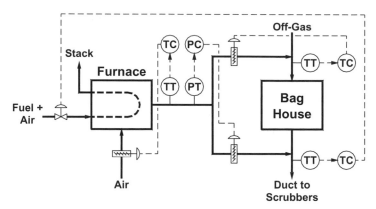

Figure 12.6. Alternate control configuration for off-gas process.

Hot gas pressure at furnace exit. The only remaining final control element is the hot gas damper to bag house outlet. This damper definitely affects the hot gas pressure.

Unfortunately, there is a concern with this configuration. The hot gas damper to bag house inlet (which is manipulated by the bag house inlet temperature controller) also affects the hot gas pressure. Therefore, any control action taken by the bag house inlet temperature controller would affect the hot gas pressure, which would lead to a response from the hot gas pressure controller. The potential exists for considerable interaction between these two loops.

The Relative Gain. There is an analytical tool that can assist in determining which control configuration is the most appropriate. This analytical tool, known as the relative gain (2), assesses the relative degree of influence of the various manipulated variables on each of the controlled variables. The relative gain is basically a quantitative measure of the steady-state interaction within the process. Based on this measure, the most appropriate single-loop control configuration (from a steady-state perspective) can be determined.

A detailed presentation of the relative gain is beyond the scope of this book. However, it will be defined later in this chapter and applied to a 2×2 process. For the off-gas process, the relative gain suggests the alternative configuration in Figure 12.6. However, the relative gain also indicates a considerable degree of interaction for that configuration.

12.3. FLOW AND PRESSURE CONTROL

The process illustrated in Figure 12.7 consists of two control valves separated by a modest length of pipe (actually, a small tubular reactor that is not shown). The flow of gas through the process is measured, and the pressure in the piping

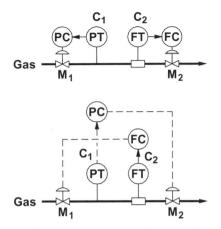

Figure 12.7. Two possible control configurations for flow-pressure process.

section between the two control valves is measured. Gas is to flow through the process at a specified flow rate. The pressure between the two control valves is to be maintained at a specified value.

This is a 2×2 multivariable process:

Controlled Variables	Manipulated Variables
Flow F	Upstream valve position M_1
Pressure P	Downstream valve position M_2

The process is definitely interacting. Opening the upstream valve increases both the flow and the pressure. Opening the downstream valve increases the flow but decreases the pressure.

For a 2×2 multivariable process, there are only two possible loop pairings. For the flow-pressure process, they are illustrated in Figure 12.7:

Configuration 1: Control pressure with upstream valve
Control flow with downstream valve
Configuration 2: Control pressure with downstream valve
Control flow with upstream valve

To most, configuration 1 initially seems to be the preferable approach. We are accustomed to seeing the flow measurement upstream of the flow control valve. However, this is due to flow measurement issues, not to control issues. Why would configuration 2 not work? There is nothing that would rule out this configuration as a potential candidate. The selection of the most appropriate configuration must be based on the degree of interaction.

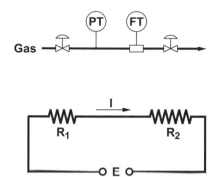

Figure 12.8. Electrical analogy to flow-pressure process.

Current Flow Analogy. An electrical analogy to the flow-pressure process can be constructed. Current flow is analogous to fluid flow. Voltage (driving force for current flow) is analogous to pressure (driving force to fluid flow). A resistor impedes current flow in the same manner as a valve impedes fluid flow. The electrical circuit in Figure 12.8 is analogous to the flow-pressure process in Figure 12.7.

In the electrical analogy, a constant voltage is maintained across two resistors in series. Suppose one resistor is larger than the other. Which resistor has the most influence on current flow? This is not a trick question; it is the larger of the two resistors.

In the flow-pressure process, which valve is providing the most resistance to fluid flow? The valve with the larger pressure drop. The valve with the larger pressure drop has more influence on flow than the other valve. Consequently, the flow controller should be manipulating the valve with the larger pressure drop. Whether it is the upstream valve or the downstream valve is immaterial.

Role of Intermediate Pressure. The appropriate control configuration is determined by the target for the intermediate pressure. This target must be between the upstream pressure and the downstream pressure. It must also be sufficiently within this range that neither control valve is driven fully open. The impact of the intermediate pressure target on the control configuration is as follows:

Intermediate pressure near upstream pressure. The major pressure drop is taken across the downstream valve, which makes the upper configuration in Figure 12.7 preferable.

Intermediate pressure near downstream pressure. The major pressure drop is taken across the upstream valve, making the lower configuration in Figure 12.7 preferable.

Intermediate pressure midway between upstream and downstream pressure. As both valves have about the same influence on the flow, there is no preference for either configuration. For this case, the interaction is most severe and neither configuration in Figure 12.7 will work very well.

Impact on Standardization. Standardization has always been popular in the process industries. For the flow-pressure process, the attitude would be to analyze the process, determine how it should be controlled, and then let us make them all the same. But how the process should be controlled depends on the operating conditions for the process. We can only standardize the control configuration provided we can standardize the process operating conditions.

This attitude is frequently encountered in distillation. The control configuration for a certain tower works very well. This leads to a proposal to implement this configuration on another tower. The results are usually disappointing. The appropriate control configuration for a tower depends on several factors, such as product purities, feed composition (relative to product compositions), ease of separation (relative volatility), external reflux ratio, and so on. The flow-pressure process is a relatively simple process; determining the appropriate control configuration can be summarized by a simple statement. Distillation is far more complex, and the dependency of the control configuration on the process characteristics reflect this.

Representations of Multivariable Processes. Applying the general approach to designating the variables in multivariable processes, the variables for the 2×2 flow-pressure process are designated as follows:

Manipulated variables	M_1	Upstream valve position
	M_2	Downstream valve position
Controlled variables	C_1	Pressure
	C_2	Flow

This notation is also indicated on Figure 12.7. The subscript that is assigned to each controlled and manipulated variable is completely arbitrary. The pressure has been designated as C_1; the upstream valve position has been designated as M_1. However, this does not imply that the upstream valve position (M_1) will be manipulated to control the pressure (C_1).

Block Diagram. The block diagram in Figure 12.9 represents a 2×2 multivariable process such as the flow-pressure process. This block diagram involves four relationships:

G_{11}	Effect of M_1 (upstream valve position) on C_1 (pressure)
G_{12}	Effect of M_2 (downstream valve position) on C_1 (pressure)
G_{21}	Effect of M_1 (upstream valve position) on C_2 (flow)
G_{22}	Effect of M_2 (downstream valve position) on C_2 (flow)

Constructing a block diagram is only manageable for a 2×2 multivariable process. For a 3×3 multivariable process, the block diagram becomes overly complex (there are nine relationships).

If any of these relationships in Figure 12.9 is zero, there will be no loop interaction. Suppose input M_2 does not affect output C_1. The consequences are as follows:

- The relationship G_{12} is zero. In such cases, that path is usually omitted when drawing the block diagram in Figure 12.9.
- C_1 can only be controlled by manipulating M_1, which means C_2 must be controlled by manipulating M_2.
- Reversing the pairing is not a viable option.
- Actions taken by the controller for C_2 do not affect the loop for C_1.

When the controller for C_1 changes M_1, these changes affect C_2. That is, control actions taken by the controller for C_1 are essentially disturbances to C_2. The controller for C_2 must respond by changing M_2 so as to maintain C_2 at target. However, the changes in M_2 do not affect C_1.

When drawing block diagrams for a 2×2 multivariable process, the representation in Figure 12.10 is frequently used to simplify constructing the block diagram. The process is represented by a block containing arrows to indicate how the inputs affect the outputs. Although this construct can be extended to higher order multivariable processes, it is still inconvenient to construct block diagrams for other than 2×2 multivariable processes.

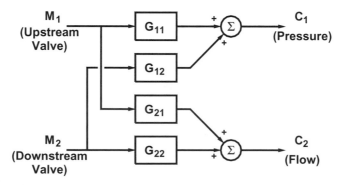

Figure 12.9. Block diagram of a 2×2 multivariable process.

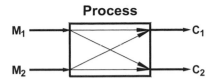

Figure 12.10. Simplified block diagram of a 2×2 multivariable process.

Equations. The block diagram in Figure 12.9 for a 2×2 multivariable process is expressed by the following equations:

$$C_1 = G_{11} M_1 + G_{12} M_2$$
$$C_2 = G_{21} M_1 + G_{22} M_2$$

For a 3×3 multivariable process, the equations would be as follows:

$$C_1 = G_{11} M_1 + G_{12} M_2 + G_{13} M_3$$
$$C_2 = G_{21} M_1 + G_{22} M_2 + G_{23} M_3$$
$$C_3 = G_{31} M_1 + G_{32} M_2 + G_{33} M_3$$

The set of equations can be easily extended to even higher order multivariable processes.

Vectors and Matrices. The equations for a multivariable process are most conveniently expressed using vectors and matrices. The following equation describes the process:

$$\begin{bmatrix} C_1 \\ C_2 \end{bmatrix} = \begin{bmatrix} G_{11} & G_{12} \\ G_{21} & G_{22} \end{bmatrix} \begin{bmatrix} M_1 \\ M_2 \end{bmatrix}$$
$$\mathbf{c} \quad = \quad \mathbf{G} \quad \mathbf{m}$$

where

$$\mathbf{c} = \begin{bmatrix} C_1 \\ C_2 \end{bmatrix} \qquad = \text{Vector of controlled variables}$$

$$\mathbf{m} = \begin{bmatrix} M_1 \\ M_2 \end{bmatrix} \qquad = \text{Vector of manipulated variables}$$

$$\mathbf{G} = \begin{bmatrix} G_{11} & G_{12} \\ G_{21} & G_{22} \end{bmatrix} = \text{matrix of relationships}$$

Most articles and books on multivariable control use such representations extensively. Engineers are exposed to vectors and matrices in school, but they use them infrequently thereafter. Consequently, few practicing engineers are comfortable with such mathematics.

12.4. GAINS AND SENSITIVITIES

For a single-input, single-output system (1×1 process), the steady-state gain is the change in the output divided by the change in the input. Using the notation for multivariable processes, the controlled variable is C_1 and the manipulated variable is M_1. The process gain K_{11} is expressed by the following derivative:

$$K_{11} = \frac{dC_1}{dM_1}$$

There is only one input and there is only one output, so there can be only one gain. For an $n \times m$ multivariable process, the number of possible gains is the product of n and m. For a 2×2 process, there are four gains. For a 3×3 process, there are nine gains. The number of gains increases rapidly with the dimensionality of the process.

But the situation is even more complex. In a 2×2 process, C_1 is a function of M_1 and M_2. Consequently, partial derivatives must be used. However, defining the process gain K_{11} as

$$K_{11} = \frac{\partial C_1}{\partial M_1}$$

is ambiguous. It is necessary to distinguish between the following two cases:

- Sensitivity of C_1 to M_1 with the other manipulated variable M_2 held constant. This gain is defined as follows:

$$K_{11} = \frac{\partial C_1}{\partial M_1}\bigg|_{M_2}$$

- Sensitivity of C_1 to M_1 with the other controlled variable C_2 held constant. This gain is defined as follows:

$$K'_{11} = \frac{\partial C_1}{\partial M_1}\bigg|_{C_2}$$

For a process with interaction, the values of these two gains will be different.

As we shall explain using the flow-pressure process as an example, the distinction between these two gains is very relevant to control. But before doing so, a word about the notation is in order. Writing partial derivatives such as those above gives the impression of extreme complexity. This is not the case. The concepts themselves are very simple. One would think that a simple concept could be expressed by a simple notation, but at least for these gains, this is unfortunately not the case. Usually the best way to bring such concepts down to earth is with a numerical example. For the flow-pressure process, we shall obtain numerical values for both of these gains, and then show that they tell us how the process should be controlled.

Pressure Controller on Manual. Let us focus on the flow loop in the upper configuration in Figure 12.7, specifically, controlling the flow C_2 by manipulating the downstream valve position M_2. As for single-input, single-output systems, we can determine the process gain by making a step change in the controller output (the downstream valve position M_2) and recording the response in the controlled variable (the flow C_2). But before making the test, we need to make a conscious decision about the pressure controller—should it be on manual or should it be on automatic?

Let us start with the pressure controller on manual, giving the test configuration illustrated in Figure 12.11. The value of the upstream valve position M_1 is fixed; that is, no control is provided for the intermediate pressure. The change in M_2 leads to changes in both the flow C_2 and the pressure C_1; however, we are not concerned with the change in pressure.

As M_1 is held constant, the sensitivity obtained using this test configuration is properly designated by the following partial derivative:

$$K_{22} = \frac{\partial C_2}{\partial M_2}\bigg|_{M_1}$$

This is the sensitivity of the flow C_2 to the downstream valve position M_2 with the upstream valve position M_1 held constant.

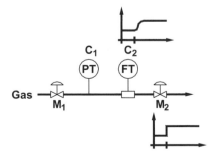

Figure 12.11. Test configuration, pressure controller on manual.

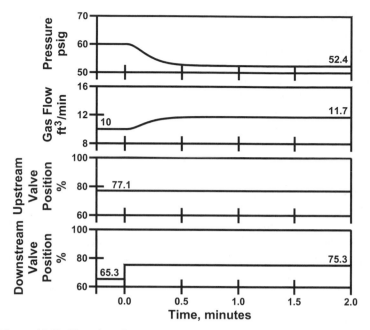

Figure 12.12. Test data for process gain, pressure controller on manual.

Figure 12.12 presents the test data obtained by making a step change in the downstream valve position from 65.3% to 75.3% with the pressure upstream valve position M_1 held constant at 77.1%. The computation of the steady-state gain is based on the change from the initial equilibrium to the final equilibrium:

	Initial	Final
Upstream pressure	80.0 psig	80.0 psig
Downstream pressure	40.0 psig	40.0 psig
Upstream valve position (M_1)	77.1%	77.1%
Downstream valve position (M_2)	65.3%	75.3%
Intermediate pressure (C_1)	60.0 psig	52.4 psig
Gas flow (C_2)	10.0 ft^3/min	11.7 ft^3/min

The value for the sensitivity K_{22} is computed as follows:

$$K_{22} = \frac{\Delta C_2}{\Delta M_2} = \frac{11.7\,\text{ft}^3/\text{min} - 10.0\,\text{ft}^3/\text{min}}{75.3\% - 65.3\%} = 0.17\,(\text{ft}^3/\text{min})/\%$$

From the responses in Figure 12.12, it would also be possible to compute the sensitivity K_{12} of the pressure C_1 to the downstream valve position M_2. But at this time, we have no interest in this value.

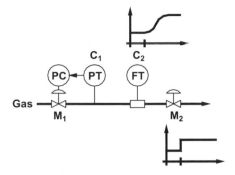

Figure 12.13. Test configuration, pressure controller on automatic.

Pressure Controller on Automatic. Let us do the same test, but with the pressure controller on automatic as illustrated by the test configuration in Figure 12.13. Starting from the same initial conditions, we make the same step change in the downstream valve position M_2 and record the response in the flow C_2. With the pressure controller in automatic, its set point SP_1 is fixed. Assuming the pressure equals its set point at steady state, the pressure C_1 is also fixed.

The sensitivity obtained using this configuration is normally designated as K'_{22}. As C_1 is held constant, it is properly designated by the following partial derivative:

$$K'_{22} = \frac{\partial C_2}{\partial M_2}\bigg|_{C_1}$$

This is the sensitivity of the flow C_2 to the downstream valve position M_2 with the pressure C_1 held constant.

Figure 12.14 presents the test data obtained by making a step change in the downstream valve position from 65.3% to 75.3% with the pressure controller set point held constant at 60 psig. The computation of the steady-state gain is based on the change from the initial equilibrium to the final equilibrium:

	Initial Value	Final Value
Upstream pressure	80.0 psig	80.0 psig
Downstream pressure	40.0 psig	40.0 psig
Upstream valve position (M_1)	77.1%	87.0%
Downstream valve position (M_2)	65.3%	75.3%
Intermediate pressure (C_1)	60.0 psig	60.0 psig
Gas flow (C_2)	$10.0\,\text{ft}^3/\text{min}$	$14.9\,\text{ft}^3/\text{min}$

During the test, the pressure initially drops below 60 psig. However, the pressure controller responds to this by opening the upstream valve from 77.1% to

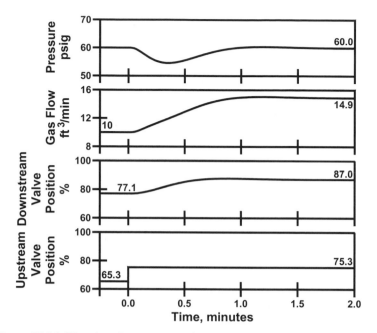

Figure 12.14. Test data for process gain, pressure controller on automatic.

87.0%. This restores the pressure to 60 psig, but also it contributes to the flow increase.

The combined effect is to increase the gas flow from 10 ft³/min to 14.9 ft³/min. The sensitivity K'_{22} is computed as follows:

$$K'_{22} = \frac{\Delta C_2}{\Delta M_2} = \frac{14.9\,\text{ft}^3/\text{min} - 10.0\,\text{ft}^3/\text{min}}{75.3\% - 65.3\%} = 0.49\,(\text{ft}^3/\text{min})/\%$$

For the flow-pressure process, K'_{22} is larger than K_{22}; however, this is not the case for all processes.

Significance of K_{22} and K'_{22}. The difference between K_{22} and K'_{22} is a measure of the degree of interaction within the process. When there is no interaction, the sensitivities K_{22} and K'_{22} will be equal. The greater the degree of interaction, the larger the difference.

The possibilities are as follows:

$K'_{22} > K_{22} > 0$ or $(-K'_{22}) > (-K_{22}) > 0$: The gains have the same sign and $|K'_{22}| > |K_{22}|$. This is the case for the flow-pressure process. Opening the downstream valve causes the flow to increase and the pressure to decrease. To bring the pressure back to the set point, the upstream valve must be opened, which further increases the flow.

$K'_{22} = K_{22}$: This is the case for no interaction.

$K_{22} > K'_{22} > 0$ or $(-K_{22}) > (-K'_{22}) > 0$: The gains have the same sign and $|K'_{22}| < |K_{22}|$. Although not the case for the flow-pressure process, there are processes that exhibit this type of behavior.

K'_{22} and K_{22} have different signs: This means that increasing the manipulated variable leads to an increase (decrease) when the other controller is in manual, but it leads to a decrease (increase) when the other controller is in automatic.

The Relative Gain. How well a loop performs in a multivariable configuration depends on the difference between K_{22} and K'_{22}, which in turn depends on the degree of interaction within the process. Therefore, one measure of interaction, known as the relative gain (2), is the ratio of these two sensitivities:

$$\lambda_{22} = \frac{K_{22}}{K'_{22}}$$

No interaction ($K_{22} = K'_{22}$) gives a relative gain $\lambda_{22} = 1.0$. As the degree of interaction increases, the relative gain differs from 1.0, with small values being as undesirable as large values. For the flow loop, the relative gain is computed as follows:

$$\lambda_{22} = \frac{K_{22}}{K'_{22}} = \frac{0.17 (\text{ft}^3/\text{min})/\%}{0.48 (\text{ft}^3/\text{min})/\%} = 0.35$$

This suggests considerable interaction. The process sensitivity of the flow loop changes by a factor of three when the pressure loop is switched between manual and automatic. As always, significant changes in the process gain lead to tuning difficulties.

We shall not do so, but the relative gain can be easily extended to higher dimensional processes. The numerator of the relative gain is the process sensitivity of a given loop with all other loops on manual. The denominator of the relative gain is the process sensitivity of that loop with all other loops on automatic.

12.5. EFFECT OF INTERACTION ON LOOP PERFORMANCE AND TUNING

Suppose we have chosen the upper configuration in Figure 12.7 for controlling the flow-pressure process. Specifically, flow is to be controlled by manipulating the downstream valve and pressure is to be controlled by manipulating the upstream valve.

We are now ready to commission the controls, which basically involves tuning the two controllers. How do we proceed? When two loops have different dynamics, one should tune the fast loop first. But for the flow-pressure process, the flow and pressure dynamics are about the same. Therefore, we should begin with the loop that is most crucial to process performance. For our process, this is the flow loop.

Tuning the Flow Loop. Since flow loops are much faster than most other loops, they can be tuned conservatively. We have recommended a controller gain of 0.2 %/% and a reset time of 3 sec. As the responses in Figure 12.15 illustrate, this tuning provides reasonable performance. For a change in the gas flow set point from 10 ft³/min to 12 ft³/min, the gas flow attains the new set point in less than 1 minute. For most processes, this would be acceptable. The flow loop could be tuned to make this happen more quickly, but in most cases, there would be no noticeable improvement in overall process performance.

Actually, there is a clue in the responses in Figure 12.15 that we may have a problem. The change in the gas flow is accompanied by a change of almost

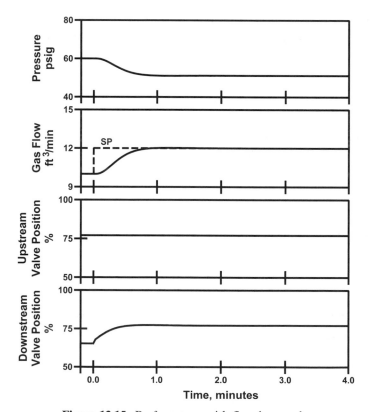

Figure 12.15. Performance with flow loop only.

10 psi in the intermediate pressure, which suggests that using the downstream valve to implement changes in the gas flow will cause significant upsets to the intermediate pressure. To have interaction, changes in the upstream valve made to control the pressure must cause significant upsets to the flow. From Figure 12.15, there is no way to determine whether this is indeed the case.

Tuning the Pressure Loop. As the dynamics of the pressure loop are essentially the same as the dynamics of the flow loop, let us try the same conservative tuning, namely a controller gain of 0.2%/% and a reset time of 3 sec.

The responses in Figure 12.16 are to a change in the gas flow set point from 10 ft³/min to 12 ft³/min (both the flow and pressure loops are on automatic). Oscillations are evident in both flow and pressure. The loops are stable, but just barely. We know that the flow loop exhibits no oscillations when used alone. Consequently, the source of the oscillations could be either of the following:

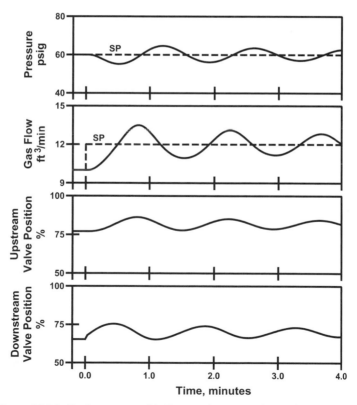

Figure 12.16. Performance with flow and pressure loops in automatic.

- Some problem within the pressure loop that leads to oscillations.
- Interaction between the flow and pressure loops.

To eliminate the first possibility, we need to examine the performance of the pressure loop when used alone.

Performance of Pressure Loop when Used Alone. The responses in Figure 12.17 illustrate the performance of the pressure loop tuned with a controller gain of 0.2%/% and a reset time of 3 sec. The response is to a change in the pressure set point from 60 psig to 70 psig. The pressure attains the new set point in less than 1 minute, which is acceptable. There is a very small overshoot of the set point but no oscillations.

The responses in Figure 12.17 complete the case for interaction. The change in the intermediate pressure is accompanied by a change of more than 2 ft^3/min in the gas flow. This suggests that changes in the upstream valve position to control the pressure will cause significant upsets to the gas flow. We now know that the upsets are significant in both directions, which means interaction.

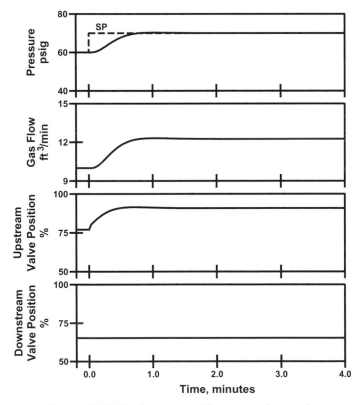

Figure 12.17. Performance with pressure loop only.

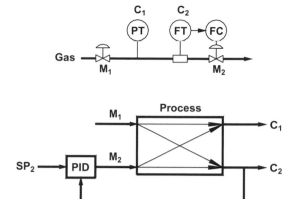

Figure 12.18. Simplified block diagram with flow loop only.

We have now demonstrated that the flow loop and the pressure loop provide very acceptable performance when used individually. But when used together, oscillations that decay very slowly are present in both the flow and pressure responses. The culprit must be interaction between the two loops.

Relationships for Flow Loop Only. The simplified block diagram in Figure 12.18 illustrates the relationships when only the flow loop is in automatic. The performance of the flow loop depends on the sensitivity of the gas flow C_2 to the downstream valve position M_2. With the pressure controller on manual, there is no change in the upstream valve position M_1. In this configuration, the sensitivity of the flow process is

$$\frac{\partial C_2}{\partial M_2}\bigg|_{M_1} = K_{22}$$

The downstream valve M_2 affects the pressure C_1 and the upstream valve M_1 affects the gas flow C_2, but these relationships have no effect on the performance of the flow loop when the pressure loop is in manual.

Impact of Pressure Loop on the Relationships for the Flow Loop. The simplified block diagram in Figure 12.19 illustrates the relationships when both loops are in automatic. The performance of the flow loop depends on the sensitivity of the gas flow C_2 to the downstream valve position M_2. Assuming that the pressure controller maintains the pressure C_1 at its set point, the steady-state sensitivity of the flow process is

$$\frac{\partial C_2}{\partial M_2}\bigg|_{C_1} = K'_{22}$$

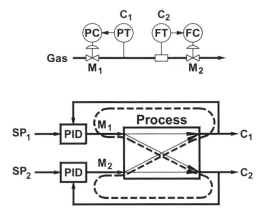

Figure 12.19. Simplified block diagram with flow and pressure loops, including the "hidden loop."

With the pressure loop in manual, the flow loop performance depends on K_{22}. With the pressure loop in automatic, the flow loop performance depends on K'_{22}. If these two sensitivities are nearly the same, the performance of the flow loop should be nearly the same with the pressure loop on automatic as with the pressure loop on manual. But to the degree that these sensitivities differ, the performance will also differ.

The Hidden Loop. When single-loop control is applied, the block diagram for a 2×2 multivariable process will contain three loops, at least when interaction is present and both controllers are on automatic. There are the usual loops for each of the controllers. But as illustrated by the dashed path in Figure 12.19, there is a third loop, sometimes referred to as the "hidden loop." Let us start at the output of the pressure controller M_1. Through the process, M_1 affects C_2. With the flow controller on automatic, C_2 affects M_2. Through the process, M_2 affects C_1. With the pressure controller on automatic, C_1 affects M_1.

Note that the hidden loop contains both controllers. When tuned with a controller gain of 0.2%/% and a reset time of 3 sec, the controllers are essentially integral-only controllers. Thus, the hidden loop essentially contains double integration, which means stability problems. The performance illustrated in Figure 12.16 is certainly consistent with this observation.

12.6. DYNAMICS

Interaction has both steady-state and dynamic aspects. If one loop is much faster than another loop, the degree of steady-state interaction is irrelevant. Tune the fast loop first, and then tune the slow loop. From the perspective of the slow loop, the fast loop always seems to be at steady state.

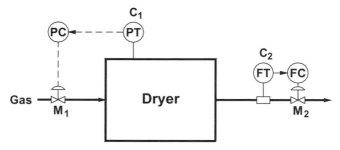

Figure 12.20. Pressure and flow control for a dryer.

Figure 12.20 illustrates flow and pressure control for a dryer. The pressure is measured within the dryer; the flow rate of the exit gas stream is measured. It is desired to control the flow of gas through the dryer and the pressure within the dryer.

Assuming the same flows, pressures, valve sizes, and so on, the steady-state characteristics of the dryer are exactly the same as for the flow-pressure process described previously. However, the dynamics are different. Specifically, the pressure will respond far more slowly to changes in the gas flow in or out. This makes the pressure loop much slower than the flow loop. The control configuration in Figure 12.20 will work regardless of the degree of steady-state interaction.

Tuning the Flow Controller. This example uses the same pressures and flows as in the previous examples for the flow-pressure process. The only difference is a large volume between the two valves. We shall examine the following control configuration:

- Control pressure C_1 using the upstream valve position M_1. The pressure C_1 is the pressure in the large volume.
- Control flow C_2 using the downstream valve position M_2. The flow C_2 is the discharge flow from the large volume.

The measurements for C_1 and C_2 are as illustrated for the dryer in Figure 12.20. One could consider this example as applying to the dryer; however, dryers do not normally operate at the pressures used in the flow-pressure process examples.

With the large volume, the flow loop is much faster than the pressure loop. Tuning should always start with the fast loop. For the flow loop, we will begin with the tuning typically recommended for flow loops, specifically, a controller gain of 0.2%/% and a reset time of 3 sec.

The responses in Figure 12.21 illustrate the performance of the flow controller with the pressure controller on manual. There is some overshoot in the

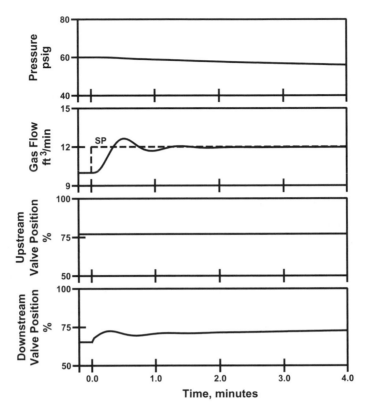

Figure 12.21. Performance with flow loop only for flow-pressure process with large volume.

flow loop, but the oscillations decay rapidly. We shall consider them to be acceptable, but if desired, the overshoot and oscillations could be eliminated by lengthening the reset time.

Because of the large volume within the dryer, the pressure now changes far more slowly than the flow. With the increase in the discharge flow, note the following two aspects of the responses in Figure 12.21:

- The pressure C_1 is slowly decreasing. This is because more gas is flowing out than is flowing in. As the pressure continues to decrease, the gas flowing in will continue to increase. This will continue until the flow in is the same as the flow out (which is C_2). It takes about an hour for the new equilibrium to be established.
- The downstream valve position M_2 is slowly increasing. As the pressure C_1 decreases, the flow controller must open the downstream valve to maintain the flow. The flow C_2 is slightly below its set point, but the resolution of the trends in Figure 12.21 is not sufficient for this to be noticeable.

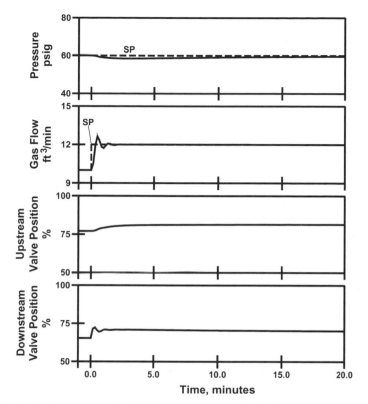

Figure 12.22. Performance with both loops on automatic for flow-pressure process with large volume.

If one waits long enough for equilibrium to be attained, the conditions will be the same as for the examples for the flow-pressure process with little volume between the control valves.

Tuning the Pressure Controller. The residence time (volume divided by gas flow rate) is approximately ten minutes. Therefore, a reasonable value for the reset time is 10 min. Suggesting a value for the controller gain is more difficult, so we shall begin with a gain of 2.0%/%.

The responses in Figure 12.22 illustrate the performance for a change in flow set point from 10 ft³/min to 12 ft³/min. Because the pressure loop is much slower, the tic marks on the time axis are 5 minutes instead of 1 minute for all previous responses. The flow controller performance is basically the same as before. The increase in discharge flow causes the pressure to drop below its set point (the maximum departure from the set point is just over 1 psig), but the pressure controller responds to return the pressure to its set point. The following tuning adjustments could be considered for the pressure controller:

1. Increase the controller gain to reduce the drop below the set point.
2. Shorten the reset time to make the pressure return to its set point more rapidly.

However, the current tuning is sufficient to illustrate the primary point—that with the large separation in the dynamics, the two loops can be on automatic and deliver acceptable performance. The value of the relative gain is the same as before, specifically, 0.35. Significant steady-state interaction is present, but the separation of dynamics makes the steady-state interaction irrelevant.

Separating Dynamics by Tuning. For the original flow-pressure process, the dynamics of the pressure loop are about the same as the dynamics of the flow loop. The steady-state interaction then led to tuning difficulties.

When the process dynamics are comparable, we can still attempt to separate the dynamics of the two loops through controller tuning. We can speed up one of the loops, preferably through increasing the controller gain. We can then slow down the other loop by reducing the controller gain and/or lengthening the reset time. For the flow-pressure process, the options are as follows:

1. Tune the flow controller to respond quickly and the pressure controller to respond slowly.
2. Tune the pressure controller to respond quickly and the flow controller to respond slowly.

Since flow is more crucial than pressure, one's initial thoughts would be that the flow controller should be the fast loop. However, it is often the noncritical loop that must be tuned to respond rapidly.

Fast Flow Loop, Slow Pressure Loop. The tuning is undertaken as follows:

1. With the pressure controller on manual, the flow controller is tuned to respond as rapidly as possible. The resulting tuning coefficients are a controller gain of 1.2%/% and a reset time of 0.15 min.
2. With both controllers on automatic, slow down the response of the pressure controller until it has only a modest influence on the performance of the flow loop. The resulting tuning coefficients are a controller gain of 0.2 %/% and a reset time of 0.5 min.

The responses in Figure 12.23 are for a flow set-point change from $10\,\text{ft}^3/\text{min}$ to $12\,\text{ft}^3/\text{min}$. On the change in the set point, the downstream valve opens quickly, which increases the flow and decreases the pressure. The pressure controller then slowly opens the upstream valve to bring the pressure back to the set point.

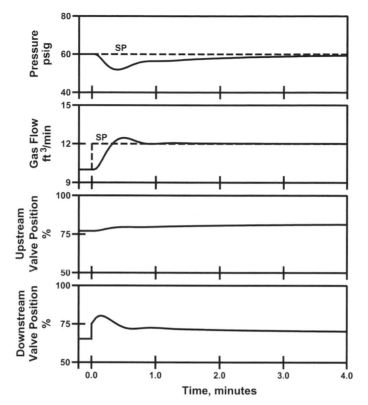

Figure 12.23. Performance with fast flow loop and slow pressure loop for original flow-pressure process.

Although tuning one loop fast and the other slow usually permits both loops to be operated on automatic, this does not mean that the performance will be acceptable under all circumstances. For the flow-pressure process with a fast flow loop and a slow pressure loop, the response to large increases in the flow set point may not be acceptable.

Figure 12.24 presents the responses to an increase in the flow set point from $10\,\text{ft}^3/\text{min}$ to $18\,\text{ft}^3/\text{min}$. The flow controller immediately drives the downstream valve fully open. But with the initial position of the upstream valve, this is not sufficient to give a flow of $18\,\text{ft}^3/\text{min}$. A flow of $18\,\text{ft}^3/\text{min}$ cannot be achieved until the pressure controller sufficiently opens the upstream valve. For the responses in Figure 12.24, this does not take too long and may be acceptable. However, the slower the pressure controller is tuned, the longer this will take.

Since the flow controller drives the downstream valve fully open, windup adversely affects the performance. For the responses in Figure 12.24, the upper output limit for the flow controller is 102%. A small overrange is tolerable, but an overly conservative setting for the upper output limit is not advisable.

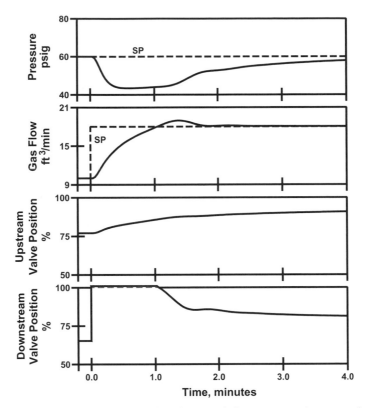

Figure 12.24. Performance with fast flow loop and slow pressure loop to a large flow set-point change.

Fast Pressure Loop, Slow Flow Loop. The tuning is undertaken as follows:

1. With the flow controller on manual, the pressure controller is tuned to respond as rapidly as possible. The resulting tuning coefficients are a controller gain of 1.8%/% and a reset time of 0.2 min.
2. With both controllers on automatic, slow down the response of the flow controller until it has only a modest influence on the performance of the pressure loop. The resulting tuning coefficients are a controller gain of 0.2%/% and a reset time of 0.1 min.

The responses in Figure 12.25 are for a flow set-point change from 10 ft^3/min to 12 ft^3/min. There is now only a small departure of the pressure from its set point. Although larger pressure changes are acceptable, the faster responding pressure loop drives the upstream valve more rapidly, enabling this tuning to respond more effectively to large changes in the flow set point.

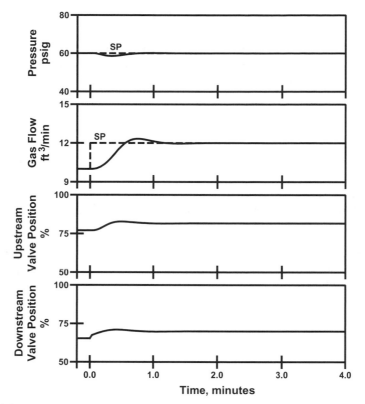

Figure 12.25. Performance with fast pressure loop and slow flow loop for original flow-pressure process.

Tuning one loop fast and the other slow can permit both loops to be oper-ated on automatic. But unless the rationale behind the tuning is understood, tuning parameter adjustments, especially in the fast loop, will not give the expected results.

Specifically, reducing the controller gain and/or increasing the reset time are normally expected to reduce overshoot and oscillations. But for processes with significant steady-state interaction, the results are as follows:

1. Reducing the controller gain in the fast loop slows down the fast loop.
2. This reduces the dynamic separation between the two loops.
3. The steady-state interaction then increases the overshoot and oscillations.

Unless one understands the need for dynamic separation between the loops, reducing the controller gain would not be expected to give these results.

12.7. ADDRESSING INTERACTION PROBLEMS

Before pursuing solutions to an interaction problem, first make sure that interaction is the problem. When tuning one loop creates cycling in another loop, the culprit could be interaction. However, it could also be a problem within the loop being tuned. When the manipulated variable from one loop is a disturbance to another loop, cycling in the loop being tuned will induce cycling into the other loop. This can occur even if the degree of interaction between the loops is zero.

To be certain that interaction is the problem, it is necessary to demonstrate that each loop will perform satisfactorily when the other loop is on manual. If two loops perform satisfactorily when used individually but cycle excessively when both are on automatic, the problem is certainly interaction between the two loops.

Leave One Loop on Manual. Although this option should probably not be considered a "solution," it is frequently done. The loop that is most crucial to process operations is tuned to provide the desired performance. The less critical loop is left on manual, usually with some instructions to the process operators to adjust manually the output of the loop should certain situations develop.

If you inquire about the problem with the second loop, you are likely to be told that the problem is tuning, the implication being is that our folks just do not know how to tune loops. Repeated attempts are often made to tune such loops. Too often these loops are candidates for automatic or self-tuning packages (we are good at throwing technology at a problem). All of this is to no avail. The tuning problems are symptoms of the interaction problem. Until the interaction problem is addressed, tuning will not be successful.

Loop Pairing. That the loop pairing is incorrect is certainly a possibility. Here we go with "understand the process" again. It is certainly possible that the basis for the current loop pairing is flawed. The first step is to review this from a qualitative perspective. If this does not suggest a different pairing to try, then consider getting quantitative values for the degree of interaction between the loops.

At this time, the only quantitative measure of interaction is the relative gain. The use of the relative gain to determine the most appropriate pairing for loops is well known but not widely practiced. Do not let the mathematics get in the way! We have already defined the relative gain. It is really a simple concept. Values for the relative gains can often be determined from process models, thus avoiding the need for process tests (although testing is often the simplest approach for fast processes). Although the relative gain has its limitations, it is unfortunately the only tool currently available.

Multivariable representations tend to suggest that all possible pairings are acceptable. For a 2×2 process, there are only two possibilities, one being the reverse of the other. Usually the pairing can be reversed, but not always and for reasons other than interaction issues. For a higher dimensional process, there are likely to be issues that make certain pairing unacceptable.

Fast Loop, Slow Loop. If two loops are dynamically separated, steady-state interaction between the loops will not prevent the loops from remaining on automatic. How much dynamic separation is needed? In cases where interaction is severe, the fast loop needs to be at least five times faster than the slow loop. Where the interaction is less severe, less dynamic separation will be required.

This approach is not always successful. Often there are consequences from one of the loops responding slowly. Often the critical loop must be tuned to respond slowly and the noncritical loop tuned to respond quickly, which is the opposite of what one would expect. And then tuning parameter changes in the fast loop will give results that are opposite of what is normally expected. Although conceptually simple, the "side effects" of this approach can be troubling.

Reduce Degree of Interaction. The objective is to reduce the degree of interaction to the point that the loops will function when all are on automatic. It is not necessary to eliminate interaction between the loops, only to reduce it sufficiently that the loops perform satisfactorily.

Let us illustrate for the flow-pressure process. What if the intermediate pressure is at target but the flow is below target? What must happen to bring the flow up to target without affecting the intermediate pressure? Both valves must open. The control configuration in Figure 12.26 does this. The output of the flow controller basically drives both valves. The position of the downstream valve is the value of the flow controller output. The position of the upstream valve is the value of the flow controller output plus a bias (that may be positive or negative). The outputs of the two controllers are as follows:

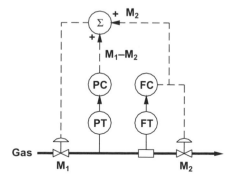

Figure 12.26. Incorporation of a summer into the control configuration.

Flow controller: Downstream valve position (M_2).

Pressure controller: Upstream valve position less downstream valve position ($M_1 - M_2$).

An increase of 1% in the flow controller output opens both valves by 1%. Does this completely eliminate the interaction between the flow controller and the pressure controller? Very unlikely. However, the objective is to reduce the interaction to the point that the two loops will perform satisfactorily.

Usually the approach is to incorporate simple computing elements (summers, ratios, characterization functions, etc.) into the control configuration with the objective of reducing the degree of interaction. There is no rigorous design methodology for developing such configurations. Sometimes the model relationships can be manipulated to suggest the proper place to insert summers, ratios, and so on. But in most cases, these configurations are based on someone's understanding of the process.

Decouplers. As illustrated in the schematic for the flow-pressure process in Figure 12.27, the decoupler is inserted between the outputs of the PID controllers and the final control elements. The decoupler contains logic that compensates for any interaction in the process. In effect, interaction is incorporated into the decoupler that cancels the interaction within the process.

Suppose the flow controller adjusts input X_2 to the decoupler. The decoupler adjusts both M_1 and M_2 such that there is a net change in the flow C_2 but no change in the pressure C_1. To the PID controllers, the decoupler–process combination seems to have no interaction.

Decouplers have been around for some time, and the theory is well developed. But by attempting to eliminate the interaction, classic decouplers are often excessively complex. Sometimes only steady-state decoupling is suffi-

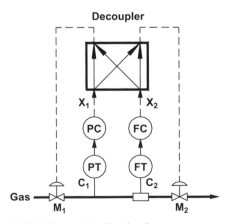

Figure 12.27. Decoupler for the flow-pressure process.

cient, which greatly reduces the complexity. But even so, the industrial applications of classic decouplers are few.

Model-Predictive Control. Pioneered by Shell Oil using a technology called dynamic matrix control, model predictive control encompasses decoupling, dead time compensation, constraint control, and other functions required to control complex, multivariable processes. Although significant performance improvements are possible, the up-front effort to implement a model predictive controller is not trivial. Before embarking on such a project, understand in complete detail the plant testing that will be required. It will be both time-consuming and expensive.

The model predictive controller is used in lieu of PID controllers. As represented in Figure 12.28 for the flow-pressure process, set points for each controlled variable are provided to the model predictive controller, which adjusts all manipulated variables to achieve these targets. A model predictive controller can be implemented to output directly to the final control elements, but more often the model predictive controller outputs to the set points of flow controllers that in turn position the final control elements. In effect, the model predictive controller becomes the outer loop of a cascade. The advantage is the same as for PID control—the flow controllers insulate the model predictive controller from the characteristics of the final control element.

As originally proposed, model predictive control was a tool that enabled the decisions of an optimization package to be implemented on a multivariable process. The benefits were from process optimization; the model predictive controller was an "enabling technology" in that it was necessary to get the benefits from optimization. When the optimization decisions were directed to single-loop controllers, process interaction led to serious control problems within the process.

Process optimization in some form is usually required to justify the effort to install a model predictive controller. The cost of installing a model predic-

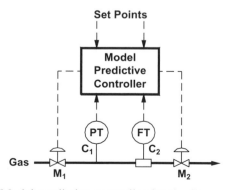

Figure 12.28. Model predictive controller for the flow-pressure process.

tive controller cannot usually be justified on processes like the flow-pressure process.

12.8. SUMMARY

Some processes will always have a certain degree of interaction. Consider a separation process such as a two-product distillation tower. Any change that affects the composition of one of the product streams will have some effect on the composition of the other product stream. This is imposed by the total material balance. A certain amount of a given component enters with the feed stream. Long term, this same amount must leave with the product streams. If the amount of the component is decreased in one of the product streams, it must increase by the same amount in the other product stream.

When developing a P&I diagram, the usual practice is to assess the degree of interaction from a qualitative perspective. This works reasonably well when there is very little effect of one variable on another. But in processes such as distillation, there is always some effect, but a question as to how much. This is difficult to resolve qualitatively, and becomes even more challenging as the complexity of the process increases. In distillation columns that separate light hydrocarbons, those familiar with distillation can assess the degree of influence of one variable on another. But for chemical towers separating complex molecules with nonideal behavior, such assessments are not so certain.

At this time, the relative gain is the only methodology for quantitatively assessing the degree of interaction. It has its limitations, and it cannot be applied blindly. The best approach is to examine each of the sensitivities that is used in the calculations, and to make sure that we understand why the sensitivities are what they are. Again, we are back to understanding the process. If the sensitivities are not consistent with what we think they should be, then either our model is incorrect or our understanding of the process is incorrect. This must be resolved, as it very likely has implications for process operations beyond loop interaction.

LITERATURE CITED

1. Trevathan, V., "Process Control in the Chemical Industry", *AIChE Symposium Series*, Vol. 72, No. 159 (1976), p. 40.
2. Bristol, E. H., "On a New Measure of Interaction for Chemical Process Control", *IEEE Transactions on Automatic Control*, Vol. 11, No. 1 (January 1966), p. 133.

Practical Process Control: Tuning and Troubleshooting, by Cecil L. Smith
Copyright © 2009 John Wiley & Sons, Inc.